Ernst Zermelo

Heinz-Dieter Ebbinghaus
In Cooperation with Volker Peckhaus

Ernst Zermelo

An Approach to His Life and Work

With 42 Illustrations

 Springer

Heinz-Dieter Ebbinghaus
Mathematisches Institut
Abteilung für Mathematische Logik
Universität Freiburg
Eckerstraße 1
79104 Freiburg, Germany
E-mail: hde@math.uni-freiburg.de

Volker Peckhaus
Kulturwissenschaftliche Fakultät
Fach Philosophie
Universität Paderborn
Warburger Straße 100
33098 Paderborn, Germany
E-mail: volker.peckhaus@upb.de

Library of Congress Control Number: 2007921876

Mathematics Subject Classification (2000): 01A70, 03-03, 03E25, 03E30, 49-03, 76-03, 82-03, 91-03

ISBN 978-3-540-49551-2 Springer Berlin Heidelberg New York

Springer is a part of Springer Science+Business Media

springer.com

© Springer-Verlag Berlin Heidelberg 2007

Typesetting by the author using a Springer TeX macro package
Production: LE-TeX Jelonek, Schmidt & Vöckler GbR, Leipzig
Cover design: WMXDesign GmbH, Heidelberg

Printed on acid-free paper   46/3100/YL - 5 4 3 2 1 0

To the memory of
Gertrud Zermelo (1902–2003)

Preface

Ernst Zermelo is best-known for the explicit statement of the axiom of choice and his axiomatization of set theory. The axiom of choice led to a methodological enrichment of mathematics, the axiomatization was the starting point of post-Cantorian set theory. His achievements, however, did not unfold in an undisputed way. They became the object of serious criticism sparked, in particular, by the inconstructive character of the axiom of choice, making it one of the most debated principles in the history of mathematics. Zermelo defended his point of view with clear insights and discerning arguments, but also with polemical formulations and sometimes hurtful sharpness. The controversial attitude shining through here has become a dominating facet of his image. Further controversies such as those with Ludwig Boltzmann about the foundations of the kinetic theory of heat and with Kurt Gödel and Thoralf Skolem about the finitary character of mathematical reasoning support this view.

Even though these features represent essential constituents of Zermelo's research and character, they fall short of providing a conclusive description. Neither is Zermelo's major scientific work limited to set theory, nor his personality to controversial traits. His scientific interests included applied mathematics and purely technical questions. His dissertation, for example, promoted the Weierstraßian direction in the calculus of variations, he wrote the first paper in what is now called the theory of games, and created the pivotal method in the theory of rating systems. The complexity of his personality shows in his striving for truth and objectivity, and in the determination with which he stood up for his convictions. Well-educated in and open-minded about philosophy, the classics, and literature, he had the ability of encountering others in a stimulating way.

Due to serious illness, which hindered and finally ended his academic career, and due to growing isolation from the dominating directions in the foundations of mathematics, he became caught in a feeling of being denied due scientific recognition, and controversy seemed to gain the upper hand. Those close to him, however, enjoyed his other sides.

The present biography attempts to shed light on all facets of Zermelo's life and achievements. In doing so, quotations from various sources play a major role. Personal and scientific aspects are kept separate as far as coherence allows, in order to enable the reader to follow the one or the other of these threads. The discussion of Zermelo's scientific work does not require detailed knowledge of the field in question. Rather than aiming at an in-depth technical analysis of his papers, the presentation is intended to explore motivations, aims, acceptance, and influence. Selected proofs and information gleaned from drafts, unpublished notes, and letters add to the analysis.

The main text is followed by a *curriculum vitae* which summarizes the main events of Zermelo's life, now in a more schematical manner. It thus provides some kind of chronological index.

All facts presented are documented by appropriate sources. Whenever possible, English versions of German texts follow a published translation. In some special cases such as axioms, the original German version is given in the main text, as well. Original German versions which have not been published previously and whose wording may be of some importance or interest, are compiled in the appendix. In particular, the appendix contains all unpublished quotations from Zermelo himself, supplemented by samples of his literary activity.

There is no claim that the biography offers a complete picture. Rather, the description of Zermelo's life and work should be considered as an approach to a rich and multifaceted personality, which may invite the reader to take an even closer look.

When Zermelo's late wife Gertrud learnt about this project, she gave all possible support. The information and the documents which she provided with warm-hearted interest contributed in essential ways to the picture of Zermelo presented here. With deep gratitude, the biography is dedicated to her memory.

Freiburg, September 2006   Heinz-Dieter Ebbinghaus

Acknowledgements

First, thanks are due to Volker Peckhaus who accompanied the project from its beginning. The (sub)sections 1.2, 2.3.1, 2.4, 2.8.4, 2.10, 2.11.4, 3.1, and 3.2 are based on drafts provided by him. He read and commented on different versions of the text. The collaboration with him has contributed to the character of the book.

Jörg Flum has looked at various drafts. His remarks and proposals provided valuable guidance.

Akihiro Kanamori read the whole manuscript with great care. His criticism and thoughtful suggestions left their mark in the final text.

Thomas Lais gave all possible support for the processing and editing of the illustrations.

Special thanks go to Anne Schwarz. With carefulness and patience, she has transferred my German English into English English.

It is a pleasure to thank all those who supported this work by giving advice on scientific and technical questions, commenting on parts treating special fields, or providing documentary material: Gerald L. Alexanderson, Hassan Aref, Massimiliano Badino, Victor Bangert, Lieselotte Bauer, Denis Blackmore, John W. Dawson, Jr., Ursula Döring, Susanne Ebbinghaus, Jürgen Elstrodt, Jens Erik Fenstad, Benjamin S. Fraenkel, Helmuth Gericke, Mark E. Glickman, Barbara Hahn, Monika Hattenbach, Markus Junker, Margret and Ronald B. Jensen, Albyn C. Jones, Lothar Klimt, Andrea Köhler, Marek Kordos, Nikola Korrodi, Egon Krause, Konrad Kunze, Ernst Kuwert, Hans Rudolf Lerche, Azriel Levy, Klaus-Dieter Lickert, Bernhard Richard Link, Moritz Müller, Paul K. Newton, Mario Ohlberger, Volker Remmert, Helmut Saurer, Gregory R. Taylor, Rüdiger Thiele, Gerhard Vollmer, Erika Wagner-Klimt.

The following institutions and persons kindly provided archival sources: Archiv der sozialen Demokratie, Bad Godesberg: Ilse Fischer; Birkhäuser Verlag; Center for American History, The University of Texas at Austin: Erin O'Brian; Deutsches Museum, Archiv: Eva A. Mayring; Geheimes Staatsarchiv Preussischer Kulturbesitz: Christiane Brandt-Salloum; Heidelberger Akademie der Wissenschaften: Hans-Günter Dosch; Institute for Advanced Study, Princeton: Marcia Tucker; Niedersächsische Staats- und Universitätsbibliothek Göttingen, Abteilung Handschriften und alte Drucke: Bärbel Mund; Österreichische Nationalbibliothek: Silke Pirolt; Staatsarchiv des Kantons Zürich: Hans Ulrich Pfister; Stadtarchiv Freiburg im Breisgau: Günther Wolf; Universitätsarchiv Freiburg: Dieter Speck and Alexander Zahoransky; Universitätarchiv Göttingen: Ulrich Hunger; Universitätsarchiv Zürich: Heinzpeter Stucki; University Archives Wrocław and the Institute of Mathematics of Wrocław University: Witold Więsław.

The Deutsche Forschungsgemeinschaft supported the project under grant no. EB 60/2.

I enjoyed the cooperation with Springer-Verlag, in particular with Ruth Allewelt and Martin Peters.

Through all the work my wife Heidi has accompanied me with warm understanding and encouragement.

Editorial Remarks

Archives and *Nachlässe* which are frequently quoted are given in an abbreviated form as follows:

ASD: Archiv der sozialen Demokratie, Bad Godesberg,
Nachlass Nelson;

CAH: Center for American History, The University of Texas
at Austin, Dehn (Max) Papers, 3.2/86-4/Box 1;

DMA: Deutsches Museum, Archiv;

GSA: Geheimes Staatsarchiv Preußischer Kulturbesitz Berlin;

SAZ: Staatsarchiv des Kantons Zürich;

SUB: Niedersächsische Staats- und Universitätsbibliothek Göttingen;

UAF: Universitätsarchiv Freiburg;

UAG: Universitätsarchiv Göttingen;

UAW: University Archives Wrocław;

UBA: University of Buffalo Archives, Marvin Farber, Papers on
Philosophy and Phenomenology, 1920s–1980s, 22/5F/768.

The abbreviation "MLF" refers to the Abteilung für Mathematische Logik, Universität Freiburg. The corresponding documents are currently being incorporated into the Zermelo *Nachlass* held in the Universitätsarchiv Freiburg under signature C 129.

Photographs reprinted without reference to a source are contained in the Zermelo photo collection held in the Abteilung für Mathematische Logik at Freiburg University. This photo collection will also be integrated into the Zermelo *Nachlass*.

References of type OV n.xy, which follow an English translation of a German text, refer to the original German version given under the same reference in the Appendix (Chapter 7). The numbers n = 1, 2, 3, 4 correspond to Chapters 1 (Berlin), 2 (Göttingen), 3 (Zurich), 4 (Freiburg), respectively.

Contents

1

Berlin *1871–1897*

1.1 Family and Youth

Ernst Friedrich Ferdinand Zermelo was born in Berlin on 27 July 1871.[1] It was the year when Georg Cantor wrote his papers on trigonometric series.[2] As Zermelo expressed it about 60 years later, these papers led with inherent necessity to the conception of transfinite ordinal numbers; they hence could be considered the birthplace of Cantorian set theory, i.e. the theory whose transformation into an axiomatic form would be an essential part of Zermelo's scientific achievements.

There exist several hypotheses about the origin of the name "Zermelo." They all seem to originate with Zermelo himself. According to the most well-known ([Rei70], 98), Zermelo is said to have replied to newcomers in Göttingen when asked about his strange name, that it resulted from the word "Wal-zermelodie" ("waltz melody") by deleting the first and the last syllable. In conversations with his wife Gertrud, Zermelo suggested the more seriously meant possibility that it was derived from "Termeulen," a form of the name "Zurmühlen" ("at the mill") current in the lower Rhine and Friesian areas. The name most probably originates from the variant "Tormählen"[3] which evolved in northeastern Germany, the region the Zermelo family comes from, during the 16th century; the Low German "T" might have been transformed into the High German "Z" in the course of one of the frequently observed hypercorrections.[4] With the death of Zermelo's widow Gertrud in December 2003 the name "Zermelo" disappeared from the German telephone listings. She seems to have been its last bearer.

Zermelo had five sisters, the one-year older Anna and the younger ones Elisabeth, Margarete, Lena, and Marie. His parents had married in Delitzsch,

---

[1]Details about Zermelo's family are taken from documents held in MLF.
[2][Can71a], [Can71b], [Can72].
[3]The "ä" is pronounced like the "ea" in "bear."
[4][Kun98], 159, and oral communication with Konrad Kunze.

Saxony, on 18 May 1869. His mother, Maria Auguste née Zieger, was born there on 11 February 1847 as the only child of the surgeon Dr. Ottomar Hugo Zieger and his wife Auguste née Meißner. Both her parents died early, her mother just one week after giving birth to her, her father one year later from tuberculosis. She then grew up in the wider family, which included the natural historian Christian Gottfried Ehrenberg (1795–1876) who was to become known for his works on microorganisms and who accompanied Alexander von Humboldt on his Russian expedition of 1829.[5] Like her parents, Maria Auguste suffered from poor health. Exhausted from the strain of multiple pregnancies, she died on 3 June 1878, soon after the birth of Marie. After her death a maid, Olga Pahlke, took care of the household.

Lena and Marie died at a young age, in 1906 and 1908, respectively. Not only they, but also the other children inherited their mother's delicate health; they all fell ill with tuberculosis. Neither of the three surviving sisters married. The letters which Zermelo wrote to them attest to a close relationship between the siblings, and care and concern on Zermelo's side.

Zermelo's father, Theodor Zermelo, was born on 9 June 1834 in the East Prussian town of Tilsit (today Sovjetsk, Russia) on the river Memel (Neman). Theodor's father Ferdinand Zermelo, a bookbinder who had graduated from the college of arts and crafts in Königsberg (today Kaliningrad, Russia), took over a shop for books, arts, and stationery in Tilsit in 1839. Documentation about him and his wife Bertha née Haberland is sparse: a carefully written diary about a trip to Paris and London, which Ferdinand undertook in the summer of 1851, and newspaper reports and obituary notices on the occasion of his death showing that he was an integral part and a supporter of cultural life in Tilsit and engaged in the town's social institutions. He served as a councillor and was a co-founder of the Tilsit Schiller Society and the Tilsit Art Society.

After finishing his secondary school education in Tilsit, Theodor Zermelo studied mainly history and geography, first in Königsberg and then in Berlin. Having attained his doctorate in 1856 with a dissertation in history, he completed his studies with the exams *pro facultate docendi*, i. e., his state exams for teaching at secondary schools, which allowed him to teach history and geography as main subjects and Greek and Latin as subsidiary ones; a later supplementary examination also permitted him to teach French. He refrained from a further examination in the natural sciences, but completed an additional diploma in mathematics. In its evaluation of 8 March 1857, the Königlich-Wissenschaftliche Prüfungskommission (Royal Academic Examination Board) found him to have "an adequate knowledge of the elements of geometry and algebra for teaching mathematics in the lower grades." It further observes: "As concerns his philosophical knowledge, he has a solid grasp of the basic notions of logic and general grammar (allgemeine Grammatik) and

---

[5]In a *curriculum vitae* written after 1933 (UAF, C 129/273) Zermelo counts not only Ehrenberg, but also the physicist Arnold Sommerfeld among his relatives.

shows an assured competence in their application." Theodor Zermelo became a teacher at the Friedrichswerder Gewerbeschule in Berlin, a secondary school that was vocationally oriented and did not have Latin and Greek as subjects like the traditional *Gymnasium*. In 1886 he was appointed *Gymnasialprofessor*.

Zermelo's parents Maria Auguste and Theodor Zermelo

Altogether, Ernst Zermelo grew up in a family with an academic background, but not necessarily with an orientation towards mathematics and the sciences. His sisters Anna and Elisabeth became painters, his sister Margarete a teacher of history and modern languages. Too little information is available about Zermelo's childhood and youth to come to reliable conclusions concerning his education. As his mother died when he was seven years old, her influence was limited to his early childhood. Zermelo described her to his wife Gertrud as a delicate and sensitive woman. She confided many of her thoughts to a diary.

The influence of his father is more obvious. As a teacher, Ferdinand Zermelo appears to have been popular and highly regarded. In a letter which former pupils wrote to congratulate him on his nomination as a professor it says:

> We all are delighted about this event, since each of us has had the good fortune to have been taught by you just a few years ago. We sincerely regret that we now have to do without. We all owe you a great part of our education, and therefore we wish you with all our heart a long and happy life.

Zermelo also respected his father and held him in high esteem. This is illustrated by a letter in which he thanks him for the presents he received on his 17th birthday:[6]

---

[6]Letter of 28 July 1888; MLF. A letter which Zermelo wrote to his father in April (copy in MLF) shows that Theodor Zermelo was not living together with his

Left: Zermelo as a baby with his mother
Right: Zermelo with his parents and his sisters Anna (right) and Elisabeth

Dear father,
Your affectionate congratulations on my birthday yesterday and your presents, which are as generous as ever, caused me great pleasure. Above all I would like to express my heartfelt thanks to you. Your very kind offer of a trip which I certainly do not particularly deserve was a complete surprise to me. I can only hope that your plan will indeed be realised, not only in my own interest, but especially in yours, as it is to depend precisely on your state of health. It goes without saying that I will gladly obey your "conditions." [. . .]

Your obedient son
Ernst. (OV 1.01)

His father's state of health to which Zermelo alludes, got worse and led to his death half a year later, on 24 January 1889. Zermelo and his sisters had become orphans. Thanks to the family's assets, their livelihood was secured for the time being.

Zermelo had a kind heart, as is attested by some of his contemporaries.[7] However, he was also prone to sharp and polemical reactions and did not refrain from trenchant irony when he was convinced of his opinions, both in

---

children at that time; perhaps he was staying in hospital because of his serious illness. Earlier letters (MLF) document that the Zermelo siblings often spent their holidays without their father.

[7]So Wilhelm Süss in a letter to Zermelo of 27 July 1951; MLF.

scientific and in political matters. He did so even if he had to suffer from the consequences, as with his criticism of the Nazis. Whereas the first trait might be attributed to the influence of his mother, his determination to speak his mind may be due to the example of his father. In 1875, Theodor Zermelo published a widely acknowledged[8] paper ([Zert75]) on the historian and philologist August Ludwig von Schlözer (1735–1809) of Göttingen, the most influential anti-absolutist German publicist of the early age of enlightenment.[9] It was his aim to raise awareness of those values expressed by Schlözer's plea for free speech, sincerity, tolerance, and humanity on which, in his opinion, the newly founded German Reich should be built. There are indications that the views of his father as reflected in this treatise and the example of von Schlözer had their effect on Zermelo. In the copy of the treatise, which he owned, many passages are underlined, among them those that concern the determined defence of one's own convictions. In a 1935 letter to his sisters, which describes his attitude towards the Nazis (4.11.3), he argues in the spirit of his father. Moreover, he used to talk about von Schlözer with his wife Gertrud, thereby arousing her interest in von Schlözer's daughter Dorothea (1770–1825), known as "the loveliest incarnation of enlightenment." In 1787, at the age of 17, she was the first woman to receive a Ph. D. at the University of Göttingen.[10]

Zermelo shared with his father an interest in poetry. As a young teacher, Theodor Zermelo had compiled a large selection of his own translations of poems from England, the US, France, and Italy under the title "Aus der Fremde" ("From Foreign Lands"). Zermelo was later to translate parts of the Homeric epics, and he enjoyed commenting on daily events in the form of poems. In 1885, at the age of thirteen, he made "a metrical translation from the first book of Virgil's Aeneid" as a birthday present for his father (MLF).

1.2 School and University Studies

In 1880 Ernst Zermelo entered the Luisenstädtisches Gymnasium in Berlin, situated in Brandenburgstraße (today Lobeckstraße).[11] Its name refers to the neighbourhood Luisenstadt honouring the Prussian Queen Luise Auguste Wilhelmine Amalie (1776–1810), the wife of King Friedrich Wilhelm III (1770–1840). The Luisenstädtisches Gymnasium had been opened on 10 October 1864, after the Berlin authorities had decided to establish new secondary schools and to extend existing ones in a reaction to the large increase in the population of Berlin.[12] Teaching started in the autumn of 1864 with 86 stu-

---

[8]Cf. *Brockhaus' Konversationslexikon*, 14th edition, 14th Volume (1895), 525.

[9]Cf., e. g., [Pet03].

[10]Cf. [Ker88] or [Kue76].

[11]The certificates relating to Zermelo's education at school and university are held in MLF.

[12]*Erster Jahresbericht über das Luisenstädtische Gymnasium in Berlin*, Berlin 1865, 33–34.

Zermelo as a schoolboy with his sisters Anna (right) and Elisabeth

dents. When Zermelo entered the school it had 587 students with 50 students in each entry class.[13] After the turn of the century the number of students in Luisenstadt decreased steadily. The school was eventually closed at its first site and transferred further to the west. The building was then destroyed in World War II.

Shortly after the death of his father, Zermelo finished school. His school leaving certificate (*Zeugnis der Reife*), dated 1 March 1889, certifies good results in religious education, German language, Latin, Greek, history, geography, mathematics, and physics. His performance in French was adequate. The comment in mathematics says that he followed the instructions with good understanding and that he has reliable knowledge and the ability to use it for skilfully solving problems. In physics it is testified that he is well-acquainted with a number of phenomena and laws. Under the heading "Behaviour and Diligence" it is remarked that he followed the lessons with reflection, but

---

[13] *Sechzehnter Jahresbericht über das Luisenstädtische Gymnasium in Berlin,* Berlin 1881, 7.

that he occasionally showed a certain passivity as a result of physical fatigue, a clear indicator that Zermelo already suffered from poor health during his school days. The certificate furthermore informs us that he was exempted from oral examinations and that he was now going to take up university studies in mathematics and physics.

In the summer semester 1889 Zermelo began his course of studies, which he did mainly at the Friedrich Wilhelm University (now Humboldt University) in Berlin, but also for one semester each at Halle-Wittenberg and Freiburg, and finally, after his Berlin examinations, at Göttingen.

When Zermelo's father died in 1889, he left behind six children who were still minors, now under the guardianship of Amtsgerichtsrat (judge of a county court) Muellner. On 24 December 1891 Muellner testified that the father's estate was needed to provide for the younger siblings, thereby helping Zermelo get a grant to finish his university studies. Zermelo was awarded grants from the Moses Mendelssohn Foundation (probably 1891/92) and the Magnus Foundation (1892/93).[14] Both foundations belonged to the Friedrich Wilhelm University, set up to support gifted students. The first was named after Moses Mendelssohn (1729–1786), the Berlin philosopher of German enlightenment, the second after the physicist Gustav Magnus (1802–1870) who is regarded as the progenitor of Berlin physics. The Magnus Foundation was the most important one in the second half of the 19th century. Gustav Magnus' widow had donated 60000 Marks to the University. The interest allowed the support of two excellent students of mathematics or the sciences with an annual grant of 1200 Marks, a significant sum at that time. Karl Weierstraß himself had outlined and defended its statutes ([Bier88], 126). As stipulated, Zermelo had to pass special examinations in order to prove that he had the level of knowledge expected at that stage of his studies. Two testimonials in mathematics, one for the Mendelssohn Foundation signed by Lazarus Fuchs and one for the Magnus Foundation signed by Hermann Amandus Schwarz, can be found among his papers.

In Berlin Zermelo enrolled for philosophy, studying mathematics, above all with Johannes Knoblauch, a student of Karl Weierstraß, and with Fuchs. Furthermore, he attended courses on experimental physics and heard a course on experimental psychology by Hermann Ebbinghaus. In the winter semester 1890/91 he changed to Halle-Wittenberg where he enrolled for mathematics and physics. There he attended Georg Cantor's courses on elliptical functions and number theory, Albert Wangerin's courses on differential equations and on spherical astronomy, but also a course given by Edmund Husserl on the philosophy of mathematics at the time when Husserl's *Philosophie der Arithmetik* ([Hus91a]) was about to be published. Zermelo also took part in the course on logic given by Benno Erdmann, one of the leading philosophical (psychologistic) logicians of that time. The next semester saw him at the University of Freiburg studying mathematical physics with Emil Warburg, analytical

---

[14]Cf. [Bier88], 126.

geometry and the method of least squares with Jakob Lüroth, experimental psychology with Hugo Münsterberg, and history of philosophy with Alois Riehl. Furthermore he attended a seminar on Heinrich von Kleist.

Zermelo then returned to Berlin. Between the winter semester 1891/92 and the summer semester 1894 he attended several courses by Max Planck on theoretical physics, among them a course on the theory of heat in the winter semester 1893/94. He also attended a course on the principle of the conservation of energy by Wilhelm Wien (summer semester 1893). In mathematics he took part in courses on differential equations by Fuchs, algebraic geometry by Georg Ferdinand Frobenius, and non-Euclidean geometry by Knoblauch. The calculus of variations, the central topic of Zermelo's later work, was taught by Hermann Amandus Schwarz in the summer semester 1892. In philosophy he attended "Philosophische Übungen" by the neo-Kantian Friedrich Paulsen and Wilhelm Dilthey's course on the history of philosophy. He also took a course on psychology, again by Hermann Ebbinghaus (winter semester 1893/94).

Zermelo's Ph. D. thesis *Untersuchungen zur Variations-Rechnung* (*Investigations On the Calculus of Variations*) (cf. 1.3) was suggested and guided by Hermann Amandus Schwarz. Schwarz, who had started his studies in chemistry, was brought over to mathematics by Weierstraß and was to become his most eminent student. In particular, he aimed at "keeping a firm hold on and handing down the state of mathematical exactness which Weierstraß had reached" ([Ham23], 9). In 1892, then a professor in Göttingen, he was appointed Weierstraß' successor in Berlin. Zermelo became his first Ph. D. student there.

On 23 March 1894 Zermelo, then 22 years old, applied to begin the Ph. D. procedure.[15] Schwarz and the second referee Fuchs delivered their reports in July 1894. The oral examination took place on 6 October 1894. It included the defence of three theses that could be proposed by the candidate. Zermelo had made the following choice ([Zer94], 98):

I. In the calculus of variations one has to attach importance to an exact definition of maximum or minimum more than has been done up to now.[16]
II. It is not justified to confront physics with the task of reducing all phenomena in nature to the mechanics of atoms.[17]
III. Measuring can be understood as the everywhere applicable means to distinguish and to compare continuously changing qualities.[18]

---

[15] Details follow the files of Zermelo's Ph. D. procedure in the archives of Humboldt University Berlin and are quoted from [Thie06], 298–303.

[16] "In der Variations-Rechnung ist auf eine genaue Definition des Maximums oder Minimums grösserer Wert als bisher zu legen."

[17] "Mit Unrecht wird der Physik die Aufgabe gestellt, alle Naturerscheinungen auf Mechanik der Atome zurückzuführen."

[18] "Die Messung ist aufzufassen als das überall anwendbare Hilfsmittel, stetig veränderliche Qualitäten zu unterscheiden und zu vergleichen."

The latter two theses, in particular the second one, are clearly aimed against early atomism in physics and the mechanical explanations coming with it. They will gain in substance within the next two years and lead to a serious debate with Ludwig Boltzmann about the scope of statistical mechanics (cf. 1.4).

After having received his doctorate, Zermelo got a position as an assistant to Max Planck at the Berlin Institute for Theoretical Physics from December 1894 to September 1897.[19] During this time he edited a German translation ([Gla97]) of Richard Tetley Glazebrook's elementary textbook on light ([Gla94]). He also prepared for his exams *pro facultate docendi*, which he passed successfully on 2 February 1897. In philosophy he wrote an essay entitled "Welche Bedeutung hat das Prinzip der Erhaltung der Energie für die Frage nach dem Verhältnis von Leib und Seele?" ("What is the Significance of the Principle of the Conservation of Energy for the Question of the Relation Between Body and Mind?").[20] According to the reports Zermelo was well-informed about the history of philosophy and systematical subjects. He was examined in religious education (showing excellent knowledge of the Holy Bible and church history) and in German language and literature. As to his exams in mathematics the report says that although he was not always aware of the methods in each domain, it was nevertheless evident that he had acquired an excellent mathematical education. He passed the physics part excellently. The exams in geography showed that he was excellent in respect to theoretical explanations, but that he was not that affected by "studying reality." Expressing its conviction that he would be able to close these gaps in the future, the committee allowed him to teach geography in higher classes. Finally it was certified that he had the knowledge for teaching mineralogy in higher classes and chemistry in the middle classes.

Already in the summer of 1896 Zermelo applied for a position as assistant at the Deutsche Seewarte in Hamburg, the central institution for maritime meteorology of the German Reich. The application was supported by Hermann Amandus Schwarz who reported on 20 July 1896 that he knew Zermelo from mathematics courses at Berlin University, from a private course ("Privatissimum"), and as supervisor of his dissertation. He expressed his conviction that Zermelo had excellent skills for investigations in theoretical mathematics. Planck confirmed on 7 July 1896 (MLF) that he was extraordinarily satisfied with Zermelo's achievements where he "made use of his special mathematical talent in an utmost conscentious manner" (OV 1.02).

Finally, however, Zermelo continued his academic career, taking up further studies at the University of Göttingen. He enrolled for mathematics on 4 November 1897. In the winter semester 1897/98 he heard David Hilbert's course on irrational numbers and attended the mathematical seminar on differential equations in mechanics directed by Hilbert and Felix Klein. He took

---

[19] Assessment from Max Planck of 3 November 1897; MLF.

[20] Draft in shorthand in UAF, C 129/270.

exercises in physics with Eduard Riecke and heard thermodynamics with Oskar Emil Meyer. The next semester he took part in a course on set theory given by Arthur Schoenflies and in Felix Klein's mathematical seminar.

Looking back to Zermelo's university studies, one may emphasize the following points:

He acquired a solid and broad knowledge in both mathematics and physics, his main subjects. His advanced studies directed him to his early specialities of research, to mathematical physics and thermodynamics, and to the calculus of variations, the topic of his dissertation.

He attended courses by Georg Cantor, but no courses on set theory, neither in Halle[21] nor at other places, at least until the summer semester 1898 when Schoenflies gave his course in Göttingen. Furthermore, he had no instruction in mathematical logic at a time, however, when German universities offered this subject only in Jena with Gottlob Frege and in Karlsruhe with Ernst Schröder. He got his logical training from Benno Erdmann, who defined logic as "the general, formal and normative science of the methodological preconditions of scientific reasoning" ([Erd92], 25).

Given his teachers in philosophy, Riehl, Paulsen, Dilthey, and Erdmann, it can be assumed that he had a broad overall knowledge of philosophical theories. In Halle he even got acquainted with Husserl's phenomenological philosophy of mathematics *in statu nascendi*. His interest in experimental psychology as taught by Hermann Ebbinghaus and Hugo Münsterberg is quite evident.

1.3 Ph. D. Thesis and the Calculus of Variations

We recall that Zermelo's Ph. D. thesis *Untersuchungen zur Variations-Rechnung* (*Investigations on the Calculus of Variations*) was guided by Hermann Amandus Schwarz. It may have been inspired by Schwarz's lectures in the summer semester of 1892. Schwarz proposed to generalize methods and results in the calculus of variations, which Weierstraß had obtained for derivations of first order, to higher derivations.[22]

The calculus of variations treats problems of the following kind: Given a functional J from the set M into the set of reals, what are the elements of M for which J has an extremal value? The transition from J to $-J$ shows that it suffices to treat either minima or maxima. In a classical example, a special form of the so-called *isoperimetric problem*, M is the set of closed curves in

---

[21]Cantor lectured on set theory only in the summer semester 1885 in a course entitled "Zahlentheorie, als Einleitung in die Theorie der Ordnungstypen" ("Number Theory, as an Introduction to the Theory of Order Types") (information by Rüdiger Thiele; cf. also [Gra70], 81, or [PurI87], 104).

[22]Also here we follow the files of Zermelo's Ph. D. procedure in the archives of Humboldt University Berlin as quoted in [Thie06], 298–303.

the plane of a fixed given perimeter, and J maps a curve in M to the area it encloses. The problem asks for the curves enclosing an area of maximal size. In Zermelo's Ph. D. thesis M is a set of curves and J the formation of integrals along curves in M for a given integrand which may now be of a more general kind. In Zermelo's own words ([Zer94], 24):

[Given] an integral

$$J = \int_{t_1}^{t_2} F\, dt,$$

[...] our task is the following: If

$$F\left(x^{(\mu)}, y^{(\mu)}\right) = F\left(x, x', .. x^{(n)}; y, y', .. y^{(n)}\right)$$

is a function which is analytical in all arguments and which has the character of an entire function in the domain under consideration and obeys the integrability conditions developed in the first section, we search in the totality A of all curves

$$x = \varphi(t), \quad y = \psi(t)$$

which satisfy certain conditions, a special curve a for which the value of the definite integral

$$J = \int_{t_1}^{t_2} F\left(x^{(\mu)}, y^{(\mu)}\right) \quad \left(x^{(\mu)} = \frac{d^\mu x}{dt^\mu}, \ y^{(\mu)} = \frac{d^\mu y}{dt^\mu}\right)$$

along the curve between certain boundaries is smaller than for all neighbouring curves $\bar{a}$ of the same totality A.

The integrability conditions to which Zermelo refers ensure that the value of the integral does not depend on the parametrization of the curve in question.

The case $n = 1$ had been solved by Weierstraß and treated in his courses, in particular in the course given in the summer semester 1879.[23] The latter one had been written up under the auspices of the Berlin Mathematical Society. Edmund Husserl had also taken part in this project.[24]

Zermelo studied the lecture notes in the summer of 1892, when he attended Schwarz's course on the calculus of variations ([Zer94], 1). Less than two years later he succeeded in solving the task Schwarz had set. In his report of 5 July 1884 Schwarz describes the subject as "very difficult;" he is convinced that Zermelo provided the very best solution and predicts a lasting influence of the methods Zermelo had developed and of the results he had obtained:

---

[23] A comprehensive version of Weierstraß' lectures has been edited as [Wei27]; cf. also [Thie06], 183–243, or [Gol80], Ch. 5.

[24] Husserl (1859–1938) studied mathematics, first in Leipzig (1876–1878) and then in Berlin (1878–1880) mainly with Weierstraß. His Ph. D. thesis *Beiträge zur Theorie der Variationsrechnung* (*Contributions to the Theory of the Calculus of Variations*) ([Hus82]) is written in the spirit of Weierstraß; cf. [Thie06], 293–295.

According to my judgement the author succeeds in generalizing the main investigations of Mr. Weierstraß [...] in the correct manner. In my opinion he thus obtained a valuable completion of our present knowledge in this part of the calculus of variations. Unless I am very much mistaken, all future researchers in this difficult area will have to take up the results of this work and the way they are deduced. (OV 1.03)

He marked the thesis with the highest degree possible, *diligentia et acuminis specimen egregium.* Co-referee Fuchs shared his evaluation.

The dissertation starts by exhibiting the integrability conditions mentioned above. Taking them as "a task of interest in itself" ([Zer94], 14), Zermelo develops them in a more general framework than needed for the later applications. In the second part he provides a careful definition of minimum (ibid., 25–29), quite in accordance with the first thesis he had chosen for the oral examination. The notion of minimum results from the conditions which he imposes on the totalities A of (parametrizations of) curves and on the relation "$\bar{a}$ is a neighbouring curve of a."[25] In the final parts he carries out the Weierstraßian programme in the framework thus created.[26]

The dissertation ends (p. 96) with "the first clear formulation and proof of the important envelope theorem,"[27] according to Gilbert A. Bliss ([Bli46], 24) "one of the most interesting and most beautiful theorems in the domain of geometrical analysis." The theorem generalizes the following proposition[28] about geodetic lines to one-parameter families of extremals, likewise providing a criterion for the non-existence of a minimum of the corresponding variational problem among the extremals under consideration:[29]

If $\mathcal{F}$ is a one-parameter family of geodetic lines issuing from a point P of a surface and E an envelope of $\mathcal{F}$, and if, moreover, $G_1 \in \mathcal{F}$ and $G_2 \in \mathcal{F}$ are tangent to E in P_1 and P_2, respectively, and P_1 precedes P_2 on E, then

$$\text{length of } PP_2 = \text{length of } PP_1 + \text{length of } P_1 P_2,$$

where, for example, PP_1 denotes the arc of G_1 between P and P_1, and $P_1 P_2$ denotes the arc on E between P_1 and P_2.

---

[25] Partly, these conditions generalize those of Weierstraß for the case "n = 1" in a natural way, demanding, for instance, that for curves in A the nth derivation exists and is continuous. Over and above that, the curves in A have to have not only the same boundary values, but also the same so-called "boundary osculation invariants," among them as a simple example the torsion.

Analogously, the definition of neighbourhood does not only refer to a natural distance between a and $\bar{a}$, but also to quantities corresponding to the osculation invariants, among them again the torsion.

[26] For technical details see [Thie06], 298–303.

[27] So Hilbert in an assessment of Zermelo of 16 January 1910; SUB, Cod. Ms. D. Hilbert 492, fols. 4/1-2. – For a sketch of the proof see [Gol80], 340–341.

[28] Attributed to Jean Gaston Darboux; cf. [Bolz04], 166, or [Gol80], 339.

[29] Namely, by providing a proof of the necessity of the so-called Jacobi condition; cf., for example, [Bli46], Sect. 10.

In the "triangle" formed by P, P_1, and P_2, the arc P_1P_2 can be replaced by a shorter curve (the envelope is not geodetic). Hence, there is a curve connecting P with P_2 that is shorter than PP_2, i.e., the (geodetic) arc PP_2 does not provide a curve of minimal length connecting P with P_2.

Later, Adolf Kneser provided generalizations of Zermelo's theorem, discussing also the existence of an envelope – a desideratum Zermelo had left open.[30] Zermelo is, however, more general than Kneser in respect to the variational integrand, allowing derivations of arbitrary order. Goldstine speaks of the Zermelo-Kneser results as of the "Zermelo-Kneser envelope theorem," giving credit to Zermelo for the "most elegant argument" ([Gol80], 340).

Various voices confirm the significance of Zermelo's dissertation for the development of the Weierstraßian direction in the calculus of variations.

Adolf Kneser's 1900 monograph on the calculus of variations recommends the dissertation as giving valuable information about the Weierstraßian methods and the case of higher derivatives. In the preface Kneser says ([Knes00], IV):

> The relationship between my work and the investigations of Weierstraß requires a special remark. By his creative activity Weierstraß opened new ways in the calculus of variations. As is well-known, his research is not available in a systematic representation; as the most productive sources I used the Ph. D. thesis of Zermelo and a paper of Kobb[31] [. . .]. In a modified and generalized form they contain all essential ideas of Weierstraß as far as they are related to our topic.

Oskar Bolza (1857–1942), one of the most influential proponents of the calculus of variations, also gives due respect to Zermelo: his epochal monograph ([Bolz09]) contains numerous quotations of Zermelo's work.[32]

Constantin Carathéodory, in his similarly influential monograph on the subject, comments that "in the beginning Weierstraß' method was known to only a few people and opened to a larger public by Zermelo's dissertation" ([Car35], 388).

The quality of the dissertation played a major role when Zermelo was considered for university positions after his *Habilitation*. In 1903 the search committee for an extraordinary professorship[33] of mathematics at the University of Breslau (now Wrocław, Poland) remarks that "[Zermelo's] doctoral dissertation, when it appeared, was taken note of considerably more than

---

[30][Knes1898], 27; [Knes00], §15. Cf. [Gol80], Sect. 7.5, or [Bolz09], §43, for details. For a further generalization of the envelope theorem obtained by Caratheodory in 1923, cf. [Car35], 292–293 and 398.

[31]The two parts of the paper are [Kob92a], [Kob92b].

[32]The monograph is a substantial extension of Bolza's *Lectures in the Calculus of Variations* ([Bolz04]). The latter is rooted in a series of eight talks about the history and recent developments of the subject which he had given at the 1901 meeting of the American Mathematical Society in Ithaca, N. Y.

[33]Corresponding to the position of an associate professor.

usually happens to such writings."[34] In 1909 Zermelo was shortlisted for the succession of Eduard von Weber at the University of Würzburg. The report of the Philosophical Faculty states that "in the calculus of variations his work had a really epoch making influence."[35] In 1913, when Zermelo was chosen for the first place in a list of three for a full professorship in mathematics at the Technical University of Breslau, the report to the minister characterizes his thesis as an investigation "basic (grundlegend) to the development of the calculus of variations."[36]

Zermelo's interest in the calculus of variations never weakened. He gave lecture courses (for example, in the summer semester 1910 at the University of Zurich and in the winter semester 1928/29 at the University of Freiburg) and continued publishing papers in the field.

In the first of these papers ([Zer02c]), he provides an intuitive description of some extensions of the problem of shortest lines on a surface, namely for the case of bounded steepness with or without bounded torsion, illustrating them with railroads and roads, respectively, in the mountains.[37] About 40 years later he will choose this topic as one of the chapters of a planned book *Mathematische Miniaturen* (*Mathematical Miniatures*) under the title "Straßenbau im Gebirge" ("Building Roads In the Mountains").

In the next paper ([Zer03]) Zermelo gives two simplified proofs of a result of Paul du Bois-Reymond ([DuB79b]) which says that, given n and an analytical F, any function y for which $y^{(n)}$ exists and is continuous and which yields an extremum of $\int_a^b F(x, y, \ldots, y^{(n)})dx$, possesses derivatives of arbitrarily high order;[38] the theorem shows that the Lagrangian method for solving the related variational problem which uses the existence of $y^{(2n)}$ and its continuity, does not exclude solutions for which $y^{(n)}$ exists and is continuous. Constantin Caratheodory characterizes Zermelo's proofs as "unsurpassable with respect to simplicity, shortness, and classical elegance."[39]

Roughly, Zermelo's proceeds as follows: Using partial integration according to Lagrange together with a suitable transformation of the first variation according to du Bois-Reymond, he obtains the first variation of the given problem in the form

$$\int_a^b \Lambda(x)\eta^{(n)}(x)dx,$$

where the variation η is supposed to possess a continuous n-th derivation and to satisfy

$$\eta^{(i)}(a) = \eta^{(i)}(b) = 0 \text{ for } 0 \leq i \leq n.$$

---

[34]UAW, signature F73, p. 114.

[35]Archives of the University of Würzburg, UWü ZV PA Emil Hilb (No. 88).

[36]UAW, signature TH 156, p. 14.

[37]Technically, the paper is concerned with so-called discontinuous solutions; cf. [Car35], 400.

[38]Zermelo adds a variant of the second proof which he attributes to Erhard Schmidt.

[39][Car10], 222; [Car54], Vol. V, 305.

The main simplification of du Bois-Reymond's proof consists in showing that Λ is a polynomial in x of a degree $\leq n - 1$.[40] In particular, Λ is arbitrarily often differentiable. It is easy to show that this property is transferred to the solutions y.

Together with Hans Hahn, one of the initiators of linear functional analysis, Zermelo writes a continuation of Adolf Kneser's contribution ([Knes04]) on the calculus of variations for the *Encyklopädie der Mathematischen Wissenschaften*, giving a clear exposition of the Weierstraßian method ([HahZ04]).

Finally, he formulates and solves "Zermelo's navigation problem" concerning optimal routes of airships under changing winds.[41]

1.4 The Boltzmann Controversy

We recall the second thesis that Zermelo defended in the oral examination of his Ph. D. procedure: "It is not justified to confront physics with the task of reducing all phenomena in nature to the mechanics of atoms." Probably he had met the questions framing his thesis in Max Planck's course on the theory of heat in the winter semester 1893/94 and had come to acknowledge Planck's critical attitude against early atomism and the mechanical explanations of natural phenomena accompanying it. During his time as an assistant to Planck (1894–1897) his criticism consolidated, now clearly aiming also against statistical mechanics, in particular against Ludwig Boltzmann's statistical explanation of the second law of thermodynamics. Having a strong mathematical argument and backed by Planck, he published it as aiming directly against Boltzmann ([Zer96a]), thus provoking a serious controversy which took place in two rounds in the *Annalen der Physik und Chemie* in 1896/97. Despite being interspersed with personal sharpness, it led to re-considering foundational questions in physics at a time when probability was showing up, competing with the paradigm of causality.[42]

1.4.1 The Situation

Around 1895 the atomistic point of view, while widely accepted in chemistry, was still under debate in physics. Looking back to this situation, Max Planck remembers ([Pla14], 87):

---

[40]For $n = 1$ this result had already been obtained by du Bois-Reymond himself in 1879 ([DuB79a]) and, with a different proof, by Hilbert in 1899 (cf. [Bolz04], §6). – It is here that Zermelo provides two arguments.

[41][Zer30c], [Zer31a]; cf. 4.2.2.

[42]For a description of the debates on the kinetic theory of gas, also in the context of 19th century cultural development, cf., for example, [Bru66] (containing reprints or English translations of the most important contributions) and [Bru78]. For an analysis of the controversy cf. also [Beh69], [JunM86], 213–217, [Kai88], 125–127, [Uff07], 983–992.

To many a cautious researcher the huge jump from the visible, directly controllable region into the invisible area, from macrocosm into microcosm, seemed to be too daring.

Discussions mainly crystallized in the theory of heat. The so-called energeticists such as Ernst Mach and Wilhelm Ostwald regarded energy as the most fundamental physical entity and the basic physical principles of heat theory as autonomous phenomenological laws that were not in need of further explanation. Among these principles we find the first law of thermodynamics[43] stating the conservation of energy and the second law of thermodynamics[44] concerning the spontaneous transition of physical systems into some state of equilibrium. The latter may be illustrated by a system A consisting of a container filled with some kind of gas and being *closed*, i. e. entirely isolated from its surroundings. As long as temperature or pressure vary inside A, there is a balancing out towards homogeneity. If the universe is considered as a closed system, the equilibrium state or *heat death* it is approaching can be characterized by a total absence of processes that come with some amount of differentiation. The parameter measuring the homogeneity, the so-called *entropy*, can be thought to represent the amount of heat energy that is no longer freely available.

From the early atomistic point of view and the mechanical conceptions accompanying it, physical systems consist of microscopic particles called *atoms* which obey the laws of mechanics. For an "atomist," the principles governing heat theory may no longer be viewed as unquestionable phenomena, but are reducible to the mechanical behaviour of the particles that constitute the system under consideration. In particular, the heat content of the system is identified with the kinetic energy of its particles. Determining its behaviour means determining the behaviour of all its constituents. However, as a rule, the number of atoms definitely excludes the possibility of calculating exactly how each of them will behave. To overcome this dilemma, atomists used statistical methods to describe the expected behaviour at least approximately. For justification they argued that the phenomena of thermodynamics, which we observe in nature, result from the global and "statistical" view, the only view we have at our disposal of the behaviour of the invisible parts lying at their root. Josiah Willard Gibbs, who shaped the mathematical theory of statistical mechanics, describes the new approach in the preface of his classical book ([Gib02], vii–viii):

> The usual point of view in the study of mechanics is that where the attention is mainly directed to the changes which take place in the course of time in a given system. The principal problem is the determination of the condition of the system with respect to configuration and velocities at any required time, when its condition in these respects has been given for some one time, and

---

[43]Henceforth also called "First Law."
[44]Henceforth also called "Second Law."

the fundamental equations are those which express the changes continually taking place in the system. [...]

The laws of thermodynamics, as empirically determined, express the approximate and probable behavior of systems of a great number of particles, or, more precisely, they express the laws of mechanics for such systems as they appear to beings who have not the fineness of perception to enable them to appreciate quantities of the order of magnitude of those which relate to single particles, and who cannot repeat their experiments often enough to obtain any but the most probable results.

According to the last part of this quotation, the statistical view includes a change of paradigm: It replaces causality by probability. In Boltzmann's presentation the probability $W(s)$ of a system A to be in state s is measured by the relative number (with respect to some suitable measure) of the configurations of A which macroscopically represent s. According to this interpretation the Second Law now says that physical systems tend toward states of maximal probability.

Coming back to the system A defined above, let us assume that the container is divided into two parts P_1 and P_2 of equal size with no wall in between. Furthermore, let s be a state where all atoms are in part P_1, and let s' be a state where the gas is homogeneously distributed over both parts. Then the probability $W(s')$ is big and $W(s)$ is low compared to $W(s')$. Therefore, if A is in a state macroscopically represented by s, it will tend toward a state macroscopically represented by s'; in other words, the atoms will tend to distribute homogeneously over the whole container. The entropy $S(s)$ of a state s may now be interpreted as to measure the probability $W(s)$. In fact, it is identified with a suitable multiple of its logarithm:

$$S = k \log W,$$

where k is the so-called Boltzmann constant. The Second Law now provides systems with a tendency towards states of increasing entropy, thereby imposing a direction on the physical processes concerned. Similarly, the distribution of the velocities of their constituents will converge to a final equilibrium distribution, the so-called Maxwell distribution.

1.4.2 The First Round

It was, in particular, Ludwig Boltzmann (1844–1906) who developed the kinetic theory of gas according to the principles of statistical mechanics, thereby providing the statistical interpretation of the Second Law sketched above.[45] In the mid 1890s, attacked already by energeticists and other non-atomists, he

---

[45]The sketch does not take into consideration the development of the respective notions in Boltzmann's work itself and the way they differ from those in [Gib02]. For respective details cf. [Uff04], [Uff07], or [Kac59], Ch. III.

Ludwig Boltzmann in 1898
Courtesy of Österreichische Nationalbibliothek

found himself confronted with a serious mathematical counterargument written up by Zermelo in December 1895 in the paper entitled "Ueber einen Satz der Dynamik und die mechanische Wärmetheorie" ("On a Theorem of Dynamics and the Mechanical Theory of Heat") ([Zer96a]), the *Wiederkehreinwand* or *recurrence objection*. For an account we return to the system A described above. Its microscopic description will only depend on the spatial coordinates of its atoms together with their velocities. Under reasonable assumptions which Boltzmann had taken for granted, the function describing these data in dependence of time falls under the so-called *recurrence theorem* proved by Poincaré in 1890:[46] System A will infinitely often approach its initial state. Hence, one may argue, also the entropy, depending only on states, will infinitely often approach the initial entropy. It thus cannot constantly increase. Moreover, the velocities of the atoms of A will not tend to a final distribution. To be fully exact, one has to assume that the initial state of A is different from some exceptional states for which Poincaré's argument does not work. These states, however, are "surrounded" by non-exceptional ones.[47] So there really seems to be a contradiction.

---

[46][Poi90], esp. Section 8, "Usage des invariants intégraux," 67–72.

[47]In precise terms the exceptional states form a set of measure zero.

Zermelo starts his paper with a lucid proof of the recurrence theorem[48] and then presents the preceding argument. To avoid the inconsistency from the statistical point of view, he continues ([Zer96a], 492), one would have to assume that nature always realizes one of the exceptional initial states that do not lead to recurrence. However, as each exceptional state has non-exceptional ones arbitrarily near, such an assumption would contradict

> the spirit of the mechanical view of nature itself which will always force us to assume that all *imaginable* mechanical initial states are physically *possible*, at least within certain boundaries.[49]

He, therefore, pleads for a modification of the mechanical model underlying the theory and *expressis verbis* claims that Boltzmann's probabilistic deduction of the Maxwell distribution[50] could no longer be maintained. He confidently concludes (ibid., 494):

> Because of the difficulties of the subject I have temporarily refrained from giving a detailed examination of the various attempts at such a deduction, in particular of those of Boltzmann and Lorentz. Rather I have explained as clearly as possible what appears to me as being rigorously provable and of fundamental importance, thereby contributing to a renewed discussion and final solution of these problems.

Where does Zermelo's self-confidence come from?

There is a first point: As he opens his paper with a clear proof of Poincaré's theorem, one may guess that the mathematical facts strengthened his conviction. According to Pierre Brémaud ([Bre98], 133) "he held a strong position" with his "striking and seemingly inescapable argument." But there was a second, likewise important point. As mentioned above, supporters of atomism such as Boltzmann together with the majority of chemists, were opposed by influential chemists and physicists such as Ostwald and Mach. Among the opponents we also find Planck, however with a different reason. Whereas Ostwald and Mach were not willing to accept a principal difference between the irreversible process of the transition of heat from a hot to a cold body and the reversible process of a body swinging from a higher level to a lower one, Planck was convinced of the fundamental difference. He clearly distinguished between the character of the First Law stating the conservation of energy and that of the Second Law involving the existence of irreversible processes. So he should have supported Boltzmann in the debate with Ostwald. But he didn't.

---

[48] Poincaré and Zermelo did not have measure theory at their disposal and performed the measure-theoretic arguments, which are required for the proof, in a naive way. The first rigorous proof of the recurrence theorem was given about 20 years later by Carathéodory ([Car19]).

[49] Quotations from [Zer96a], [Bol96], [Zer96b], and [Bol97a] are based on the English translations given in [Bru66].

[50] Apparently Zermelo does not refer to one specific deduction, but to various ones Boltzmann performed, for example, in [Bol68], [Bol72], [Bol77].

In his "Personal Reminiscences" he makes this disagreement clear ([Pla49b], 12–13):

> For Boltzmann knew very well that my point of view was in fact rather different from his. In particular he got angry that I was not only indifferent towards the atomistic theory which was the basis of all of his research, but even a little bit negative. The reason was that during this time I attributed the same validity without exception to the principle of the increase of entropy as to the principle of the conservation of energy, whereas with Boltzmann the former principle appears only as a probabilistic law which as such also admits exceptions.

Zermelo being his assistant, Planck of course got acquainted with the recurrence argument and – as we shall see in a moment – supported (or maybe even initiated) its publication.[51] This is evidence that Zermelo should not only be seen in the role of a capable mathematician who forwarded a strong mathematical argument, but also in the role of Planck's mouthpiece against Boltzmann's atomistic theories.

Boltzmann knew about this situation. Therefore he took Zermelo's paper seriously and felt himself compelled to give an immediate answer ([Bol96]).[52] In his reply he defends his theory by arguing that the Second Law as well as the Maxwell law about the distribution of velocities would in itself be subject to probability (an argument that, as just described, asked for Planck's criticism). The increase of entropy was not absolutely certain and a decrease or a return to the initial state was possible, but so improbable that one was fully justified in excluding it. Zermelo's paper would demonstrate that this point had not been understood. Boltzmann's résumé is definite (ibid., 773): "Poincaré's theorem, which Zermelo explains at the beginning of his paper, is clearly correct, but its application to the theory of heat is not." The contradiction Zermelo thought he had found could only apply if he had been able to prove that a return to the initial state (or its neighbourhood) had to happen within an observable time. He compares Zermelo with a dice player who calculates that the probability of throwing one thousand consecutive ones is different from zero and, faced with the fact that he never met such an event, concludes that his dice is loaded. Boltzmann summarizes (ibid., 779-780):

> If one considers heat to be molecular motion which takes place according to the general equations of mechanics, and assumes that the complexes of

---

[51]For the debate between Boltzmann and Planck himself cf. [HolZ85], 19–28.

[52]Three years earlier Poincaré had already formulated the recurrence objection as one of the major difficulties when bringing together experimental results and mechanism ([Poi93a], 537). However, there had not been a reaction by physicists. Also Zermelo is not aware of Poincaré's argument; in his paper he says ([Zer96a], 485) that "apparently Poincaré did not notice the applicability [of the recurrence theorem] to systems of molecules or atoms and, hence, to the mechanical theory of heat." More than a decade later the encyclopedic contributions [BolN07] and [EE11] still attribute the recurrence objection only to Zermelo.

bodies that we observe are at present in very improbable states, then one can obtain a theorem which agrees with the Second Law for all phenomena observed up to now.

Full of self-confidence like Zermelo, but more stridently, he ends (ibid., 781–782):

> All the objections raised against the mechanical viewpoint are therefore meaningless and based on errors. Those however, who cannot overcome the difficulties offered by the clear comprehension of the gas-theoretic theorems should indeed follow the suggestion of Mr. Zermelo and decide to give up the theory entirely.

In his "Personal Reminiscences" Planck describes the impression he got from Boltzmann's reply ([Pla49b], 13):

> In any case he answered the young Zermelo with scathing sharpness a part of which also hit me because actually Zermelo's paper had appeared with my permission.

As a result of this sharpness the exchange of arguments which followed saw a more polemical tone also on Zermelo's side.

1.4.3 The Second Round

As explained above, Planck had rejected Boltzmann's theory because of the resulting probabilistic character of the Second Law. So he did not agree with his arguments either. As Zermelo shared Planck's opinion, it is not surprising to see him publishing a reply to Boltzmann's reply ([Zer96b]). To begin with, he considers Boltzmann's comments rather as a confirmation of his view than as a refutation. For Boltzmann had acknowledged that in a strict mathematical sense the behaviour of a closed system of gas molecules was essentially periodic and, hence, not irreversible. He thus had accepted just the point which he, Zermelo, had had in mind when writing his paper. Boltzmann's claim to be in correspondence with the Second Law if the system under consideration was in an improbable state suffered from not providing a *mechanical* explanation for the fact that the systems one observes are always in a state of growing entropy (ibid., 795):

> It is not admissible to accept this property simply as a fact for the initial states that we can observe at present, for it is not a certain unique variable we have to deal with (as, for example, the eccentricity of the earth's orbit which is just decreasing for a still very long time) but the entropy of *any arbitrary* system free of external influences. How does it happen, then, that in such a system there always occurs only an *increase* of entropy and *equalization* of temperature and concentration differences, but never the reverse? And to what extent are we justified in expecting that this behaviour will continue, at least for the immediate future? A satisfactory answer to these questions must be given in order to be allowed to speak of a truly mechanical analogue of the Second Law.

Among further objections, he puts forward as "*a priori* clear" (ibid., 799) that
the notion of probability did not refer to time and, hence, could not be used to
impose a direction on physical processes as is the case with the Second Law. He
thereby touches Josef Loschmidt's *Umkehreinwand* or *reversibility objection*
([Los76], 139): As the mechanical laws are invariant under the inversion of
time, a deduction based on these laws and yielding an increase of entropy in
time t should also yield an increase in time $-t$, i. e. a *decrease* in time t.[53]

Faced with the Second Law as a well-established empirical fact and the in-
sufficiencies of the mechanical theory of gas he had exhibited, he condenses the
essence of his criticism in the following methodological conclusion ([Zer96b],
793–794):

> As for me (and probably I am not alone in this opinion), I believe that a
> single universally valid principle summarizing an abundance of established
> experimental facts according to the rules of induction, is more reliable than
> a *theory* which by its nature can never be directly verified; so I prefer to
> give up the *theory* rather than the *principle*, if the two are incompatible.

Already on 16 December 1896 Boltzmann completes a reply ([Bol97a]) where
he systematically answers the questions Zermelo had raised, among them the
request to explain *mechanically* why the systems observed should be in an
improbable state. Boltzmann's explanation is based on what he calls "Assumption A" (ibid., 392): Viewed as a mechanical system, the universe (or
at least an extended part of it surrounding us) has started in a highly improbable state and still is in such a state. Therefore, any system which is
separated from the rest of the world, as a rule will be in a highly improbable
state as well, and as long as it is closed, will tend to more probable states.
Hence, if the mechanical theory of heat is applied not to *arbitrary* systems,
but to systems which are part of our actual world, it will be in accordance
with our experience. Concerning a justification of "the of course unprovable"
Assumption A, he basically repeats an argument from his first answer (ibid.,
392–393):

> Assumption A is the physical explanation, comprehended according to the
> laws of mechanics, of the peculiarity of the initial states, or better, it is a
> uniform viewpoint corresponding to these laws, which allows one to predict
> the type of peculiarity of the initial state in any special case; for nobody
> will demand an explanation of the final explanatory principle itself.

At the end he varies his model, thereby taking care of the reversibility objection as well: The universe may already be in a state of equilibrium, of heat
death; however, there might be probabilistic fluctuations, i. e, relatively small
areas, possibly as large as our "space of stars" – *single worlds* (*Einzelwelten*)
in his terminology – which are in an improbable state, among them worlds
which presently tend to more probable states and worlds which tend to still

---

[53]For time reversal cf., for example, [Gre04], 146–149 and 506–507.

less probable ones. Thus the universe might realize both directions of time and in this sense does not prefer one of them; in other words ([BolN07], 522):

> The statistical method shows that the irreversibility of processes in a relatively small part of the world or, under special initial conditions also in the whole world, certainly is compatible with the symmetry of the mechanical equations with respect to the positive and the negative direction of time.

Faced with the possibility of successful mechanical explanations of a series of phenomena, certain discrepancies between the phenomenological view and the mechanical method (for example, with respect to "uncontrollable questions" such as that for the development of the world for a very long time) should not be taken to justify Zermelo's demand for a methodological change. For just such discrepancies would support the view that ([Bol97a], 397)

> the universality of our mental pictures will be improved by studying not only the consequences of the Second Law in the Carnot-Clausius version, but also the consequences in the mechanical version.

There was no further reply by Zermelo.

1.4.4 After the Debate

For ten more years, Zermelo maintained interest in statistical mechanics. In his *Habilitation* address ([Zer00a]), given 1899 in Göttingen, he offered a solution of a problem which he had mentioned at the end of his rejoinder to Boltzmann ([Zer96b], 800): On the one hand, Boltzmann applied his statistical view to each state of a system developing in time, whereas, on the other hand, these states were not independent from each other, but were intimately linked by the laws of mechanics. In order to obtain results which are "mathematically reliable at least in the sense of probability theory" ([Zer00a], 1), Zermelo pleads for a notion of probability that takes this problem into account,[54]

---

[54] Let S be a system of n particles in, say, 3-dimensional space whose possible states are given by the coordinates $x = (x_i)_{i \leq 3n}$ and by the components $m = (m_i)_{i \leq 3n}$ of the momenta of its particles and whose behaviour can be described by Hamiltonian equations. The totality of states $s = (x, m)$ fills a body Q of finite contents $|Q|$. Let s be a state of S and q a cuboid in Q of contents $|q|$. Zermelo considers the probability $p(s, q) := |q|/|Q|$ for s belonging to q. This notion of probability is independent of physical developments in the following sense: If $t_0 < t_1$ and s is considered as an initial state at time t_0 which develops into state $s(t_1)$ at time t_1 and if, moreover, q is considered as a set of initial states at time t_0 which develops into $q(t_1)$ at time t_1, then, by Liouville's theorem, $|q| = |q(t_1)|$ and, hence, $p(s, q) = |q|/|Q| = |q(t_1)|/|Q| = p(s(t_1), q(t_1))$. – Similar relations between probability and phase space volume were well-known at that time, originating, e. g., in [Bol68]. However, so Massimiliano Badino (written communication), Zermelo's paper "is terser and much clearer than Boltzmann's" and "expresses an important critique of the use of probability in physics."

and ends by treating mean values and promising a deduction of the Maxwell distribution of velocities "without any further hypothesis and solely based on the definition of probability given here" (ibid., 4). The promise remained unfulfilled, perhaps for the reason that soon statistical mechanics would find a representation along similar lines in Gibbs' monograph from which we have quoted above. As described below, Zermelo appreciated it to such an extent that he provided a German translation.

Some of the members of the Göttingen *Habilitation* committee comment also on Zermelo's anti-Boltzmann papers.[55] Hilbert states that Zermelo's recurrence objection is "very remarkable and found the attention it deserves." Felix Klein is more critical: In connection with "a certain tendency to one-sided criticism" which he sees in Zermelo, he emphasizes

> that the two notes directed against Boltzmann are very astute and, in particular, that one has to acknowledge the courage by which the young author opposes a man like Boltzmann, but that, finally, it was Boltzmann who was right; it must be said, however, that Boltzmann's deep and original way of looking at these things was difficult to understand. (OV 1.04)

Woldemar Voigt, professor of mathematical physics, who follows Klein's assessment, adds that

> already the intensive preoccupation with the highly interesting papers of Boltzmann, which are not really taken into account by the majority of physicists, represents some kind of achievement, and that Dr. Zermelo's criticism undoubtedly attributed to a clarification of the difficult subject. (OV 1.05)

Six years later Zermelo published his German translation ([Gib05]) of Gibbs' *Elementary Principles* ([Gib02]) which "played an important role in making Gibbs' work known in Germany" ([Uff04], 1.2). In the preface he appreciates Gibbs' book ([Gib05], iii) as being

> the first attempt to develop strictly and on a secure mathematical basis the statistical and the probability theoretical considerations in mechanics, as they are indispensable in various areas of physics, in particular in the kinetic theory of gas, independent of their applications.

However, despite the success of the statistical method, his reservations concerning the range of statistical mechanics are not settled (ibid., iii):

> I intend to work out soon and elsewhere the objections I have, in particular, against the considerations in the twelfth chapter as far as they attribute to the mechanical systems a tendency towards a state of statistical balance and, based on this, a full analogy to thermodynamical systems in the sense of the Second Law.

---

[55]UAG, Philosophische Fakultät 184a, 149seq.

The elaboration appeared in his review of Gibbs' book, where he repeats his objections from 1896, in particular pointing to the irreversibility problem ([Zer06a], 241):

> Neither Boltzmann's nor Gibbs' deductions were able to shake my conviction that – also in the sense of probability theory – a kinetic theory of heat can be brought together with the Second Law only if one decides to base it not on Hamilton's equations of movement, but on differential equations already containing the principles of irreversibility.

Contrary to Zermelo's scepticism, Boltzmann's views gained general acknowledgement. Already at the end of 1900 even Planck *expressis verbis* used Boltzmann's probabilistic approach (without instantly accepting the atomistic hypothesis). He deduced an improved radiation formula by working with a probabilistic interpretation of the distribution of oscillator frequencies, i.e. "by introducing probability considerations into the electromagnetic theory of radiation, whose importance for the second law of thermodynamics was first discovered by Mr. L. *Boltzmann*" ([Pla00], 238). Finally Planck fully accepted Boltzmann's atomism ([Pla14], 87):

> It was L. Boltzmann who reduced the contents of the Second Law and, hence, the totality of irreversible processes, whose properties had caused unsurmountable difficulties for a general dynamic explanation, to their true root by introducing the atomistic point of view.

As a matter of course, it was Boltzmann who was asked by Felix Klein to contribute an article on the kinetic theory in the *Encyklopädie der Mathematischen Wissenschaften*. Boltzmann remembers ([Bol05b], 406):

> When Klein asked me for an encyclopedic article, I refused for a long time. Finally he wrote: 'If you do not do it, I will give it to Zermelo.' Zermelo represents a view that is just diametrically opposed to mine. That view should not become the dominant one in the encyclopedia. Hence, I answered immediately: 'Rather than this is done by that Pestalutz, I'll do it.'[56]

The finished article ([BolN07]) discusses also the Boltzmann-Zermelo controversy, repeating essential parts of Boltzmann's arguments (ibid., 519–522). Boltzmann himself did not see it appear; mentally weakened he committed suicide in 1906.[57]

Already in his first answer to Zermelo, Boltzmann had illustrated by an example that the recurrence time in Poincaré's theorem should be far greater than the age of the universe even for systems comprising only a small number

---

[56]Boltzmann refers to a minor figure in Friedrich Schiller's drama *Wallenstein's Tod* (*The Death of Wallenstein*). Pestalutz does not appear on the stage. When three conspirators are haggling over who is to assassinate Wallenstein, their ringleader persuades the other ones to perform the assassination by threatening to turn to Pestalutz in case they would not.

[57]Cf., e. g., [Lind01] for Boltzmann's life.

of particles. Paul and Tatiana Ehrenfest were able to support his arguments and substantially contribute to a more precise elaboration ([EE07]). They conceived a simple model consisting of two adjacent containers connected by an opening and filled with gas atoms, per time unit allowing one atom, chosen by chance, to change its container. By easy calculations they obtained the somewhat paradoxical facts that this model meets Boltzmann's assertions, at the same time being invariant under a reversion of time. They thus showed (ibid., 314) that neither the recurrence objection nor the reversibility objection are sufficient or suitable to refute Boltzmann's theory, in other words, that "it is always possible to reconcile both time reversibility and recurrence with 'observable' irreversible behaviour" ([Kac59], 73).[58] So in 1908 Planck could write ([Pla08], 41):

> Nature simply prefers more probable states to less probable ones by performing only transitions in the direction of larger probability. [...] By this point of view all at once the second law of the theory of heat is moved out of its isolated position, what is mysterious with the preference of nature vanishes, and the principle of entropy as a well-founded theorem of probability theory gets connected with the introduction of atomism into the physical conception of the world.

Nevertheless, problems remained, among them, for example, Zermelo's question for a physical explanation why, say, our universe might have been in a state of low entropy when it came to exist, or the question about how Boltzmann's *Einzelwelten* would fit the cosmological picture. They initiated a development leading to difficult conceptual challenges and ever more sophisticated technical resources ([Skl93], xiii), finally reaching the limits of scientific research at the roots of the cosmological phenomena, the big bang.[59]

After 1906, having contributed to the growing effort to clarify the basis of the theory of heat, Zermelo left the lively scene. His scientific energy focused on the foundations of mathematics. Already in 1904 he had formulated the axiom of choice and used it to prove the well-ordering theorem, thereby opening a new era in set theory and at the same time one of its most serious debates.

---

[58]For the recurrence problem see also [EE11], 22–33, and [Bru66], 15–18; for an extended treatment of the irreversibility problem cf. [HolZ85] and [Zeh89].

[59]For a detailed discussion cf., e. g., [Gre04], in particular Chs. 6 and 11.

2

Göttingen *1897–1910*

2.1 Introduction

On 19 July 1897 Zermelo, still in Berlin, turned to Felix Klein in Göttingen for advice:[1]

> Since my doctorate in October 1894 I have been an assistant to Herr Prof. Planck for almost three years at the "Institute for Theoretical Physics" here in Berlin. However, I intend to give up this position at the beginning of the winter semester in order to continue my scientific work in a smaller town and prepare myself for a possible later *Habilitation* in theoretical physics, mechanics, *etc.* at some university or technical university. At the same time I plan to complete my practical training in physics.
>
> As I am well aware of the rich stimulations and support which you, Herr Professor, offer to younger mathematicians, and also of the vivid interest which you continue to show in problems of mathematical physics, I would be particularly honoured if I could make use of your appreciated scientific advice in pursuing my plans, and I would feel deeply obliged to you. (OV 2.01)

Klein's answer has not survived. But evidently Zermelo was encouraged to come to Göttingen. He enrolled there for mathematics on 4 November 1897. At that time Berlin and Göttingen were still struggling for supremacy in mathematics in Germany. Göttingen was a good choice for Zermelo, given his special field of competence, because applied mathematics played a dominant role there.[2] It was still to become, however, the leading centre for research in the foundations of mathematics, a development that began after the turn of the century, in the end giving birth to a new mathematical subdiscipline, proof theory or metamathematics. Furthermore, the main impulses for an institutionalization of mathematical logic in the German academic system were to have their origins in Göttingen. The driving force for these developments was

---

[1]SUB, Cod. Ms. F. Klein, 12:443 A.

[2]For Klein's support of applied mathematics cf., e. g., [Schu89].

David Hilbert. On the initiative of Klein he had come from Königsberg University two years before Zermelo arrived. In 1897 he was still chiefly working in algebraic number theory, having just published his famous "Zahlbericht" ([Hil1897]). He subsequently changed his main field of research, concentrating for some years on the foundations of geometry and the axiomatic method. Zermelo became deeply involved from the first hour on. He grew into the role of Hilbert's most important collaborator in foundational studies in this important early period, slowly moving from applied mathematics and mathematical physics to foundational studies, set theory, and logic. Decades later, in a report to the Deutsche Forschungsgemeinschaft, he reflected on this move and on the role of Hilbert, his "first and sole teacher in science,"[3] with the following words ([Zer30d]):

> As many as thirty years ago, when I was a *Privatdozent*[4] in Göttingen, I began, under the influence of D. Hilbert, to whom I owe more than to anybody else with regard to my scientific development, to occupy myself with questions concerning the foundations of mathematics, especially with the basic problems of Cantorian *set theory,* which only came to my full consciousness in the productive cooperation among Göttingen mathematicians.

2.2 Working in Applied Mathematics

For the time being and according to his plans, applied mathematics and physics became Zermelo's main subjects. We have learnt already that during his first Göttingen semester he attended the mathematical seminar on differential equations in mechanics directed by Hilbert and Klein and that he took exercises in physics with Eduard Riecke and heard thermodynamics with Oskar Emil Meyer. His first Göttingen publication about a year later ([Zer99a]) was concerned with a problem in applied mathematics. He succeeded in generalizing a result of Adolph Mayer ([May99]) on the uniqueness of the solutions of a differential equation that represents the accelerations of the points of a frictionless system in terms of their coordinates and their velocities where, in addition, the coordinates are restricted by inequalities. Whereas Mayer had proved uniqueness only in the case of at most two inequalities, Zermelo was able to solve the general case. He thereby followed a suggestion of Hilbert in making essential use of the fact that certain side conditions which Mayer had derived in his proof, were linear in the (unknown) accelerations.

---

[3]So in a letter to Courant of 4 February 1932; UAF, C 129/22.

[4]A position in the German academic system with the right to lecture, but without a regular salary.

2.2.1 *Habilitation*

At the same time Zermelo completed his *Habilitation* thesis.[5] As the letter to Klein indicates, work on it had already been started with Planck in Berlin. The aim was to develop a systematic theory of incompressible and frictionless fluids streaming in the sphere, as this had been done already for the plane and had been presented in particular in Poincaré's *Théorie de tourbillons* ([Poi93b]). In the introduction Zermelo describes his motivations and procedure as follows ([Zer02b], 201):

> It may be possible that one can in this way succeed in getting informa-
> tion about many a process in the development of atmospheric cyclones and
> ocean currents as far as they concern the whole of the earth and as far
> as the vertical component of the current can be neglected against its hor-
> izontal one. This geo-physical point of view gave me the first idea for this
> work. However, during the elaboration it took second place after the purely
> geometric-analytic problems and methods. It was my endeavour here to give
> a presentation as unified as possible and independent of external assump-
> tions. The whole development is solely based on the main hydrodynamic
> equations in orthogonal surface coordinates. They are introduced right at
> the beginning, and all quotations of the literature I shall give merely serve
> the purpose of reference or comparison.

In particular, Zermelo does not make systematic use of the stereographic projection of the sphere onto the plane, as this method "refers only to the respective momentary vector field, but not to the movement of the vortices, to the temporal development of the phenomenon" (ibid., 201).[6]

The contents are described in detail in Hilbert's assessment of 9 February 1899:[7] We quote some parts which also expound the structure of the thesis.

> The [...] thesis is purely mathematical, although it is related to the physi-
> cal-meteorological problem of the movement of cyclones on the surface of
> the earth.
>
> The *first* chapter deals with the movement of a fluid on an arbitrary
> surface in space. Based on the Lagrangian differential equations of 2nd kind
> [...] the author develops the form of the general laws of the movement of
> a fluid in the present case. [...] Using the Helmholtz theorem stating the
> constancy of the moments of the vortices in the case of an incompressible

---

[5] A post-doctorate thesis which had to be submitted to initiate the *Habilitation* procedure, then the only way in the German academic system to become eligible for a professorship.

[6] Zermelo refers to Gustav Robert Kirchhoff (probably [Kir76]) and to Horace Lamb ([Lam95], 114 and 253). He may have been influenced also by the pioneering work of Hermann von Helmholtz ([Hel58]).

[7] The quotations given here stem from UAG, Philosophische Fakultät 184a, 149seq.

fluid, the author obtains a partial differential equation of 3rd order for the current which completely determines its temporal course.[8]

The *second* chapter applies this general theory to the sphere. For the [corresponding] partial differential equation [. . .] a special basic solution is introduced which has only one singular point, a so-called "vortex point," and otherwise represents a constant density of vortices on the whole sphere. Using this basic solution instead of the usual Green function in the plane, the author obtains the general integral of the partial differential equation above: the so-called spheric potential [. . .] where the density [. . .] of vortices takes over the role of mass density. [. . .]

In the *third* chapter the author calculates the velocity of a vortex point and sets up the equation of its movement in case there is [. . .] a finite number of further vortex points. [. . .]

The *last* chapter treats the case of three vortex points. Essentially this amounts to the investigation of the change of the form of the triangle which is formed by them. [. . .] (OV 2.02)

Zermelo does not acknowledge earlier relevant literature on the topic. He does not mention the work of Walter Gröbli ([Gro77a], [Gro77b]) on the three vortex problem in the plane. He is evidently unaware of Ippolit Stepanovich Gromeka's work on vortex motions on a sphere ([Gro85]).[9]

The members of the *Habilitation* committee, among them Hilbert himself, Klein, and the physicists Woldemar Voigt, then Dean of the Philosophical Faculty, Walther Nernst, and Eduard Riecke, give positive assessments of both the thesis and Zermelo's abilities. Hilbert states that

the thesis represents a careful and thorough investigation and has led to new and remarkable results; it attests not only knowledge and abilities, but also the author's scientific inclination and his ideal pursuit. [. . .][10]

Dr. Zermelo is well-known to me through personal contacts; while staying here, he has always shown the most vivid scientific interest. In particular, he has demonstrated by talks and scientific remarks in the Mathematical Society[11] that he is well-suited for an academic profession. (OV 2.03)

---

[8]Among others, Zermelo proves the theorem ([Zer02b], 211, Satz III) that a closed or bounded simply connected surface does not admit a movement with continuous velocities which is free of vortices.

[9]Gröbli's dissertation was guided by Heinrich Weber; for details cf. [AreRT92]. For the history of vortex dynamics see [AreM06].

[10]OV 2.02, last paragraph.

[11]The Mathematische Gesellschaft in Göttingen (Mathematical Society in Göttingen) was an institution where the Göttingen mathematicians and physicists as well as invited guests reported both on their own work and on recent developments in mathematics and physics. According to the protocols (SUB, Cod. Ms. D. Hilbert 741) Zermelo had given the following talks up to that time:

9 November 1897: On Boltzmann's *Prinzipien der Mechanik* ([Bol97b]);

1 February 1898: On Kneser's work on the calculus of variations;

10 May 1898: On the state of his *Habilitation* thesis;

Felix Klein adds that "the *Habilitation* thesis makes rather a favourable impression" and appreciates that it is written "in quite a productive way" ("durchaus produktiv"). Voigt states that "the question treated is well put and the way to its solution chosen skilfully." He characterizes Zermelo as "knowledgable and a scholar particularly astute in his criticisms."

On 4 March Zermelo gave his *Habilitation* address on the applicability of probability theory to dynamical systems ([Zer00a]) which we discussed in connection with the Boltzmann controversy, and was granted the *venia legendi* for mathematics.[12]

Only the first two chapters of the *Habilitation* thesis appeared in print ([Zer02b]), with the addition "Erste Mitteilung" ("First Part") added between author's name and main text. The *Nachlass* contains the carefully handwritten version of Chapters 3 and 4 with the addition "Zweite Mitteilung" ("Second Part").[13] The introduction of the published part covers all four chapters. There are no documents known that might explain why the last two chapters remained unpublished and whether this was due to a conscious decision.[14] But there are complaints: In a letter to Max Dehn of 25 September 1900 (CAH) when describing his progress during the summer, Zermelo comments:

> However, I have still a serious problem with [. . .] the preparations of my lecture course, not to mention the disastrous vortex paper. Its completion and printing is still to come. I wonder how this will finally end. (OV 2.04)

In any case, after the *Habilitation*, Zermelo no longer pursued questions of hydrodynamics.

The review in the *Jahrbuch der Fortschritte der Mathematik*"[15] essentially quotes from the introduction. In later reports on Zermelo's achievements in connection with university positions, his *Habilitation* thesis did not play a major role, whereas his Ph. D. thesis was highly praised. Perhaps the overall reason for this modest reception may be seen in the fact that for the time being the field of fluid dynamics was shaped by Ludwig Prandtl in another direc-

---

2 August 1898: On a report of Wilhelm Wien on the movement of the ether, probably Wien's talk "Über die Fragen, welche die translatorische Bewegung des Lichtäthers betreffen" ("On Questions which Concern the Translatorial Movement of the Ether") at the 70th Versammlung Deutscher Naturforscher und Ärzte at Düsseldorf in the same year;

10 January 1899: On a paper of Leo Koenigsberger (1837–1921) concerning Weber's law; furthermore, he had put forward there the problem of distributing points on a sphere in such a way that the product of their mutual distances is minimal.

[12]Notification of the Dean of the Philosophical Faculty to the Royal Curator, No. 68, 15 March 1899; UAG, Files of the Kgl. Kuratorium, 4/Vc 229. Moreover, letter from the Dean Voigt to Zermelo of 6 March 1899; MLF.

[13]With page numbers 61–117; UAF, C 129/225.

[14]In later years, when Zermelo made plans for an edition of his collected works, he counted the last two chapters among his papers.

[15]Vol. **33**, 0781.01.

tion.[16] Prandtl's influence resulted from far-reaching achievements such as his boundary layer theory from 1904, his research on turbulent movements from 1910, and the wing theory from 1918/19.[17] A systematic treatment of fluid dynamics on the sphere started about 50 years later, in contrast to Zermelo strongly emphasizing numerical methods for obtaining solutions.[18] Vortex dynamics on the sphere corresponding to Zermelo's unpublished chapters turned up again only in the 1970s.[19]

Zermelo's *Habilitation* thesis may be considered one of the early essential papers in the field. However, unlike the pioneering work of, say, Gröbli and Gromeka, even the published part remained undiscovered when fluid dynamics developed into a fascinating theory.

2.2.2 Applied Mathematics After the *Habilitation*

For the time being Zermelo continued his research in applied mathematics. Hilbert's *Nachlass* contains a manuscript ([Zer99b]) where he treats the movement of an unstretchable material thread in a potential field. Soon he returned to the calculus of variations, publishing two shorter papers ([Zer02c], [Zer03]; cf. 1.3). At this time his shift to set theory and the foundations of mathematics had already started: The year 1902 saw his first paper in set theory ([Zer02a]) and the year 1904 his sensational paper on the well-ordering theorem and the axiom of choice ([Zer04]). Nevertheless, his turning away from applied mathematics and physics was a gradual process. As we have already learnt, he still co-authored a report with Hans Hahn about recent developments in the calculus of variations ([HahZ04]). Furthermore, he co-published the second edition of the second and third volume of Joseph Alfred Serret's *Lehrbuch der Differential- und Integralrechnung* ([Ser99], [Ser04].) Even after the proof of the well-ordering theorem had appeared, he was not yet fully occupied with foundational questions, but was also working on the translation of Josiah Willard Gibbs' *Elementary Principles in Statistical Mechanics* ([Gib02]), which was published in 1905 ([Gib05]). As late as 1909 and together with Ernst Riesenfeld, a former student of Nernst, he studied the limit concentration of solvents which are brought together in a horizontal cylinder ([RieZ09]).

Thanks to his interest in the calculus of variations, Zermelo got into closer contact with Constantin Carathéodory, who had also studied in Berlin with Hermann Amandus Schwarz and Max Planck, but was now working in Göttingen together with Klein, Hilbert, and Hermann Minkowski, writing his Ph. D.

---

[16]Ludwig Prandtl (1875–1953) obtained a professorship in Göttingen in 1904, becoming also co-chairman of the *Aerodynamische Versuchsanstalt* there. In 1908 he built the first wind-tunnel in Germany.

[17]Cf., for example, [Tie29] and [Tie31] or [Tie57].

[18]Cf., for example, [Ped79].

[19]Influential papers are, for example, [Bog77] and [Bog79]. For a systematic treatment see [New01], in particular Ch. 4.

thesis ([Car04]) under the final guidance of Minkowski. It was completed in 1904 and became a highly acknowledged masterpiece in the calculus of variations. In the preface Carathéodory thanked not only Minkowski, but also Zermelo for the stimulating interactions ([Geo04], 35). His mathematical interests, which extended also into statistical mechanics, led to a long period of mutual scientific exchange with Zermelo and developed into a deep, lifelong friendship. In 1906 both were working together on a book on the calculus of variations, a book "which in any case promises to become the best in this field."[20] The book remained unfinished. Reasons may be found in Zermelo's poor state of health then and Carathéodory's move to Bonn in 1908. In 1907, when Zermelo's plans to get a position at the Academy of Agriculture in Poppelsdorf (now part of Bonn) had not been realized (2.10.1) and Carathéodory intended to change to the University of Bonn, we see him writing to Zermelo[21] that

> it is still undecided whether I will go to Bonn. Now, since Hergl[otz] and you will stay, I would almost prefer G[öttingen] again.

Later, he dedicated his *Vorlesungen über reelle Funktionen* ([Car18]) to Erhard Schmidt and to Zermelo. In 1910 the Carathéodory family and Zermelo spent their holidays together in the Swiss Alps.[22] On 12 October 1943 "Cara," then living in Munich, complained that Zermelo "is so far away." His last letter to Zermelo was written on 4 January 1950, four weeks before his death on 2 February.

2.3 First Years as a *Privatdozent* in Göttingen

2.3.1 *Privatdozenten* Grant and Titular Professorship

Two years after his *Habilitation* Zermelo applied for a *Privatdozenten* grant in order to take care of his livelihood.[23] To justify his neediness he claimed that he had no income besides the students' fees. He was mainly living on the interest from a small inheritance which he had come to after the death of his parents, but was also continually forced to draw on the inheritance itself. His application was supported by Hilbert:[24]

---

[20]Letter from Hermann Minkowski to Wilhelm Wien of 12 October 1906; DMA, prel. sign. NL 056-009. Already in 1903 Hilbert writes that Zermelo is working on such a monograph (assessment for the University of Breslau; SUB, Cod. Ms. D. Hilbert 490).

[21]Postcard of 30 August 1907; UAF, C 129/19.

[22]Letter from Carathéodory to Zermelo of 14 July 1935; together with the following two letters in UAF, C 129/19.

[23]Letter to the Prussian Minister for Cultural Affairs Konrad Studt of 13 March 1901; UAG, 4/Vc 229.

[24]Letter from Hilbert to the University Curator of 20 March 1901; ibid.

Dr. Zermelo is a gifted scholar with a sharp judgement and a quick intellectual grasp; he shows a lively interest and open understanding for the questions of our science and has furthermore comprehensive theoretical knowledge in the domain of mathematical physics. I am in continuous scientific
exchange with him. (OV 2.05)

Hilbert recommended

with utmost warmth taking Zermelo into consideration for the award of the
Privatdozenten grant, because he has become an estimable member of our
teaching staff. (OV 2.06)

Zermelo's application was successful. As the grants for 1901 had already been
made, support started in 1902 and amounted to double the size of the income
from his inheritance (6 000 RM per year).[25] By a skilful manipulation the
usual period of five years could be extended to six years, thus ensuring his
economic situation, albeit modestly, until 1907.[26]

Zermelo's academic career progressed slowly. Only after nearly six years
as a *Privatdozent* did he receive the title "Professor."[27] The appointment
resulted from an application filed a year earlier by Hilbert, who was then
acting director of the Seminar for Mathematics and Physics.[28] Hilbert had
especially emphasized Zermelo's penetrating grasp of the problems of mathematical physics. The University Curator had supported the application with
some supplementary comments regarding Zermelo's teaching skills. According
to him, Zermelo's lecture courses were "more deep than pedagogically skilful,"
resulting in less success for younger students than for mature ones. Nevertheless, "the lecture courses are among the best [...] that are currently delivered
in Göttingen."[29]

Very likely, the long time which passed between Hilbert's application and
Zermelo's appointment, resulted from severe problems of Zermelo's health.
In the winter of 1904/05 he fell seriously ill. In order to recover, he spent
the spring and early summer of 1905 in Italy. The next two years brought
further health problems, forcing his absence from Göttingen and adding to
his difficulties in obtaining a permanent position (2.11.1).

2.3.2 Under Consideration in Breslau 1903

Because of his work in applied mathematics and physics, Zermelo was under
consideration for a position at the University of Breslau. On its third meeting
on 15 June 1903 the search committee for a new extraordinary professorship of

---

[25]Ministry to Curator, UI No. 751, 9 May 1901; ibid.

[26]UAG, 4/Vc 229 and 4/Vb 267a. – For details cf. [Pec90a], 22seq.

[27]Certificate of 21 December 1905; MLF.

[28]Hilbert's application, 14 December 1904; GSA, Rep. 76 Va Sekt. 6 Tit. IV No.
4, Vol. 4, fols. 165–166.

[29]Curator's report, 9 January 1905; ibid., fols. 160–163.

mathematics at the Faculty of Philosophy decided to recommend the *Privatdozent* at the University of Breslau Franz London, the extraordinary professor at the University of Greifswald Gerhard Kowalewski, the extraordinary professor at the University of Gießen Josef Wellstein, and Zermelo *aequo loco* as candidates.[30]

In the draft of the report the committee emphasizes the quality of Zermelo's dissertation and gives a short description of his papers, concluding as follows:

> His literary achievements are still few in number; however, it has to be appreciated that they are concerned with difficult fields. We learn from Göttingen that he proves himself as an academic teacher and that he has already given lecture courses on many different disciplines.

Two days later the chairman of the committee asked for an additional meeting on 18 June in order "to discuss some changes in the draft of the letter to the minister." The essential change was to name London, Kowalewski, and Wellstein *aequo loco* in the first place and Zermelo in the second place. So Zermelo was out.

The files give no indication why the Faculty made this sudden change. Hermann Minkowski, who had come to Göttingen in 1902, wrote in 1906 that the main reason for Zermelo's repeated failure to obtain a professorship was his "nervous haste" (2.11.2). Probably the Breslau case is one he had in mind.

In moving Zermelo's name to the end of the list, the Faculty ignored Hilbert's staunch recommendation. Hilbert had been asked by them to report on some candidates, among them London, Kowalewski, and Wellstein, but not Zermelo. In his answer of 31 May[31] he characterizes Kowalewski and Wellstein as "good," but is extremely negative about London ("I would consider an appointment a disaster for the Breslau Faculty"). He then adds:

> Now, concerning further names, I immmediately start with the one whom I consider the real candidate for the Breslau Faculty, namely Zermelo. (OV 2.07)

Having emphasized Zermelo's "extremely varied" lecturing activity and shortly characterized Zermelo's papers, he continues, thereby weaving in some interesting remarks about the calculus of variations:

> Zermelo is a modern mathematician who combines versatility with depth in a rare way.[32] He is an expert in the calculus of variations (and working on a comprehensive monograph about it). I regard the calculus of variations as a branch of mathematics which will belong to the most important ones in the

---

[30]The presentation follows the documents in UAW, signature F73, pages 99 and 102–115.

[31]Outline in SUB, Cod. Ms. D. Hilbert 490.

[32]This phrase will return in later assessments, for example, in the one which Hilbert wrote for Zurich in 1910.

future. I hope that by developing this branch and by using it for the good of the neighbouring areas, one can contribute to giving mathematics back the reputation which it possessed 100 years ago and has been threatening to forfeit in recent times. (OV 2.07)

Having praised Zermelo's knowledge in set theory and physics, he concludes with some personal remarks:

> You will not presume that I intend to praise Zermelo away. Before Minkowski came here and before Blumenthal had matured,[33] Zermelo was my main mathematical company, and I have learnt a lot from him, for example, the Weierstraßian calculus of variations, and so I would miss him here most of all. (OV 2.07)

2.4 The Context: Hilbert's Research on Foundations

Soon after his arrival in Göttingen and under the influence of David Hilbert, Zermelo turned to foundational questions. In fact it was his foundational work, in particular his work on set theory, which led to his greatest and most influential scientific achievements, but in the end also to a depressing failure. It is concentrated in two periods which roughly coincide with his decade in Göttingen and the years around 1930 in Freiburg and corresponds to two specific periods of research on the foundations of mathematics by Hilbert and his collaborators in Göttingen. In the first period Hilbert elaborated his early axiomatic program, and Zermelo's work is clearly along the lines proposed by Hilbert, it is *pro* Hilbert. The second period falls into the time when Hilbert developed his proof theory which was, because of its finitistic character, flatly rejected by Zermelo.[34] His work at that time is *contra* Hilbert. The present section is intended to provide the backdrop for the description of the first period.

2.4.1 Mathematical Problems and Foundations

The Göttingen claim for leadership in mathematics found its expression in Hilbert's address at the Second International Congress of Mathematicians which took place early in August 1900 on the occasion of the World Fair in Paris (cf. [Gra00b]). Hilbert presented his famous list of 23 unsolved problems to the mathematical community, discussing for reasons of time only 10 problems. He thereby set the mathematical agenda for the new century. According to Constance Reid ([Rei70], 84) it soon became quite clear

---

[33] Minkowski came to Göttingen in 1902. Otto Blumenthal took his doctorate with Hilbert in 1898 and became a *Privatdozent* in Göttingen in 1901.

[34] For the development of Hilbert's programme cf., for example, [Zac06].

David Hilbert in the early 1900s
Courtesy of Niedersächsische Staats- und Universitätsbibliothek Göttingen
Abteilung Handschriften und alte Drucke
Sammlung Voit: D. Hilbert, Nr. 9

that David Hilbert had captured the imagination of the mathematical world with his list of problems for the twentieth century. His practical experience seemed to guarantee that they met the criteria which he had set up in his lecture; his judgement, that they could actually be solved in the years to come. His rapidly growing fame – exceeded now only by that of Poincaré – promised that a mathematician could make a reputation for himself by solving one of the problems.

In his introductory part Hilbert considers the general preconditions for the solution of a mathematical problem. For Hilbert it is a requirement of mathematical rigour in reasoning that the correctness of a solution can be established "by means of a finite number of inferences based on a finite number of hypotheses," i. e., using "logical deduction by means of a finite number of inferences."[35] Thus Hilbert's later proverbial finitistic attitude was clearly in evidence early on. In his 1900 address it is combined with an unrestricted optimism, when speaking of "the axiom of the solvability of every problem" in the sense of the conviction ([Hil96b], 1102)

---

[35][Hil96b], 1099 (modified).

that every definite mathematical problem must necessarily be susceptible of an exact settlement, either in the form of an actual answer to the question asked, or by the proof of the impossibility of its solution and therewith the necessary failure of all attempts.

It is this conviction which gives the incentive for mathematical research (ibid., 1102):

> We hear within us the perpetual call: There is the problem. Search for the solution. You can find it by pure reason, for in mathematics there is no ignorabimus![36]

The first six problems deal with foundations, some of them with special questions about geometrical axiomatics. They reflect the state of philosophical insights Hilbert had reached during his foundational studies. They can also be read as a programme for future foundational research in Göttingen.

The first problem concerns "Cantor's problem of the cardinal number of the continuum,"[37] i. e., to prove Cantor's continuum hypothesis, in Hilbert's formulation:

> Every system of infinitely many real numbers, i. e., every infinite set of numbers (or points), is either equivalent to the set of natural numbers, $1, 2, 3, \ldots$ or to the set of real numbers and therefore to the continuum, that is, to the points of a line; *as regards equivalence there are, therefore, only two [infinite] sets of numbers, the countable set and the continuum.*

In this context Hilbert mentions another "very remarkable" problem "which stands in the closest connection with the theorem mentioned and which, perhaps, offers the key to its proof," the question whether

> the totality of all numbers may not be arranged in another manner so that every subset may have a first element, i. e., whether the continuum cannot be considered as a well-ordered set – a question which Cantor thinks must be answered in the affirmative.

Hilbert asks for "a direct proof, perhaps by actually giving an arrangement of numbers such that in every partial system a first number can be pointed out."

The second problem concerns "the consistency of the arithmetical axioms," i. e., the consistency of the axioms for the real numbers, a central problem for subsequent Göttingen research on foundations. In his comment Hilbert characterizes the basic features of his axiomatic method, stressing for the first time the central role of consistency proofs: The investigation of the foundations of a scientific discipline starts with setting up an axiom system which contains

---

[36]Hilbert refers to Emil du Bois-Reymond's conviction in respect to scientific knowledge: "Ignoramus et ignorabimus" ("We do not know and we shall never know") in the closing words of his lecture "Ueber die Grenzen des naturwissenschaftlichen Erkennens" of 1872 ([DuB71], cf. [Mcc04]).

[37]Essentially quotations here follow [Hil96b], 1103–1105.

"an exact and complete description of the relations subsisting between the elementary ideas" of that discipline. The axioms thus set up are at the same time the definitions of those elementary ideas,

> and no statement within the realm of the science whose foundation we are testing is held to be correct unless it can be derived from those axioms by means of a finite number of logical steps.

Having addressed the issue of independence of the axiom system, he formulates

> the most important among the numerous questions which can be asked with regard to the axioms: *To prove that they are not contradictory, that is, that a finite number of logical steps based upon them can never lead to contradictory results.*

Concerning mathematics, the consistency of the axioms of geometry can be reduced to the consistency of the arithmetical axioms via a suitable model. But "a direct method is needed for the proof of the consistency of the arithmetical axioms."

The feature of consistency stands out, because it becomes a criterion for mathematical existence:

> If contradictory attributes be assigned to a concept, I say, that *mathematically the concept does not exist.* [...] But if it can be proved that the attributes assigned to the concept can never lead to a contradiction by the application of a finite number of logical inferences, I say that the mathematical existence of the concept [...] is thereby proved.

In other words, mathematical existence means consistent possibility (cf. [Pec04c]).

It is significant that Hilbert illustrates the perspectives of his method by examples from set theory. The proof of the consistency of the arithmetical axioms, he writes, would at the same time prove the mathematical existence of the complete system of the real numbers ("Inbegriff der reellen Zahlen"), i. e., the continuum, and he continues that such an existence proof via a consistent axiomatization should also be possible, for example, for Cantor's number classes and cardinals, but not for the system of all cardinal numbers or all Cantorian alephs,

> for which, as may be shown, a system of axioms, consistent in my sense, cannot be set up. Either of these systems is, therefore, according to my terminology, mathematically non-existent.

So there clearly is a desideratum: A consistent axiomatization of set theory.

Hilbert's comments have two roots, his development of an axiomatic system for Euclidean geometry (2.4.2) and certain problems in Cantor's set theory related to inconsistent sets (2.4.3). Both got a new quality when the publication of the set-theoretic paradoxes in 1903 suddenly questioned the foundations of mathematics, the discipline with the reputation of being the most certain among the sciences.

2.4.2 Foundations of Geometry and Analysis

Since his school days Hilbert was interested in geometry.[38] This interest found its expression in the *Grundlagen der Geometrie* ([Hil1899]), published as part of a *Festschrift* on the occasion of the unveiling of the Gauss-Weber monument in Göttingen on 17 June 1899.[39] A French edition appeared in 1900 ([Hil00c]), an English translation followed in 1902 ([Hil02a]). The first German separate book edition with new attachments was published in 1903 ([Hil03]) and is now in its 14th edition ([Hil1999]).

In his *Grundlagen* Hilbert does not really reflect on mathematical methodology. He actually *gives* a foundation of Euclidean geometry which he calls "a new attempt" at establishing a simple and complete system of independent axioms in such a way that "the significance of the different groups of axioms can be recognized and the consequences of the individual axioms become as clear as possible" ([Hil1899], 4). With this book modern axiomatics was created. Later, Hilbert defined the axiomatic method as follows ([Hil02c], 50):

> I understand under the *axiomatic* exploration of a mathematical truth [or theorem] an investigation which does not aim at finding new or more general theorems connected to this truth, but which seeks to determine the position of this theorem within the system of known truths together with their logical connections in such a way that it can be clearly said which conditions are necessary and sufficient for giving a foundation of this truth.

He thereby introduced what was called a formalist position in the philosophy of mathematics. This *formalism* is usually set against *logicism* and *intuitionism*. Logicism, going back to Gottlob Frege's philosophy of mathematics, maintains that mathematical objects are logical objects, i. e., they can be given by logical means alone. Intuitionism, on the other hand, maintains that all mathematical knowledge originates in (mental) constructions. Its main representative was Luitzen Egbertus Jan Brouwer (1881–1966). As to formalism, Hilbert himself gave a metaphorical definition in a letter to Gottlob Frege of 29 December 1899:[40]

> But it is surely obvious that every theory is only a scaffolding or schema of concepts together with their necessary mutual relations, and that the basic elements can be thought [or imagined] in any way one likes. E. g., instead of points, think [or imagine] a system of love, law, chimney-sweep . . . which satisfies all axioms; then Pythagoras' theorem also applies to these things. In other words, any theory can always be applied to infinitely many systems of basic elements.

---

[38] On the pre-history of Hilbert's *Grundlagen der Geometrie* cf. [Toe86], [Toe99].

[39] The monument celebrates the Göttingen professors Carl Friedrich Gauß and Wilhelm Weber. It symbolizes the close connection between mathematics and physics in Göttingen, showing Gauss and Weber with the first electromagnetic telegraph which both had built in 1833.

[40] [Fre76], 67; [Fre80], 42.

Unlike Euclid, Hilbert does not try to determine the underlying concepts in their intension or their extension. Using a Kantian phrase, he speaks of "thought-things," i. e., mental creations independent of non-mental reality ([Hil1899], 4):

> We think [or imagine] three different systems of things: we call the things of the *first* system *points* [...], the things of the *second* system *lines* [...], the things of the *third* system *planes* [...]. We think [or imagine] the points, lines, planes in certain relations and designate these relations with words like "lying," "between," "congruent;" the exact and for mathematical purposes complete description of these relations is done by the *axioms of geometry*.

In the *Grundlagen* Hilbert gives a description by 20 axioms. He also investigates this system of axioms as an object in itself, treating questions of completeness for mathematical purposes, independence of the axioms, and consistency. The consistency is reduced to the consistency of the arithmetical axioms. Therefore, an absolute consistency proof is, in fact, postponed, and a new task is set: to find a consistent set of axioms for the real numbers.

Hilbert presented his ideas concerning a foundation of analysis – in his opinion the fundamental discipline of mathematics – at the annual meeting of the Deutsche Mathematiker-Vereinigung in Munich in September 1899. In his address "Über den Zahlbegriff" ([Hil00a]) he provided an axiom system for the real numbers independent of set-theoretic considerations. Due to the sketchy character of the address, he did not attempt to go into detail about completeness, independence, and consistency. With regard to the latter he did, however, assert that the "necessary task" of providing consistency required "only a suitable modification of known methods of inference" (ibid., 184). These optimistic words from September 1899 clearly indicate that Hilbert initially underestimated the enormity of the task at hand. But he soon changed his views, and in August 1900, less than one year later, he included the consistency proof for the arithmetical axioms as the second among his Paris problems. Three years later, on 27 October 1903, he again emphasized the distinguished role of the consistency proof in a lecture delivered before the Göttingen Mathematical Society, coining the brief formula: "The principle of contradiction the pièce de résistance."[41]

2.4.3 The Paradoxes

Until 1897 Hilbert had largely ignored set theory.[42] This changed when in the late 1890s Hilbert and Cantor corresponded on set-theoretic topics,[43] a

---

[41] *Jahresbericht der Deutschen Mathematiker-Vereinigung* **12** (1903), 592.

[42] For Hilbert's changing attitude towards set theory cf. [Moo02a].

[43] For a comprehensive discussion of this correspondence cf. [Purl87], 147–166. Extracts are published in [Can1991], 387seq., and, in English translation, in [Ewa96], 923seq. Cf. also [Ferr99], 290–294.

special topic being the problem of the existence of paradoxical sets such as the set of all alephs, or more generally, in Cantorian terminology, the problem of inconsistent multiplicities. First viewed in Göttingen as a matter concerning merely set theory as a mathematical discipline and even enriched by further examples, only the publication of Russell's paradox in 1903 led to the insight that the paradoxes meant a severe threat to Hilbert's early consistency programme.

Unrestricted Comprehension

In his first letter to Hilbert, dated 26 September 1897,[44] Cantor argues that the totality of all alephs does not exist:[45]

> For the totality of all alephs is one that cannot be conceived as a determinate, well-defined, *finished* set. If this were the case, then this totality would be *followed* in size by a *determinate aleph*, which would therefore both *belong* to this totality (as an element) and *not belong*, which would be a contradiction.[46]

On 2 October 1897, he answers criticism by Hilbert quoted by him as saying that "the totality of alephs can be conceived as a determinate, well-defined set, since if any thing is given, it must always be possible to decide whether this thing is an aleph or not; and nothing more belongs to a well-defined set:"[47]

> The totality of all alephs cannot be conceived as a determinate, a well-defined, and *also a finished* set. This is the *punctum saliens* [...]. I say of a set that it can be thought of as *finished* (and call such a set, if it contains infinitely many elements, "transfinite" or "super-finite") if it is possible without contradictions [...] to think of *all its elements as existing together* [...] or (in other words) if it is *possible* to imagine the set as *actually existing* with the totality of its elements.
>
>   In contrast, infinite sets such that the *totality* of their elements cannot be thought of as "existing together" or as a "thing for itself" [...] I call "*absolutely infinite* sets,"[48] and to them belongs the "set of all alephs."

Hilbert's responses in correspondence have not been preserved, but he published his opinion in prominent places. In his Munich address "Über den Zahlbegriff" he commented, thereby anticipating crucial features of his Paris talk:[49]

---

[44][Can1991], no. 156, 388-389.

[45]Quotations here follow [Ewa96], 926–928.

[46]Using this consideration, Cantor then argues that every cardinality is an aleph.

[47][Can1991], 390, also published in [Purl87], no. 44, 226–27.

[48]Later, so in a letter to Dedekind of 3 August 1899 ([Can99]), absolutely infinite sets are called "inconsistente Vielheiten" ("inconsistent multiplicities") by Cantor.

[49][Hil00a], 184; [Hil96a], 1095.

[In the proof of the consistency of the arithmetical axioms] I see the proof of the existence of the totality of real numbers, or – in the terminology of G. Cantor – the proof that the system of real numbers is a consistent (finished) set. [...]

   If we should wish to prove in a similar manner the existence of a totality of all powers (or of all Cantorian alephs), this attempt would fail; for in fact the totality of all powers does not exist, or – in Cantor's terminology – the system of all powers is an inconsistent (unfinished) set.

Obviously Hilbert had accepted Cantor's distinction between finished or consistent sets and unfinished sets or inconsistent multiplicities and, hence, recognized that the inconsistent multiplicity of all cardinals served only as a paradigmatic example for other inconsistent multiplicities resulting from unrestricted comprehension. A suitable axiomatization, however, would exclude these problematic totalities from the outset.

   The need for a safe axiomatization was enhanced by the disturbing experience that methods used by mathematicians as a matter of course had proved to lead to contradictory results. As discussed below, both Hilbert himself and Zermelo had found further paradoxes, Hilbert's paradox being quite mathematical in nature. Nevertheless it took some time until the full significance of these paradoxes was understood in Göttingen. In particular, it was a longer process to grasp the role of the paradoxes in logic and, thus, in mathematical proofs. The insight was forced on the Göttingen mathematicians in 1903 when Bertrand Russell published his paradox in Chapter 10 of his *Principles of Mathematics* ([Rus03]) under the heading "The Contradiction," and when almost at the same time Gottlob Frege addressed it in the postscript of the second volume of his *Grundgesetze der Arithmetik* ([Fre03]), admitting that the foundational system of the *Grundgesetze* had proved to be inconsistent. The insight had serious consequences for Hilbert's early programme, because the consistency proof for the arithmetical axioms could not be carried out in a logic which had been shown to be inconsistent. Therefore, logic had to be moved into the focus of the Göttingen foundational interests – a process which was still to last for some years.[50]

Hilbert's Paradox

Hilbert never published his paradox, "Hilbert's paradox," as it was referred to in Göttingen.[51] He discussed it, however, in a Göttingen lecture course on "Logische Principien des mathematischen Denkens" ("Logical Principles of Mathematical Thinking") in the summer semester 1905. The course is preserved in two sets of notes elaborated by Ernst Hellinger and Max Born.[52]

---

[50]Cf. also [Ber35].

[51]It was mentioned, e. g., in Otto Blumenthal's biographical note in the third volume of Hilbert's collected works ([Blu35], 421–422). On its history cf. [Pec90b], [PecK02], [Pec04b].

[52][Hil05b], [Hil05c]. For an intensive discussion of this course cf. [Pec90b], 58–75.

Part B of these notes, on "The Logical Foundations," starts with a comprehensive discussion of the paradoxes of set theory. It begins with metaphorical considerations on the general development of science ([Hil05c], 122):

> It was, indeed, usual practice in the historical development of science that we began cultivating a discipline without many scruples, pressing onwards as far as possible, that we in doing so, then ran into difficulties (often only after a long time) that forced us to turn back and reflect on the foundations of the discipline. The house of knowledge is not erected like a dwelling where the foundation is first well laid-out before the erection of the living quarters begins. Science prefers to obtain comfortable rooms as quickly as possible in which it can rule, and only subsequently, when it becomes clear that, here and there, the loosely joined foundations are unable to support the completion of the rooms, science proceeds in propping up and securing them. This is no shortcoming but rather a correct and healthy development. (OV 2.08)

Although contradictions are quite common in science, Hilbert continued, in the case of set theory they seemed to be of a special kind because of a tendency towards theoretical philosophy. In set theory, the common Aristotelian logic and its standard methods of concept formation, hitherto used without hesitation, had proved to be responsible for the new contradictions. He elucidated his considerations by three examples,[53] the liar paradox, "Zermelo's paradox," as Russell's paradox was called in Göttingen, and the "purely mathematical" paradox of his own ([Hil05b], 210), commenting[54] that the latter

> appears to be especially important; when I found it, I thought in the beginning that it causes invincible problems for set theory that would finally lead to the latter's eventual failure; now I firmly believe, however, that everything essential can be kept after a revision of the foundations, as always in science up to now.

Hilbert starts with two set formation principles: the *addition principle* (*Additionsprinzip*) allowing the formation of the union of "several sets and even infinitely many ones," and the *mapping principle* (*Belegungsprinzip*) allowing the formation, for a given set S, of the set of *self-mappings* (*Selbstbelegungen*) of S, i.e., the set S^S of mappings of S into itself. He then considers all sets which result from the set of natural numbers "by applying the operations of addition and self-mapping an arbitrary number of times." Using the two principles again, he obtains the union U of all these sets and the corresponding set U^U of self-mappings of U. By definition of U, the set U^U is a subset of U. However, as U^U has larger cardinality than U, this yields a contradiction.[55]

---

[53]There was a further paradox discussed in Göttingen, Grelling's paradox. Cf. [Pec95] for its genesis and [Pec04b] for its role in these discussions.

[54]Ibid., 204; original text reprinted in [PecK02], 170.

[55]In order to obtain this contradiction, Hilbert follows Cantor's proof that the power set of a set S has larger cardinality than S ([Can90/91]). Essentially, he

As this paradox is based on operations which are part of the everyday practice of working mathematicians, Hilbert views it as of purely mathematical nature and, therefore, as particularly more serious for mathematics than Cantor's inconsistent multiplicities.

The Zermelo-Russell Paradox

In his letter to Gottlob Frege of 16 June 1902 ([Rus02]), Bertrand Russell gave two versions of a new paradox, the second one saying that

> there is no class (as a totality) of those classes which, each taken as a totality, do not belong to themselves.[56]

Such a class had to belong to itself and at the same time not to belong to itself.

Hilbert was informed about the paradox by a copy of the second volume of Frege's *Grundgesetze* which Frege had sent him. It is instructive to study Hilbert's reaction. He wrote to Frege that "this example" had been known before in Göttingen, adding in a footnote that, as he believed, "Dr. Zermelo discovered it three or four years ago after I had communicated my examples to him," and continuing with respect to these examples:[57]

> I found other, even more convincing contradictions as long as four or five years ago; they led me to the conviction that traditional logic is inadequate and that the theory of concept formation needs to be sharpened and refined.
>
> Up to now all logicians and mathematicians have believed that a notion has come into being as soon as can be determined for any object whether it falls under it or not. I regard this assumption as the most essential gap in the traditional building up of logic. It appears to me as insufficient. Rather, the essential point consists in recognizing that the axioms defining the notion are consistent.

These quotations give evidence of a discourse between Hilbert and Zermelo on set theory that must already have taken place before the turn of the century. They also indicate that the revolutionary impact of the paradoxes was only seen in Göttingen when their effect on Frege's logic had become evident. Two years later, in his 1905 lecture, Hilbert was to present Russell's paradox as a "purely logical" one[58] and as a paradox which was probably more convincing

---

proceeds as follows: There are 'not more' elements in U^U than there are in U. So let f be a function from U onto U^U. Choose $g \in U^U$ such that for all $x \in U$ the image $g(x)$ is different from $(f(x))(x)$. For any $h \in U^U$, say, $h = f(x)$ with suitable $x \in U$, one has $h(x) = (f(x))(x) \neq g(x)$; hence, g is different from h. Therefore, $g \notin U^U$, a contradiction. – For an English translation of Hilbert's original argument cf. [PecK02], 168–170.

[56] "[Es gibt] keine Klasse (als Ganzes) derjenigen Klassen[,] die als Ganze sich selber nicht angehören." English translation given here according to [vanH67], 125.

[57] [Fre80], 51; [Fre76], 79–80.

[58] Its purely logical character being due to the fact that it was solely based on concept formation.

for non-mathematicians, claiming that his own paradox seemed to be more decisive for mathematicians ([Hil05b], 210).

The Göttingen custom of attributing Russell's paradox to Zermelo was supported by Zermelo himself: In his 1908 paper on the well-ordering theorem Zermelo claims to have formulated the paradox independently of Russell and already reported it to Hilbert and other people before 1903.[59] Moreover, in recollections which he communicated to Heinrich Scholz[60] in 1936, he says:[61]

> Around 1900 the set-theoretic antinomies[62] were intensively discussed in the Hilbert circle. It was then that I gave the antinomy of the largest cardinality the precise form (as "the set of all sets that do not contain themselves") which was later named after Russell. When Russell's work (Principles of Mathematics 1903?) appeared, we already knew about that. (OV 2.09)

At this time Scholz was working on Frege's papers which he had acquired for his department at the University of Münster. He had found Hilbert's letter to Frege mentioned above, and now was curious what paradoxes Hilbert referred to in his letter.

Independent evidence of what exactly Zermelo had discovered, was found among the papers of Edmund Husserl in a note from Husserl's hand saying that Zermelo had informed him on 16 April 1902 about a paradox in set theory.[63] Zermelo's message was a comment on a review which Husserl had written of the first volume of Ernst Schröder's *Vorlesungen über die Algebra der Logik* ([Schr90]). There Schröder had criticized (ibid., 245) George Boole's interpretation ([Boo54], 42–43) of the symbol 1 as the class of everything that can be a subject of discourse. Husserl had dismissed Schröder's argumentation as sophistical ([Hus91b], 272) and was now advised by Zermelo that Schröder

---

[59] [Zer08a], footnote p. 118–119; [Zer67b], 191.

[60] Originally a professor for theology in Breslau from 1917 to 1919, Heinrich Scholz (1884–1956) studied mathematics while holding a professorship for philosophy in Kiel in order to lay a sound foundation for his awakening interest in mathematical logic. Having obtained a professorship for philosophy at the University of Münster in 1928, he started to build up what was later called the Münster School of Mathematical Logic. In 1943 he had his professorship renamed into a professorship for mathematical logic and foundational research. He established an institute under the same name in 1944; it was the first institute for mathematical logic in Germany. For Scholz's life and work cf. [Scholh61].

[61] Letter from Zermelo to Heinrich Scholz of 10 April 1936; *Nachlass* of Heinrich Scholz, Institut für Mathematische Logik und Grundlagenforschung, University of Münster.

[62] Also in his axiomatization paper ([Zer08b], 261; [Zer67c], 200) Zermelo prefers "antinomy" to "paradox," as the latter word denotes a statement which merely "conflicts the general opinion without containing an interior contradiction" (postcards to Leonard Nelson of 22 December 1907 and 1 January 1908; ASD, 1/LNAA000399); cf. 2.10.2. For the use of "antinomy" *vs.* "paradox" in the foundational discussion see [Ferr99], 309.

[63] [Hus1979], 399; English translation in [RanT81].

was right "in the issue," but not "in the method of proof." According to Husserl's note Zermelo claimed that "a set S which contains each of its subsets is an inconsistent set, i. e., such a set, if at all treated as a set, leads to contradictions." To prove this assertion, Zermelo argued that the subsets of S which do not contain themselves as elements, "constitute in their totality a set S_0" which "(1) does not contain itself as an element and (2) contains itself as an element." [64]

Taking into consideration Zermelo's role as just described, we follow Abraham A. Fraenkel and henceforth speak of Russell's paradox as the *Zermelo-Russell paradox*. [65]

2.5 Starting Set Theory

2.5.1 Zermelo's First Period of Foundational Research

Before discussing Zermelo's achievements during his first period of foundational investigations, we will briefly draw the main lines in order to sketch the extent to which Zermelo followed Hilbert's directives, thus confirming Zermelo's judgement quoted at the beginning of this chapter, that it was his teacher Hilbert and the questions concerning the foundations of mathematics together with "the productive cooperation among Göttingen mathematicians" which shaped his scientific development.

As remarked above, Hilbert and Zermelo may have discussed set-theoretic questions already before 1900 at a time when Zermelo's primary interests were in applied mathematics. After the turn of the century, however, set theory moved into the centre of Zermelo's scientific endeavour. With two epochal achievements he became the founder of modern axiomatic set theory, at the same time solving two essential desiderata Hilbert had formulated.

First, in 1904 he proved that every set can be well-ordered, i. e., admits a total ordering in which every non-empty subset has a least element ([Zer04]). He thus showed that the size of every infinite set can be measured by a Cantorian aleph, and at the same time solved a part of Hilbert's first Paris problem. His proof, however, did not fulfil Hilbert's hope of opening the way to a confirmation of Cantor's continuum hypothesis. But it had striking methodological consequences: By formulating the axiom of choice as a prerequisite and emphasizing its necessity and methodological power, he initiated a debate which finally led to the acceptance of this axiom, a development which resulted in a remarkable enrichment of mathematics.

Then, in 1908 he published the first set-theoretic axiom system ([Zer08b]). He thereby succeeded in realizing a part of Hilbert's axiomatization programme. In doing so, he followed Hilbert's idealistic attitude towards mathe-

---

[64]For a discussion of the question to what extent Husserl's representation may meet Zermelo's original formulation, cf. [RanT81], 20.

[65]Fraenkel ([Fra27], 21) uses the term "(ZERMELO-)RUSSELLsches Paradoxon."

matical objects. As mentioned earlier, Hilbert regarded mathematical objects as "thought-things," creations of the mind whose ontological status was simply left open, identifying the existence of mathematical objects with consistent possibility within a given axiomatic theory. His ontology was therefore without any realistic commitment and his axiomatics epistemologically and ontologically neutral.[66] Zermelo shared this attitude of keeping epistemological and ontological questions open. He started the axiomatization of set theory in set theory itself as it was historically given. He then established the principles "necessary to found this mathematical discipline," but with regard to deeper philosophical considerations remarked that "the further, more philosophical, question about the origin of these principles and the extent to which they are valid will not be discussed here."[67] So, like Hilbert in the pre-war era, he simply avoided dealing with philosophical problems.

Between his striking achievements of 1904 and 1908 Zermelo turned to the theory of finite sets. In particular, he explored the relationship between different versions of finiteness going back to Dedekind's treatise on the natural numbers ([Ded88]). He thereby recognized a decisive role of the axiom of choice. This insight must surely have added to his decision to advocate the axiom as sharply as he finally did in the paper defending his well-ordering proof from 1904 ([Zer08a]). It may also have strengthened the self-confidence he exhibited there.

But Zermelo did not confine his foundational work to set theory. After the significance of the paradoxes had been understood in Göttingen, a new field of research was opened: logic. In 1903, faced with the consequences of the Zermelo-Russell paradox, it had become evident that a consistency proof for arithmetic could not be found by slightly revising existing methods of inference, as Hilbert had initially assumed ([Hil00a], 184). For these methods had turned out inconsistent. First attempts at initiating a new logic for his projects were made by Hilbert himself in 1905 ([Hil05a]) and elaborated in the lecture course on the logical principles of mathematical thinking ([Hil05c]). Zermelo then took over the task of creating logical competence in Göttingen with his lecture course on mathematical logic in the summer term of 1908 ([Zer08c]), the first one given in Germany that was based on an official assignment (2.10.1).

2.5.2 Early Work

During the winter semester 1900/01 Zermelo already gave a course on set theory for seven students.[68] It is the first course ever given which is entirely

---

[66] For details see [Pec05b].

[67] [Zer08b], 262; [Zer67c], 200.

[68] The number of listeners can be derived from an attachment of Zermelo's application for the *Privatdozenten* grant (2.3.1), containing a list of his lecture courses; Zermelo to Minister Konrad Studt, 13 March 1901, UAG, 4/Vc 229.

devoted to set theory.[69] The notebook ([Zer00b]) contains additional remarks in Zermelo's hand, perhaps from the time between 1904 and 1906, which anticipate his later axiom system for set theory.

A first result of his set-theoretic research was presented to the Königliche Gesellschaft der Wissenschaften zu Göttingen at the meeting of 9 March 1901. The main theorem says ([Zer02a], 35):

> If a cardinal $\mathfrak{m}$ remains unchanged under the addition of any cardinal from the infinite series $\mathfrak{p}_1, \mathfrak{p}_2, \mathfrak{p}_3, \ldots$, then it remains unchanged if all of them are added at once.

Zermelo's argument makes tacit use of the axiom of choice – less than four years before he will initiate a serious discussion about this axiom by formulating it for the first time explicitly and in full generality. One of the corollaries of his main theorem is the Schröder-Bernstein equivalence theorem[70] which asserts that two sets are equivalent (i.e that they can be mapped onto each other by one-to-one functions) if each set is equivalent to a subset of the other one.[71]

About five years later Zermelo will give a new proof of the equivalence theorem. Whereas the present proof employs cardinal addition and cardinal multiplication,[72] the new one will be based on Dedekind's chain theory. We conclude this subsection by sketching the present proof together with that of the main theorem; for the new proof and the history of the equivalence theorem, cf. 2.9.5.

Under the assumption of the main theorem, let M be a set of cardinality $\mathfrak{m}$ and (using the axiom of choice) $\varphi_1, \varphi_2, \varphi_3, \ldots$ a sequence of one-to-one mappings from M onto subsets $M_1, M_2, M_3, \ldots$ of M such that with $P_i := M \setminus M_i$ for $i = 1, 2, 3, \ldots$ and with "$\sqcup$" meaning disjoint union,

$$|P_i| = \mathfrak{p}_i \text{ for } i = 1, 2, 3 \ldots \text{ and } M = M_1 \sqcup P_1 = M_2 \sqcup P_2 = \cdots.$$

Then, with $\varphi_i[S] := \{\varphi_i(x) \mid x \in S\}$ for S a set and $i = 1, 2, 3, \ldots$,

$$M = \varphi_1[M] \sqcup P_1$$

$$= \varphi_1[\varphi_2[M] \sqcup P_2] \sqcup P_1 = (\varphi_1\varphi_2)[M] \sqcup P_1 \sqcup \varphi_1[P_2]$$

$$= (\varphi_1\varphi_2\varphi_3)[M] \sqcup P_1 \sqcup \varphi_1[P_2] \sqcup (\varphi_1\varphi_2)[P_3] = \cdots$$

and, therefore, with a suitable cardinal $\mathfrak{m}'$,

---

[69]Cf. [Moo02a], 44. It is, however, not known to what extent Cantor's 1885 course on number theory (cf. Ch. 1, fn. 21) was devoted to set theory.

[70]The theorem is a special case of Zermelo's result and does not need the axiom of choice; see the proof given below.

[71]For additional information cf. [Dei05].

[72]It uses cardinals in Cantor's sense ([Can95], 481-482; [Can32], 282-283); each set S is assigned a cardinal $|S|$. Two sets are assigned the same cardinal, they have the same cardinality, if and only if they are equivalent.

$$(*) \qquad \mathfrak{m} = \mathfrak{m}' + \sum_{i=1}^{\infty} \mathfrak{p}_i.$$

In particular, given i and setting $\mathfrak{p}_j := \mathfrak{p}_i$ and $\varphi_j := \varphi_i$ for $j \neq i$, there is a cardinal $\mathfrak{m}'_i$ such that

$$\mathfrak{m} = \mathfrak{m}'_i + \aleph_0 \mathfrak{p}_i$$

and, hence,

$$\mathfrak{m} = \mathfrak{m}'_i + 2\aleph_0 \mathfrak{p}_i = \mathfrak{m} + \aleph_0 \mathfrak{p}_i.$$

Taking now $\aleph_0 \mathfrak{p}_i$ instead of $\mathfrak{p}_i$ in $(*)$, there is a cardinal $\mathfrak{m}''$ such that

$$\mathfrak{m} = \mathfrak{m}'' + \sum_{i=1}^{\infty} \aleph_0 \mathfrak{p}_i$$

and, therefore,

$$\mathfrak{m} = \mathfrak{m}'' + (\aleph_0 + 1) \sum_{i=1}^{\infty} \mathfrak{p}_i = \mathfrak{m} + \sum_{i=1}^{\infty} \mathfrak{p}_i$$

as claimed.

To prove the Schröder-Bernstein equivalence theorem, let

$$\mathfrak{m} = \mathfrak{n} + \mathfrak{p} \text{ and } \mathfrak{n} = \mathfrak{m} + \mathfrak{q}.$$

Then

$$(\dagger) \qquad \mathfrak{m} = \mathfrak{m} + \mathfrak{p} + \mathfrak{q}.$$

By the preceding proof, letting $\mathfrak{p}_i := \mathfrak{p} + \mathfrak{q}$ for $i = 1, 2, 3, \ldots$ and taking $\varphi_2 := \varphi_1$, $\varphi_3 := \varphi_1, \ldots$ (thus making no use of the axiom of choice), one obtains

$$\mathfrak{m} = \mathfrak{m} + \aleph_0(\mathfrak{p} + \mathfrak{q})$$

and, therefore,

$$\mathfrak{m} = \mathfrak{m} + \aleph_0 \mathfrak{p} + (\aleph_0 + 1)\mathfrak{q} = \mathfrak{m} + \mathfrak{q} = \mathfrak{n}.$$

In his new proof of the Schröder-Bernstein equivalence theorem (cf. 2.9.5), Zermelo shows first that for any cardinals $\mathfrak{m}, \mathfrak{p}, \mathfrak{q}$:

$$\text{If } \mathfrak{m} = \mathfrak{m} + \mathfrak{p} + \mathfrak{q} \text{ then } \mathfrak{m} = \mathfrak{m} + \mathfrak{q}.$$

Then $(\dagger)$ yields immediately that $\mathfrak{m} = \mathfrak{m} + \mathfrak{q} = \mathfrak{n}$.

2.5.3 The Heidelberg Congress 1904

Zermelo's striking entry into the development of set theory was heralded in by a sensational event during the Third International Congress of Mathematicians in Heidelberg from 8 to 13 August 1904. It concerned Cantor's continuum hypothesis which states that the cardinality of the continuum – i.e. the cardinality of the set of real numbers – equals the first uncountable cardinal $\aleph_1$. A vivid report was given by Gerhard Kowalewski ([Kow50], 198–203).[73] On 9

---

[73]Cf. also [Dau79], 247–250.

August Julius König delivered a lecture "Zum Kontinuum-Problem"[74] where he claimed to have proved that Cantor's conjecture was false, and more, that the cardinality of the continuum was not an aleph and not the cardinality of a well-ordered set. He thus also claimed to have refuted the well-ordering principle, Cantor's "fundamental law of thought" ([Can83], 550), according to which every set could be given a well-ordering. About the effects of König's lecture we read ([Kow50], 202):

> After the lecture Cantor took the floor with deep emotion. Among other things he thanked God that he had the privilege of living to see the refutation of his errors. The newspapers brought reports on König's important lecture. The grand duke of Baden had Felix Klein inform him about this sensation.

The proceedings report ([Kra05], 42) that Cantor, Hilbert, and Schoenflies contributed to the discussion which followed the address.

On the very next day, Kowalewski continues, Zermelo, an "extremely astute and quick-working thinker" ([Kow50], 202), was able to point out an error in König's argument. In the first part of his lecture König had proved what is sometimes called König's inequality: For a cardinal $\aleph_\gamma$ which is the supremum of a countable set of smaller cardinals, i. e., for a cardinal $\aleph_\gamma$ of cofinality ω, the cardinal $\aleph_\gamma^{\aleph_0}$ of the set of functions from $\aleph_0$ into $\aleph_\gamma$ is bigger than $\aleph_\gamma$. Assuming that the power $2^{\aleph_0}$ of the continuum is an aleph, say $\aleph_\beta$, and taking as $\aleph_\gamma$ the cardinal $\aleph_{\beta+\omega}$, i. e. the supremum of $\aleph_\beta, \aleph_{\beta+1}, \aleph_{\beta+2}, \ldots$ of cofinality ω, he used the result

$$(*) \qquad\qquad \aleph_\alpha^{\aleph_0} = \aleph_\alpha \cdot 2^{\aleph_0} \text{ for every ordinal } \alpha$$

from Felix Bernstein's dissertation ([Bern01]) to obtain the equation $\aleph_{\beta+\omega}^{\aleph_0} = \aleph_{\beta+\omega} \cdot \aleph_\beta$ and, hence, the equation $\aleph_{\beta+\omega}^{\aleph_0} = \aleph_{\beta+\omega}$ contradicting his inequality.[75]

One might conclude from Kowalewski's report that Zermelo had spotted the place where König's proof breaks down: Bernstein's argument for $(*)$ did not work for cardinals of cofinality ω. Doubts have been put forward that Zermelo played such a decisive role in this discussion. As argued by Ivor Grattan-Guinness ([Gra00a], 334) and Walter Purkert ([Pur02], 9–12), it was Felix Hausdorff who discovered the gap some time after the congress, the following passage in Hausdorff's letter to Hilbert from 29 September 1904[76] giving evidence for this:[77]

---

[74]Published in revised form in the congress proceedings ([Koej05a]) and in *Mathematische Annalen* ([Koej05b]).

[75]A detailed analysis is given in [Moo82], 86–88.

[76]SUB, Cod. Ms. D. Hilbert 136. For the German original see [Pur02] or [Ebb05]. For a discussion of the letter in the context of Hausdorff's work on cofinalities, see [Moo89], 108–109.

[77]See also [Haus04], 571.

After the continuum problem had tormented me at Wengen[78] nearly like
a monomania, my first glance here of course turned to Bernstein's dissertation. The worm is sitting at just the place I expected. [...] Bernstein's
consideration gives a recursion from $\aleph_{\alpha+1}$ to $\aleph_\alpha$, but it fails for such $\aleph_\alpha$
which do not have a predecessor, hence exactly for those alephs which Mr.
J. König necessarily needs. I had written in this sense to Mr. König already
while on my way, as far as I could do so without Bernstein's work, but
have received no answer. So, I all the more tend to take König's proof to
be wrong and König's theorem to be the summit of the improbable. [...] I
am really looking forward to the printed proceedings of the congress.

This version is supported by Schoenflies, who attributes a critical attitude
also to Cantor ([Scho22], 101–102):

It is characteristic for Cantor that he right away did not believe that König's
result was valid, despite the exact proof. (He used to joke that he didn't
distrust the king, but the minister.)[79] [...] The Heidelberg congress was
followed by some kind of successor congress at Wengen. Hilbert, Hensel,
Hausdorff, and I myself happened to meet there. Cantor, who had initially
stayed nearby, joined us immediately, because he was constantly eager for
discussion. Again and again, König's theorem was the centre of our conversations. It was really a dramatic moment when one morning Cantor showed
up very early in the hotel where Hilbert and I were staying, and, having
waited for us for a long time in the breakfast room, couldn't help but instantly present us and those around us with a new refutation of König's
theorem. It is well known that the domain of validity of Bernstein's aleph
relation was only checked by Hausdorff: his result, which he already obtained
at Wengen, leads to [its restriction ...] and, hence, to an undermining of
the consequences for the continuum.

Nevertheless, Kowalewski's version may be correct to the effect that also Zermelo (and even König himself) found the error in König's proof. In the postcard which Zermelo wrote to Max Dehn, then a *Privatdozent* at the University
of Münster, on 27 October 1904[80] it says:

So you still do not know the fate of König's talk? Innocent person! Actually, nothing else was discussed all during the holidays. K[önig] solemnly
informed both Hilbert and me that he revoked the Heidelberg proof.
B[ernstein] did the same with his theorem on powers. K[önig] can be happy
that the library in Heidelberg had closed so early; otherwise he might have
made a fool of himself. Thus I had to wait for checking until my return
to G[öttingen]. Then it was instantly obvious. In the meantime, however,
K[önig] himself had found that out when elaborating his proof. So, please,
try to apply the Ph. D. thesis p. 50(?) [two unreadable letters] to the case

---

[78] A resort in the Swiss Alps where some in the audience of König's Heidelberg talk, among them Cantor, Hausdorff, Hilbert, and Schoenflies, met after the
congress.

[79] "König" is German for "king." – The text in brackets is a footnote.

[80] CAH. Reprinted and with English translation in [Ebb05].

$\mathfrak{s} = \mathfrak{m}_1 + \mathfrak{m}_2 + \dots$ treated by K[önig]. You will see the mistake at once which is the only result K[önig] uses. Blument[h]al[81] says that this is a real "Bernstein ruin." (OV 2.10)

In the proceedings of the Heidelberg congress König referred to the objections ([Koej05a], 147):

> Unfortunately [the] proof [of Bernstein's theorem] had an essential gap [...]. I mention this above all in order to withdraw explicitly the conclusion I had made in my lecture at the congress assuming that the Bernsteinian theorem was correct.[82]

The refutation of the continuum hypothesis had thus been itself refuted, and the status of the hypothesis together with that of the related well-ordering principle remained open.[83]

2.6 Zermelo's First Well-Ordering Proof

2.6.1 The Proof and its Motivation

About five weeks after the Heidelberg incident the status of the well-ordering principle was clarified: On 24 September 1904, while staying in the small town of Münden about 25 kilometres southwest of Göttingen, Zermelo wrote a letter to Hilbert which contains a proof of it. Being aware of the significance of the result, Hilbert had the relevant parts of the letter immediately printed in the *Mathematische Annalen* ([Zer04]). Acccording to Zermelo, the proof resulted from conversations which he had had with Erhard Schmidt the week before.

The paper started an "unforeseen and dramatic development" ([Fra66], 132). The lively discussions[84] were more concerned with the method of proof than with its result. Zermelo based the proof on and for the first time explicitly formulated a "logical principle" that "cannot be reduced to a still simpler one, but is used everywhere in mathematical deductions without hesitation,"[85] the *principle of choice*, later called the *axiom of choice*. It guarantees for any set

---

[81]At that time Blumenthal was still a *Privatdozent* in Göttingen.

[82]König's remark forced Bernstein to comment publicly on the validity of his theorem ([Bern05b], 464): "A theorem being as such correct, correctly proved and applied, was in an *auxiliary consideration* mistakenly formulated by me to too great an extent, without having made further application of it, or without leading to any further error. [...] By the way, it is not very probable that the continuum problem can be solved by the notions and methods developed so far."

[83]For the history of the continuum hypothesis cf. [Moo89].

[84]Cf. [Moo82], esp. 85–141. Part of the contributions to the debate were edited by Gerhard Heinzmann ([Hein86]).

[85][Zer04], 516; [Zer67a], 141.

S of non-empty sets the existence of a so-called *choice function* on S,[86] i.e. a function which is defined on S, assigning to each set in S one of its elements. In Zermelo's words,[87]

> the present proof rests upon the assumption that even for an infinite totality of sets there always exist mappings by which each set corresponds to one of its elements, or formally expressed, that the product of an infinite totality of sets each of which contains at least one element is different from zero.[88]

Zermelo gave Erhard Schmidt priority for the idea of basing the well-ordering on an arbitrary choice function. In notes on set theory which fill an only partially conserved writing pad[89] together with notes on vortices and moving fluids and may have been written around 1905, he even attributes the well-ordering theorem to both Schmidt and himself; for in the first line of p. 31 it says: "*Satz*. Jede Menge M ist wohlgeordnet. (E. Schmidt u. E. Z.)."

As a matter of fact, the axiom of choice had already been used earlier without arousing attention, and Zermelo was right in speaking of a usage "everywhere in mathematical deductions without hesitation." Applications ranged from analysis to set theory itself, some of them avoidable, some of them not. Applications of the latter kind include[90] the proof of a theorem of Cantor, published by Eduard Heine in 1872 ([Hei72], 182–183), according to which the continuity of a real function f at some point x equals the sequential continuity of f at x.[91] Moreover, frequent use both in analysis and set theory was made of the fact originating with Cantor[92] that the union of countably many countable sets $S_0, S_1, \ldots$ is countable; the usual argument consisted in choosing enumerations φ_i of the S_i which then were combined into an enumeration of the union. Zermelo himself proceeded similarly in his 1902 paper without referring to the axiom of choice (2.5.2). Both Cantor[93] and Borel ([Bor98], 12–14) proved that every infinite set contains a countably infinite subset. Also without mentioning the axiom, the essence of their argument consists in choosing different elements $x_0, x_1, x_2 \ldots$ from the infinite set

---

[86] A *covering* (*Belegung*) of S in Zermelo's terminology. Also Cantor (for example [Can95], 486; [Can32], 287) and Hilbert (for example [Hil05b], 205) speak of functions as coverings (Belegungen). José Ferreirós ([Ferr99], 317) regards Zermelo's use of "covering" as giving evidence that he bases his work on Cantor's.

[87] [Zer04], 516; [Zer67a], 141.

[88] "Auch für eine unendliche Gesamtheit von Mengen [gibt es] immer Zuordnungen, bei denen jeder Menge eines ihrer Elemente entspricht, oder formal ausgedrückt, daß das Produkt einer unendlichen Gesamtheit von Mengen, deren jede mindestens ein Element enthält, selbst von Null verschieden ist."

[89] UAF, C 129/247.

[90] Cf., e.g., [Jec73].

[91] The real function f is sequentially continuous at x if for any sequence converging to x the corresponding sequence of f-values converges to $f(x)$.

[92] [Can78], 243; [Can32], 120.

[93] [Can95], 493; [Can32], 293.

in question.[94] Later ([Can32], 352) Zermelo will characterize this procedure as

> unsatisfying both intuitively (rein anschaulich) and logically, it calls to mind the well-known primitive attempt to get a *well-ordering* of a given set by successively choosing arbitrary elements. We obtain a correct proof only if we *start* with a *well-ordered* set; then the smallest transfinite section of it will indeed have the required cardinality $\aleph_0$.

In his letter to Dedekind of 3 August 1899 ([Can99]) Cantor sketches an intuitive proof of the well-ordering principle, where he generalizes the argument of successive choices along all ordinals, thereby speaking of a "projection" of ordinals into sets. The key of the proof consists in showing that the cardinal of an infinite set is an aleph. (Then infinite sets are equivalent to well-ordered sets and so are well-orderable as well.) If the set S were a counterexample, i. e. a set without an aleph, then a projection of the inconsistent multiplicity of all ordinals into S (by choosing different elements $s_0, s_1, \ldots, s_\omega, \ldots$) would show that S itself encompassed an inconsistent multiplicity and, hence, could not be a set – a contradiction.[95]

Later ([Can32], 451) Zermelo would likewise criticize this procedure as being based on a "vague intuition" ("vage Anschauung"). By successively choosing new elements of the set S in question along all ordinals, Cantor would apply the idea of time to a process that transcended any intuition. Concerns of just this kind would have let him, Zermelo, base the 1904 proof of the well-ordering principle solely on the axiom of choice which

> postulates the existence of a *simultaneous* selection and which Cantor has used everywhere instinctively without being conscious of it, but which he nowhere formulates explicitly.

In fact, Zermelo's argument can be seen as a corresponding modification of Cantor's procedure. In order to enumerate a set S by ordinals without choosing elements successively, he first applies the axiom of choice to get a choice function that maps every non-empty subset S' of S onto an element $\gamma(S') \in S'$. Then the enumeration can be pictured as

$$(*) \qquad s_0 := \gamma(S), \ s_1 := \gamma(S \setminus \{s_0\}), \ s_2 := \gamma(S \setminus \{s_0, s_1\}), \ldots .$$

It obviously yields a well-ordering on S. The proof itself brings in some further ingredients that serve to avoid the use of ordinals and of Cantor's inconsistent multiplicities. The central point is a definition of the set of all initial sections

---

[94] Cantor proves the theorem "Every transfinite set T has subsets with the cardinal number $\aleph_0$" as follows: "If, by any rule, we have taken away a finite number of elements $t_1, t_2, \ldots t_{\nu-1}$ from T, there always remains the possibility of taking away a further element t_ν. The set $\{t_\nu\}$ where ν denotes an arbitrary finite cardinal number, is a subset of T with the cardinal number $\aleph_0$, because $\{t_\nu\} \sim \{\nu\}$."

[95] Already two years earlier Cantor had informed Hilbert about his argument (cf. 2.4.3, in particular fn. 46).

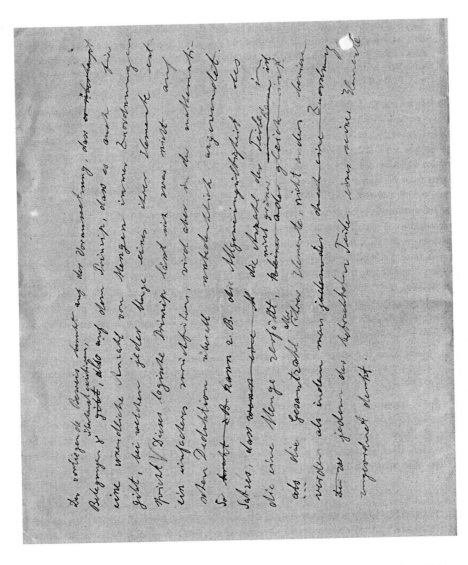

A page from a draft of Zermelo's letter to David Hilbert of 24 September 1904,
upper half, showing the formulation of the Axiom of Choice
Universitätsarchiv Freiburg, Zermelo Nachlass, C 129/255

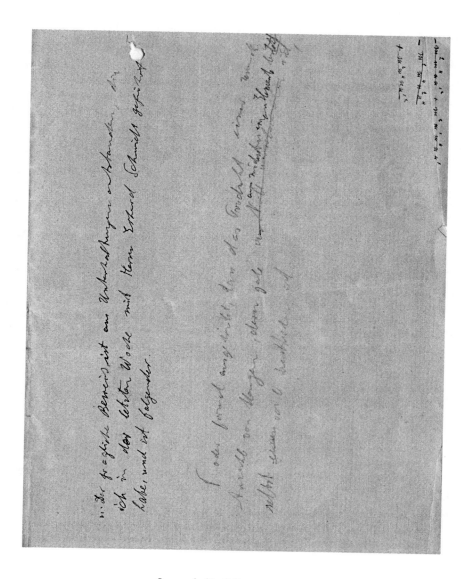

Lower half of the same page
with a reformulation of the Axiom of Choice (in pencil)[96]

---

[96] "oder formal ausgedrückt, dass das Produkt einer unendl[ichen] Anzahl von Mengen, deren jede aus mindestens einem Element besteht, selbst von 0 verschieden ist."

of (∗), Zermelo's γ-sets, and their natural well-orderings. The union L_γ of all γ-sets is again a γ-set. L_γ has to coincide with S, because otherwise it could be extended by $\gamma(S \setminus L_\gamma)$ to a larger γ-set. It therefore yields a well-ordering of S.

We conclude this subsection by giving the proof in detail, thereby closely following Zermelo's argumentation, but exhibiting the well-orderings explicitly.

Let S and γ be given as above. With the initial sections of (∗) in mind, define $S' \subseteq S$ to be a γ-*set* if and only if there is a relation $\prec'$ such that $(S', \prec')$ is a well-ordering and

$$\text{for all } a \in S', \ \gamma(S \setminus \{b \mid b \prec' a\}) = a.$$

We call $\prec'$ a *witnessing relation* of S'.

Some examples (Zermelo's item 4)): If S contains more than one element, the sets $\{s_0\}$ and $\{s_0, s_1\}$ (with s_0, s_1 as in (∗)) are γ-sets; the corresponding witnessing relations are uniquely determined and reflect the ordering given by (∗).

The main observation (Zermelo's claim 5)) says:

(†) *If S_1' and S_2' are γ-sets with witnessing relations $\prec_1'$ and $\prec_2'$, respectively, then $(S_1', \prec_1')$ is an initial section of $(S_2', \prec_2')$ or vice versa.*

To prove (†), let $S_1', S_2', \prec_1'$, and $\prec_2'$ be given according to the assumption. By the comparability of well-orderings, $(S_1', \prec_1')$ is isomorphic to an initial section of $(S_2', \prec_2')$ or *vice versa*. Without loss of generality, let $\iota : S_1' \to S_2'$ be an isomorphism from $(S_1', \prec_1')$ onto an initial section of $(S_2', \prec_2')$. An *reductio ad absurdum* argument shows that ι is the identity on S_1' and, hence, $(S_1', \prec_1')$ an initial section of $(S_2', \prec_2')$. Namely, let $c \in S_1'$ be $\prec_1'$-minimal such that $\iota(c) \neq c$. Then

$$\{b \mid b \prec_1' c\} \ = \ \{b \mid b \prec_2' \iota(c)\}$$

and, hence,

$$c \ = \ \gamma(S \setminus \{b \mid b \prec_1' c\}) \ = \ \gamma(S \setminus \{b \mid b \prec_2' \iota(c)\}) \ = \ \iota(c),$$

a contradiction.

By taking $S_1' = S_2'$, (†) yields that a γ-set has exactly one witnessing relation (which we called the "natural" one in the preceding sketch of the well-ordering proof). As a further immediate consequence (Zermelo's item 6)) we get:

(‡) *Let S_1' and S_2' be γ-sets with witnessing relations $\prec_1'$ and $\prec_2'$, respectively, and let $a, b \in S_1' \cap S_2'$. Then $a \prec_1' b$ if and only if $a \prec_2' b$, and the initial section of $(S_1', \prec_1')$ with domain $\{c \mid c \prec_1' a\}$ coincides with the initial section of $(S_2', \prec_2')$ with domain $\{c \mid c \prec_2' a\}$.*

The union of all witnessing relations well-orders S (Zermelo's claim 7)):

The union L_γ of all γ-sets is a γ-set which equals S, the corresponding witnessing relation $\prec$ being the union of all witnessing relations, i. e.,

$a \prec b$ if and only if
there is a γ-set S' with witnessing relation $\prec'$ such that $a \prec' b$.[97]

---

[97] Zermelo uses (iii) below to define $\prec$.

Proof. First, $(L_\gamma, \prec)$ is an ordering.

(i) By definition, $\prec$ is irreflexiv. (This point is not explicitly mentioned by Zermelo.)

(ii) To show that $\prec$ is transitive, let $a \prec b$ via $(S_1', \prec_1')$ and $b \prec c$ via $(S_2', \prec_2')$. By (†) one may assume that, say, $(S_1', \prec_1')$ is an initial section of $(S_2', \prec_2')$. Then, $a \prec_2' b$ and $b \prec_2' c$. The transitivity of $\prec_2'$ yields $a \prec_2' c$ and, hence, $a \prec c$.

(iii) If $a, b \in L_\gamma$ with $a \neq b$, then, again by (†), there is a γ-set S' such that $a, b \in S'$, and (‡) yields that either $a \prec b$ or $b \prec a$.

Next, $(L_\gamma, \prec)$ is a well-ordering. For assume that $L' \subseteq L_\gamma$ is non-empty, $a \in L'$, and S' is a γ-set with witnessing relation $\prec'$ such that $a \in S'$. By (‡), $\{b \mid b \prec a\} \subseteq S'$, and the $\prec'$-smallest element of $L' \cap \{b \mid b \prec a \text{ or } b = a\}$ is the $\prec$-smallest element of L'.

Furthermore, L_γ is a γ-set with witnessing relation $\prec$. Namely, if $a \in L_\gamma$, say, $a \in S'$ where S' is a γ-set with witnessing relation $\prec'$, then (‡) yields

$$\gamma(S \setminus \{b \mid b \prec a\}) = \gamma(S \setminus \{b \mid b \prec' a\}) = a.$$

Finally, L_γ coincides with S. Otherwise, $L_\gamma \cup \{\gamma(S \setminus L_\gamma)\}$ is a γ-set, its witnessing relation arising from $\prec$ by adding $\gamma(S \setminus L_\gamma)$ at the end, a contradiction.

The proof is complete. Note that, apart from the existence of γ, it does not make use of the axiom of choice.[98]

2.6.2 Reactions

Despite its lucidity, the proof became an object of intensive criticism. In particular, the formulation of the axiom of choice and its use started a debate that is unsurpassed in the modern history of classical mathematics and subsided only after decades. As early as on 15 November 1904, at a meeting of the Göttingen Mathematical Society, Zermelo, in a talk dealing in particular with "the question of the well-orderedness or non-well-orderedness of the continuum," already had to defend his proof against immediate objections by Émile Borel, Julius König, Felix Bernstein, and Arthur Schoenflies.[99] Three years later he would decide to publish a justification in *Mathematische Annalen* where he would defend his point of view in a self-conscious and somewhat polemical way ([Zer08a]). In the postcard to Dehn from 27 October 1904 mentioned above[100] he writes:

> Best thanks for your friendly assent. Two days ago I also received such an assent from the "younger Berlin school," i.e., apparently from Landau and Schur. But if you knew how I am bombarded by the experts of set theory, by Bernstein, Schoenflies, and König, with objections, for the time being in letter form! Some fine polemics will still develop in the Annalen; but I am not afraid of them. [...] König [...] fantasizes about the set W of *all* order types which *could be* contained in many a set so that my $L_\gamma = W$ and then

---

[98] For a detailed analysis of the proof cf. [Kan97], 292seq., or [Kan04], 492seq.
[99] *Jahresbericht der Deutschen Mathematiker-Vereinigung* **14** (1905), 61.
[100] In 2.5.3; cf. fn. 52.

my conclusion $L\gamma = S^{101}$ would be wrong. Strangely enough, he is backed up in this matter by Bernstein and Schoenflies, by the latter of course in quite a confused and misunderstood manner. The W-believers already deny the def[inition] of subset, perhaps even the principle of the excluded middle. (OV 2.10)

As indicated here, the doubts about the well-ordering proof were partly due to lack of understanding and document that Zermelo was ahead of many of his critics with respect to handling the set-theoretic notions in a clear manner.

The essential point of concern was the axiom of choice, stating the existence of choice functions without providing a rule of how to perform the selections. A typical statement is given by Borel ([Bor05], 194):

> Any argument where one supposes an *arbitrary choice* to be made a non-denumerably infinite number of times [...] is outside the domain of mathematics.[102]

Already more than a decade earlier, Guiseppe Peano ([Pea90], 210) had claimed that "one cannot apply infinitely many times an *arbitrary* rule by which one assigns to a class an individual of this class."[103] He therefore diligently specified a deterministic rule in the case under consideration. Rodolpho Bettazzi ([Bett96], 512) had criticised infinitely many choices, too, but had been close to justifying them by an axiom. He had, however, drawn back, perhaps under the influence of Peano: with respect to corresponding choices of Dedekind in the *Zahlen* treatise he remarks that such a procedure "does not seem rigorous, unless one wishes to accept as a postulate that such a choice can be carried out – something, however, which seems ill-advised to us."[104]

Nevertheless, explicit criticism of infinitely many choices had been rare, and tacit use of the axiom of choice had been abundant. Why then such an opposition, when the axiom appeared *expressis verbis* on the stage? The opposition was so strong that Fraenkel and Bar-Hillel ([FraB58], 47) characterize the axiom as "probably the most interesting and, in spite of its late appearance, the most discussed axiom of mathematics, second only to Euclid's axiom of parallels which was introduced more than two thousand years ago," and Hilbert ([Hil26], 178) calls it "one of the most attacked axioms up to now in mathematical literature."

---

[101]Zermelo writes "M" instead of "S".

[102]"[...] tout raisonnement oú l'on suppose un *choix arbitraire* fait une infinité non dénombrable de fois; de tels raisonnements sont en dehors du domaine des mathématiques." – Borel remarks in addition that the axiom of choice is equivalent to the well-ordering principle, as the principle trivially implied the axiom.

[103]"Mais comme on ne peut pas appliquer une infinité de fois une loi *arbitraire* avec laquelle à une classe a on fait correspondre un individu de cette class." Cf. also [Ferr99], 315seq.

[104]Citation from [Moo82], 26–27.

As argued by Moore,[105] the situation was characterized by two features: First, there was no sharp boundary between constructive and non-constructive methods,[106] and second, those who used the axiom of choice tacitly were not aware of its deductive strength. So "like a stroke of lightning, Zermelo's proof suddenly illuminated the landscape and caused mathematicians to scrutinize the assumptions upon which they had been resting unawares."

Speaking of functions as of *laws* – so Cantor[107] and even Zermelo still in 1908[108] with respect to choice functions – calls to mind the view that functions came with some kind of explicitness.[109] And now Zermelo emphasized that any *arbitrary* choice function worked for his proof of the well-ordering principle, couching his view in the ostentatious form: *"Imagine that with every* [non-empty] *subset S′ there is associated an arbitrary element s′ that occurs in S′ itself and that may be called the 'distinguished' element of S′."*[110] According to Akihiro Kanamori this really meant a turning point for and a conceptual shift in mathematics,[111] at the same time marking the beginning of modern abstract set theory ([Kan97], 291).[112]

2.7 Finite Sets

In May 1905 Zermelo spent part of a longer spring cure in Italy in the health resort of Casamicciola on the island of Ischia near Naples (2.11.1). Not only was he reading the galley proofs of his Gibbs' translation ([Gib05]), but, as he wrote to Hilbert,[113] he was "once again well occupied" ("flüssig beschäftigt") with set theory, too. He concentrated mainly on Dedekind's concept of number and on finite sets. About two months later, then staying in the town of Tremezzo on Lake Como, he informed Hilbert in detail about his results:[114]

> [I succeeded in] completely and with clarity solving the problem that I am going to call "Dedekind's problem of the concept of number." In doing so,

---

[105][Moo82], 83–84.

[106]As late as 1905 Bernstein ([Bern05c]) explicitly restricts the notion of a decimal representation of a real number (conceived as a function on the set of integers) in a certain constructive manner and concludes that there are only $\aleph_1$ many of them, thus claiming to have proved the continuum hypothesis.

[107]For example, [Can95], 486; [Can32], 287.

[108][Zer08a], 108; [Zer67b], 184.

[109]See, for example, [Fel02a] for the development of the notion of function.

[110][Zer04], 514: "*Jeder Teilmenge M′ denke man sich ein beliebiges Element m′_1 zugeordnet, das in M′ selbst vorkommt und das 'ausgezeichnete' Element von M′ genannt werden möge.*"

[111][Kan96], 10; [Kan04], 498.

[112]Not only conceptually, but also technically Zermelo's proof anticipates important developments, among them von Neumann's transfinite recursion theorem ([Kan97], 292; [Kan04], 8).

[113]Letter of 7 May 1905; SUB, Cod. Ms. D. Hilbert 447, fol. 1.

[114]Letter of 29 June 1905; SUB, Cod. Ms. D. Hilbert 447, fols. 2/1–2/3, p. 2.

I was able to make Dedekind's deduction, which is not very appealing and hard to understand, obsolete. (OV 2.11)

Zermelo refers to Dedekind's Theorem 160 of the *Zahlen* treatise ([Ded88]) which essentially states that sets are finite if and only if they are – in modern terminology – Dedekind finite. Here, sets are *finite* if they are equivalent to a proper initial section of the natural numbers; they are *Dedekind finite* if they are not equivalent to a proper subset. By slightly reformulating Dedekind's definition of the series of natural numbers (ibid., Erklärung 71), he can incorporate the method of his 1904 well-ordering proof to show directly that the set of natural numbers can be well-ordered of type ω,[115] the underlying successor function also being the successor function with respect to the well-ordering. The critical direction of Theorem 160 consists in showing that Dedekind finiteness implies finiteness. Without comment Dedekind uses the axiom of choice.[116] Zermelo sketches a proof using his well-ordering theorem:[117] Let S be a Dedekind finite set. In order to show that S is finite,

> imagine that it is *well-ordered*. If then some initial section or the whole set had *no last element*, one could associate with any of its elements the immediately following element, where, of course, the *first* element is *not* used, and the set would be mapped on a [proper] part of itself, against the assumption. However, a "sequence" which together with all its initial sections contains a last element, is similar to an initial section of the "number series."[118] (OV 2.12)

The use of the axiom of choice would be neither accidental nor bad; indeed, the axiom was (ibid., 6–7)

> a *necessary presupposition* of the theorem which could *not at all* be proved *without* "choice."[119] *Cantor* as well proves the theorem that "every infinite set contains a subset of power $\aleph_0$" by first choosing an element, then another one, and thereupon using the notorious "etc."[120] Made precise in the right manner, it comes down exactly to my well-ordering theorem. One can also *not* argue that in the case of *finite* sets one has to "choose" only a *finite* number of times, which is supposed to be still allowed according to the

---

[115]I. e., in such a way that each proper initial section possesses a last element, but the whole set does not.

[116]Namely for the critical direction of the preparatory theorem 159 which essentially says that a set is not Dedekind finite if and only if for each finite set S it contains a subset equivalent to S. As is known today, the axiom of choice is necessary; cf., e. g. [Jec73], Ch. 5.

[117]Letter of 29 June 1905; SUB, Cod. Ms. D. Hilbert 447, fols. 2/1–2/3, p. 5.

[118]In his new proof paper ([Zer08a], 114; [Zer67b], 188) Zermelo formulates this direction as follows: "A set that is not equivalent to any of its parts [i. e., its proper subsets] can always be ordered in such a way that every subset possesses a first as well as a last element.

[119]Russell ([Rus06], 49) expresses a similar opinion.

[120][Can95], 493; [Can32], 293.

opinion of those "empiricists." For we know from the set considered *only* the *negative* feature that it is not equivalent to any of its parts; the "successive selection" in the empiric sense, however, requires a finite *order type* whose existence cannot be presupposed, but just should be *proved*.[121] (OV 2.13)

Zermelo intended to talk about his results at the 77th meeting of the Gesellschaft deutscher Naturforscher und Ärzte at Merano in September 1905 ("'about the theory of finite number' or something similar"). In the end, he did not do so. However, the investigations strengthened his conviction that the well-ordering principle and the axiom of choice were of fundamental importance. These insights might have triggered off his decision to confront sharply the opponents of his well-ordering proof:[122]

> The theory of finite sets is impossible without the "principle of choice," and the "well-ordering theorem" is the true fundament of the whole theory of number. This would be a valuable weapon against my "empiristic-sceptical" opponents (Borel, Enriques, Peano, *et al.*), whereas the "dogmatic" opponents (Bernstein, Schoenflies, and pals) in general are not taken seriously in this question, as I could convince myself recently in a conversation with Enriques (in Florence). The latter consequently doubted almost the whole of set theory including the concept of power, so that it is actually no longer possible to quarrel with him.[123] (OV 2.14)

The cancellation of the talk in Merano did not mean that he would give up the topic. In his axiomatization paper ([Zer08b]) he announces:[124]

> A second paper, which will develop the theory of well-ordering together with its application to finite sets and the principles of arithmetic, is in preparation.

The "second paper" announced here was not written. However, already earlier, in May 1907, Zermelo had finished a paper on the intended topic. He had

---

[121]Zermelo refers to Philip Jourdain's theory of choice in [Jou05], esp. 468. He also discusses it critically in [Zer08a], 120–121.

[122]Letter to Hilbert of 29 June 1905; SUB, Cod. Ms. D. Hilbert 447, fols. 2/1–2/3, p. 7–8.

[123]Federigo Enriques (1871–1946), a professor at the universities of Bologna and Rome; he worked in projective and algebraic geometry, but was also strongly interested in logic and the history and philosophy of mathematics. Zermelo's judgement cannot be claimed anymore for the later Enriques. For example, in his *Per la storia della logica* ([Enr22]; German [Enr27]) he writes on Cantor's theory of the infinite ([Enr27], 126): "[Cantor investigates] without prejudice the properties of the relations which may exist between infinite sets and thereby arrives at further remarkable and nevertheless sensible features, so that all semblance of contradiction, to speak with Bolzano, can now be recognized as semblance. The result obtained between 1878 and 1883 means a landmark in the history of human thought and it is, like the analysis of the infinite in the Ancient World, an event that closely touches the interests of logic."

[124][Zer08b], 161; [Zer67c], 201.

sent it to Poincaré for publication in the *Revue de métaphysique et de morale*,
as it addressed Poincaré's "empiristic-sceptical" point of view. Because of its
mathematical character Poincaré recommended it to Magnus Gustav Mittag-
Leffler for publication in *Acta mathematica*[125] where it appeared two years
later ([Zer09a]). Moreover, Zermelo gave a corresponding talk at the Fourth
International Congress of Mathematicians 1908 in Rome ([Zer09b]). Both papers are early documents of Zermelo's set-theoretic reductionism, the reduction of mathematical concepts to set-theoretic ones, exemplified here by the
notion of number.[126] At the beginning of the first paper ([Zer09a]) we read
that "every theorem stated about finite numbers is nothing but a theorem
about *finite sets*." The introduction to the second paper ([Zer09b]) repeats
this point of view, at the same time also describing the contents:

> If one intends to base arithmetic on the theory of natural numbers as *finite
> cardinals*, one has to deal mainly with the definition of *finite set*; for the
> cardinal is, according to its nature, a property of a set, and any proposition
> about finite cardinals can always be expressed as a proposition about finite
> sets. In the following I will try to deduce the most important property
> of natural numbers, namely the *principle of complete induction*, from a
> definition of finite set which is as simple as possible, at the same time
> showing that the different definitions [of *finite set*] given so far are equivalent
> to the one given here. When proceeding in this way, we will have to make
> use of the basic notions and methods of general set theory as created by G.
> Cantor and R. Dedekind. However, unlike Dedekind [...] we will *not* use
> the assumption that there is an "infinite set," i. e. a set which is equivalent
> to one of its proper parts.

Both papers end with "a sharp point against Poincaré's scepticism"[127] by
emphasizing that in contrast to the notion of natural number the field of
analysis needs the existence of infinite sets ([Zer09b], 11):

> As a consequence, those who are really serious about the rejection of the
> actual infinite in mathematics should stop at general set theory and the
> lower number theory and do without the whole of modern analysis.[128]

Zermelo's concern about avoiding the assumption of the existence of infinite
sets clearly follows Hilbert's axiomatic programme which included a discussion
of the axioms needed in a given proof. His "simple definition" of finiteness says
that a set S is, say, *Zermelo finite* if it has at most one element[129] or there is
a non-empty proper subset P of S and a one-to-one function f from P onto a

---

[125]Letter from Poincaré to Zermelo of 19 June 1907; UAF, C 129/90.

[126]Cf. [Hal84], 244seq., and [Tay93] for a detailed discussion of Zermelo's reductionism.

[127]Zermelo in a letter to Cantor ([Zer08e]).

[128]Zermelo is not quite right here. As shown, for example, by Gaisi Takeuti
([Tak78]), analytic number theory is not stronger than elementary number theory,
in technical terms: it is a conservative extension of Peano arithmetic.

[129]This case is missing in Zermelo's paper.

proper subset $P' \neq P$ of S such that f "shifts" P to P', i. e., for any partition of S into two non-empty subsets at least one of these subsets contains an element s of P, but does not contain $f(s)$. Having proved by use of the well-ordering theorem that Zermelo finiteness coincides with Dedekind finiteness, he once more emphasizes the indispensability of the axiom of choice: Without it one probably would have to admit a third kind of finite set as being possible without being able to say "something positive" about them.[130] The "basic notions and methods of set theory as created by Cantor and Dedekind" which he mentions near the end of the quotation, refer to his axiom system of set theory that was nearly completed when he wrote the earlier paper ([Zer09a]). His principle of complete induction says that any property which holds for the empty set and which, for any Zermelo finite set S and any $s \in S$, is passed on from $S \setminus \{s\}$ to S, holds for all Zermelo finite sets.

Zermelo did not extend his investigations by giving a set-theoretic model of arithmetic. This was done by Kurt Grelling in his dissertation ([Grel10]) which was officially supervised by Hilbert, but *de facto* by Zermelo.[131]

Altogether, Zermelo's considerations on finite sets may be viewed as illuminating the road to the axiomatization of set theory and adding to his decision to strongly advocate the axiom of choice.

2.8 The New Proof Paper

Four years after he had shed light on the axiom of choice, Zermelo gave a comprehensive answer to the critics of his well-ordering proof. The paper ([Zer08a]) pursues several lines of defence and also provides a new proof of the well-ordering principle. He speaks restrainedly of his intention to bring up some objections; however, besides all their brilliance, some of his remarks demonstrate a high degree of engagement and polemical sharpness both in argumentation and tone that we have met in the Boltzmann debate and that we will meet again about 25 years later in discussions with Gödel and Skolem when he was confronted with ideas and opinions that totally contradicted his point of view. At the same time, his "spirited" response "provided his first extended articulation of his expansive view about mathematics" ([Kan04], 499).

2.8.1 Zermelo's Defence

In a clear and definite way Zermelo refutes criticism concerning the technical aspects of his 1904 proof. Those who criticized the axiom of choice are reminded of their former tacit use of it. In particular, Zermelo argues against

---

[130] A systematic comparison of different notions of finiteness, including that of Zermelo, was later given by Alfred Tarski ([Tar24]).

[131] For details cf. [Par87]. Zermelo's assessment is in UAF, C129/281.

criticism put forward, for instance, by Émile Borel ([Bor05], 194) and Guiseppe Peano[132] which blamed him for using the axiom without having proved it. In his answer to these "empiristic-sceptical" opponents he strongly pleads for evidence as a source of mathematics:[133]

> I just cannot *prove* this postulate, as I expressly emphasized at the end of my note,[134] and therefore cannot compel anyone to accept it apodictically. [...] Now even in mathematics *unprovability*, as is well known, is in no way equivalent to *nonvalidity*, since, after all, not everything can be proved, but every proof in turn presupposes unproved principles. Thus, in order to reject such a fundamental principle, one would have to ascertain that in some particular case it did not hold or to derive contradictory consequences from it; but none of my opponents has made any attempt to do this. [...] How does Peano arrive at his own fundamental principles?[135] [...] Evidently by analyzing the modes of inference that in the course of history have come to be recognized as valid and by pointing out that the principles are intuitively evident and necessary for science – considerations that can all be urged equally well in favor of the disputed principle. [...] That this axiom, even though it was never formulated in textbook style, has frequently been used, and successfully at that, in the most diverse fields of mathematics, especially in set theory, by Dedekind, Cantor, F. Bernstein, Schoenflies, J. König, and others is an indisputable fact. [...] Such an extensive use of a principle can be explained only by its *self-evidence*,[136] which, of course, must not be confused with its provability. No matter, if this self-evidence is to a certain degree subjective – it is surely a necessary source of mathematical principles, even if it is not a tool of mathematical proofs.

And he continues:[137]

> Banishing fundamental facts or problems from science merely because they cannot be dealt with by means of certain prescribed principles would be like forbidding the further extension of the theory of parallels in geometry because the axiom upon which this theory rests has been shown to be unprovable. Actually, principles must be judged from the point of view of science, and not science from the point of view of principles fixed once and for all. Geometry existed before Euclid's *Elements*, just as arithmetic and set theory did before Peano's *Formulaire*, and both of them will no doubt survive all further attempts to systematize them in such a textbook manner.

To demonstrate the indispensability of the axiom of choice, he lists seven theorems ranging from set theory to analysis which would be well acknowledged, whereas the known proofs could not dispense with the axiom. Among them

---

[132][Pea06b], 145seq.; [Pea73], 207seq.

[133][Zer08a], 111seq.; [Zer67b], 186seq.

[134]I. e., [Zer04].

[135]Zermelo refers to Peano's project of systematizing mathematics in his *Formulaire de mathématiques* ([Pea95]).

[136]For a discussion of Zermelo's notion of self-evidence, cf. [Sha06], 25seq.

[137][Zer08a], 115; [Zer67b], 189.

are the statement that a countable union of countable sets is countable and the theorems about the existence of a Hamel basis[138] and about the existence of non-continuous solutions of the functional equation $f(x+y) = f(x) + f(y)$. The latter examples go back to existence proofs by Georg Hamel ([Ham05]) which are based on a well-ordering of the reals.

Part of Zermelo's defence is concerned with Poincaré who doubted the 1904 well-ordering proof because of the use of an impredicative definition, a definition which implicitly involved the object it was to define, thus giving rise to a *vicious circle*. Poincaré had in mind the union L_γ of all initial sections of the series

$$(*) \qquad s_0 := \gamma(S), \ s_1 := \gamma(S \setminus \{s_0\}), \ s_2 := \gamma(S \setminus \{s_0, s_1\}), \dots$$

as defined in 2.6.1. It plays a central role because it yields the desired well-ordering of the set S concerned. Zermelo defends himself by pointing to ample use of impredicative definitions in set theory and analysis and summarizes:[139]

> A definition may very well rely upon notions that are equivalent to the one to be defined; indeed, in every definition *definiens* and *definiendum* are equivalent notions, and the strict observance of Poincaré's demand would make every definition, hence all of science, impossible.

However, he acknowledges that Poincaré did justice to the axiom of choice by characterizing it[140] as unprovable but indispensable for the theory of cardinalities, even if denying the existence of $\aleph_1$ (ibid., 315; 1069). Poincaré's discussion in Kantian terms about the character of mathematical propositions and the characterization of the axiom of choice as a synthetic *a priori* judgement is dismissed by Zermelo as being primarily philosophical and not touching mathematics as such.

A last type of criticism raised about the 1904 proof, for example by Felix Bernstein ([Bern05a]) and Arthur Schoenflies ([Scho05]), likewise originated in the maximal section L_γ, but now rooted in concerns that Zermelo's procedure might lead to an analogue of the Burali-Forti paradox.[141] The paradox arises from the assumption that the ordinals form a set W, because then the natural order type of W or, even more, the order type of a one-element extension of W yields an ordinal not in W, a contradiction. Schoenflies, for example, apparently believed that L_γ might be order-isomorphic to W; so there might be the possibility that the well-ordering on L_γ could not be extended, a point that was central for Zermelo's proof. It is, in particular, with the refutation of this type of criticism that Zermelo documents a deep and sharp understanding

---

[138] I. e., a basis of the vector space of the reals over the field of the rationals.

[139] [Zer08a], 118; [Zer67b], 191. – For details cf. [Fef05], in particular 591–599.

[140] [Poi06a], 313; [Ewa96], 1067–1068.

[141] The criticism had been encouraged by Hilbert, although he did not find the objections valid; cf. [Moo82], 109.

of the set-theoretic picture.[142] Apart from discussing the impact of Bernstein's criticism in 2.8.4, we will not go into further details.[143]

In the end, Zermelo sums up:[144]

> The relatively large number of criticisms directed against my small note testifies to the fact that, apparently, strong prejudices stand in the way of the theorem that any arbitrary set can be well-ordered. But the fact that in spite of a searching examination, for which I am indebted to all the critics, no *mathematical error* could be demonstrated in my proof and the objections raised against my *principles* are mutually contradictory and thus in a sense cancel each other, allows me to hope that in time all of this resistance can be overcome through adequate clarification.

The polemic style of his defence found a varied echo. A medium stance is represented by Hessenberg:[145]

> Zermelo has found a new proof of the well-ordering theorem. I held its proof sheets in my own hands yesterday. The proof is marvellous and very simple. I myself had failed to go that way. He also has 8 pages of polemics with Poincaré, Bernstein, Jourdain, Peano, Hardy, Schoenflies etc., which are very funny and hit the logicists hard. He stresses the synthetic character of mathematics everywhere and accuses Poincaré of confusing set theory and logicism. (OV 2.15)

A sharp judgement is given by Fraenkel who speaks of "a treatise which has in respect to its sarcasm nothing like it in mathematical literature."[146]

A different opinion is represented by the board of directors of the Göttingen Seminar of Mathematics and Physics. They argue that the criticism of the 1904 well-ordering proof was unjustified according to the judgement of the experts among them and might have hindered Zermelo's academic career. They then continue:[147]

> [This criticism] confused public opinion which had hardly been won for the necessity and utility of set theory. We hope that Zermelo's formally well thought-out and moderate anti-criticism, which appears to us as entirely convincing, will soon redeem the minds. (OV 2.16)

---

[142] In this connection Zermelo criticizes Julius König's paper [Koej05c] where König tries to prove along the lines of Richard's paradox that the set of real numbers cannot be well-ordered ([Zer08a], 119, first footnote; [Zer67b], 191, footnote 8). Probably Zermelo herewith follows a request of Hilbert in spite of his opinion that König's arguments were "nowhere else taken seriously, as far as I know" (letter to Hilbert, dated Saas-Fee 27 August 1907; SUB, Cod. Ms. D. Hilbert 447, fol. 6, 3-4). For a discussion of König's paper and its successor ([Koej07]), cf. [Moo82], 119–120.

[143] For thorough information see [Moo82], Chs. 2 and 3.

[144] [Zer08a], 128; [Zer67b], 198.

[145] Letter to Leonard Nelson of 7 September 1907; ASD, 1/LNAA000272.

[146] [Fra67], 149, fn. 55. The footnote will be quoted in 3.1.1.

[147] Outline of a letter to the Prussian Minister of Cultural Affairs of 24 January 1910; SUB, Cd. Ms. D. Hilbert 492, fol. 3/2.

2.8.2 The New Proof

Zermelo's 1904 proof of the well-ordering principle together with the formulation of the axiom of choice marks a shift to abstract set theory. The new proof paper emphasizes the abstract character of both the proof and the axiom even more seriously, thereby perhaps losing some of the transparency of the 1904 proof.

Zermelo had not accepted Cantor's procedure of well-ordering a set or an inconsistent multiplicity by successively choosing new elements. On the contrary, he claimed, that he had been led to accept the indispensability of the axiom of choice just by the endeavour to state more precisely this intuitive procedure. As he had learnt from the reactions, the point of successive choices had been a condensation point of misunderstanding and criticism – despite the fact that successive choices do not appear in his 1904 paper, but have been replaced by choice functions. Therefore, the new proof is designed[148]

> [to bring out] more clearly than the first proof did, the purely formal character of the well-ordering, which has nothing at all to do with spatio-temporal arrangement.

Furthermore, it might be intended to meet the criticism against the formation of the union L_γ of all initial sections of the series

$$(*) \qquad s_0 := \gamma(S), \ s_1 := \gamma(S \setminus \{s_0\}), \ s_2 := \gamma(S \setminus \{s_0, s_1\}), \dots$$

given above: Instead of getting the well-ordering of the set S in question "from below" by forming the union L_γ (cf. 2.6.1), he now obtains it "from above" by forming the intersection of all γ-chains:[149] Having in mind the set of final sections of $(*)$, he defines a γ-chain Θ as a set of subsets of S which satisfies the following conditions:

(i) $S \in \Theta$,
(ii) $S' \setminus \{\gamma(S')\} \in \Theta$ for nonempty $S' \in \Theta$,
(iii) $\bigcap T \in \Theta$ for nonempty $T \subseteq \Theta$.

The power set of S is a γ-chain. Furthermore, the intersection of a non-empty set of γ-chains is a γ-chain. In particular, the intersection Θ_γ of the set of all γ-chains is a γ-chain.

Zermelo does not mention explicitly, but proves implicitly, that Θ_γ represents the set of final sections of the series $(*)$, i.e. with $\prec$ being the witnessing relation for L_γ, that

$$\Theta_\gamma = \big\{ \{b \mid a \prec b \text{ or } a = b\} \mid a \in S \big\}.$$

With this correspondence in mind, the definition of the intended well-ordering is immediate:

---

[148] [Zer08a], 107; [Zer67b], 183.
[149] Zermelo uses Dedekind's chain theory and its generalization into the transfinite; for an analysis of the chain argument cf., e. g., [Ferr99], 319.

For all elements $s, s' \in S$, $s < s'$ if and only if there is a set in Θ_γ which contains s' but not s.[150]

Zermelo then proves that $(S, <)$ is a well-ordering, showing as a first essential step that Θ_γ is an $\supseteq$-chain such that for any $S_1, S_2 \in \Theta_\gamma$ either $S_1 \supseteq S_2$ or S_2 is not empty and $S_2 \setminus \{\gamma(S_2)\} \supseteq S_1$.

In order to support his view that the well-ordering he has obtained "has nothing to do with spatio-temporal arrangement," Zermelo also reformulates the axiom of choice. The new version was to exhibit more clearly its "formal" character and to exclude any idea of successive choices and even the idea of simultaneous choices – an idea still suggested by choice functions and "still subjectively coloured and exposed to misinterpretations:"

> A set S which is decomposed into a set of disjoint parts $A, B, C, \ldots$ each containing at least one element, possesses at least one subset S_1 having exactly one element in common with each of the parts $A, B, C, \ldots$ considered.[151]

Without being aware of it, he had thus arrived at Russell's multiplicative axiom ([Rus06], 48–49):

> Given a mutually exclusive set of classes k, no one of which is null, there is at least one class composed of one term out of each member of k.

It was only in 1908 that Zermelo became informed about this fact and about Russell's proof of the equivalence of the multiplicative axiom with the 1904 version ([Rus08], 259).

2.8.3 Toward Acceptance of the Axiom of Choice

After 1908 objections against the axiom of choice gradually ceased, in particular due to the work of Wacław Sierpiński[152] and the Polish school starting in the mid 1910s.[153] As an early witness who forsaw the development which would finally lead from the insight into the usefulness of the axiom and its necessity to its acceptance, we quote Ernst Steinitz ([Ste10], 170–171):

---

[150]Zermelo chooses an equivalent variant.

[151]"Eine Menge S, welche in eine Menge getrennter Teile $A, B, C, \ldots$ zerfällt, deren jeder mindestens ein Element enthält, besitzt mindestens eine Untermenge S_1, welche mit jedem der betrachteten Teile $A, B, C, \ldots$ genau ein Element gemein hat." In the axiomatization paper ([Zer08b], 266; [Zer67c], 294) Zermelo adds the following comment which seems to contradict his aims: "We can also express this axiom by saying that it is always possible to *choose* a single element from each [part $A, B, C, \ldots$ of S] and to combine all the chosen elements, $m, n, r, \ldots$, into a set S_1." According to Fraenkel ([Fra22b], 232) Zermelo considers it "an annotation that does not touch the theory." For more information on this point see 3.5.

[152]For example, by his survey [Sie18].

[153]For a detailed description of this process cf. [Moo82], Ch. 4.

Many mathematicians still oppose the axiom of choice. With the increasing recognition *that there are questions in mathematics which cannot be decided without this axiom,* the resistance to it may increasingly disappear.

This process culminated in the mid 1930s when Max Zorn demonstrated the usefulness of his maximum principle ([Zor35], 667) that is equivalent to the axiom of choice and in fact goes back to a maximum principle of Hausdorff ([Haus09], 300–301). Slightly generalized by Bourbaki ([Bou39], 36–37) it became what is now Zorn's lemma: "Any inductively ordered set [i. e. any set which is partially ordered such that every ordered subset has an upper bound in it], contains a maximal element."[154]

Zorn's lemma revealed a convincing power in algebra and topology, at the same time enabling mathematicians to replace the use of the well-ordering principle and of transfinite induction by a method of greater convenience. Additional support was provided by Gödel's proof of relative consistency ([Goe38]): If set theory without the axiom of choice is consistent, then so is set theory with the axiom. In 1963 Paul J. Cohen proved ([Coh66]) that the same is true for the negation of the axiom. So altogether, the axiom of choice is independent of the other axioms of set theory.

Coming back to Zermelo, it seems that after 1908 he lost interest in further discussions of the axiom. The matter seemed to be definitely settled for him. On 16 March 1912 Russell invited him to the Fifth International Congress of Mathematicians in Cambridge:[155]

As the representative of the philosophical section of the coming mathematical congress I urgently ask you to come to Cambridge and give a talk to us. If you would lecture, for example, about the reasons for your principle of choice, this would be of highest interest. Moreover, it would be a great personal pleasure for me to renew your acquaintance.[156]

Zermelo accepted the invitation, proposing a talk about "the theory of the game of chess."[157] Following Russell's repeated assurance that "it would be of highest interest if you gave a talk on the foundations of set theory," he finally decided to give two talks, the proposed talk on the theory of chess, and an additional one entitled "On axiomatic and genetic methods in the foundation of mathematical disciplines."[158] The latter talk probably did not involve the axiom of choice either (4.3.1). In any case, Zermelo chose only the first talk for publication in the proceedings. The resulting paper ([Zer13]) marks the beginning of the theory of games (3.4.1).

---

[154] "Tout ensemble ordonné inductif possède au moins un élément maximal."

[155] Letter from Russell to Zermelo of 16 March 1912; UAF, C 129/99.

[156] Russell and Zermelo had met each other at the Fourth International Congress of Mathematicians 1908 in Rome.

[157] Letter from Russell to Zermelo, dated 8 April 1912; UAF, C 129/99.

[158] "Ueber axiomatische und genetische Methoden bei der Grundlegung mathematischer Disziplinen."

Felix Bernstein in the 1920s
Courtesy of Niedersächsische Staats- und Universitätsbibliothek Göttingen
Abteilung für Handschriften und seltene Drucke

2.8.4 Postscriptum: Zermelo and Bernstein

From the time between the completion of his 1908 papers and their publication, when Zermelo was staying in Glion near Montreux in the Swiss Alps to cure his tuberculosis, some of his correspondence with the Göttingen philosopher Leonard Nelson (1882–1927) have been preserved.[159] In cooperation with Gerhard Hessenberg, Nelson worked on a programme of "Critical Mathematics" ("Kritische Mathematik") as a philosophical alternative to Frege's logicism and Poincaré's anti-logicism. Already in 1904, when Nelson was still a student, both had become co-founders of the new series of *Abhandlungen der Friesschen Schule* which was to promote Kant's critical philosophy in the continuation of the philosopher Jakob Friedrich Fries (1778–1843).[160] Nelson regarded his contributions to the questions of the epistemological base of mathematics and of the nature of mathematical axioms as a philosophical foundation of Hilbert's axiomatic programme.[161] There were lively personal

---

[159] ASD, 1/LNAA000399. Letters of Zermelo quoted in this section stem from this source. The correspondence is edited in [Pec90a], 46–51.

[160] The old *Abhandlungen* had been published by Fries' disciple Ernst Friedrich Apelt (1815–1859) from 1847 to 1859.

[161] Cf., e. g., [Nel28]. For details see [Pec90b], 123seq.

contacts between his followers and members of the Hilbert group.[162] Zermelo and Nelson first met, when Nelson began his studies in Göttingen. In 1904 Zermelo participated, at least for some time, in Nelson's private philosophical discussion circle.[163] Nelson remarks on this to Hessenberg:[164]

> Unfortunately, after we had made many time-consuming, indescribable efforts to make pure intuition clear to him, Zermelo suddenly stayed away from our evenings; I do not know why.

The Nelson correspondence provides information about the reactions Zermelo's colleagues showed after they became acquainted with the 1908 papers through proof sheets which Zermelo had circulated, and about Zermelo's response. It thereby reveals something about what Moore calls the "polemical aspect" of Zermelo's personality, concluding that "controversy spurred [him] to his greatest efforts" ([Moo82], 159). Gericke who got to know Zermelo as a research assistant in Freiburg around 1934, comments ([Ger55], 73) that "sometimes [he] even insulted his friends."

As the correspondence illustrates, Zermelo's personal resentment was particularly directed against his younger colleague Felix Bernstein (1878–1956). Bernstein, whose studies were also focussed on set theory at that time, was hoping to become permanently employed in Göttingen. He had studied philosophy, archaeology, and especially art history in Pisa and Rome and then, between 1896 and 1901, mathematics in Munich, Halle, Berlin, and Göttingen. As a grammar school pupil he had already participated in the seminars of Georg Cantor in Halle. While reading the corrections of one of Cantor's publications in the winter semester 1896/97, he had found a proof of the Schröder-Bernstein equivalence theorem. Cantor let Émile Borel know about it, and Borel published it in 1898.[165] Bernstein took his doctorate on set theory in Göttingen in 1901 ([Bern01]). Two years later he made his *Habilitation* in Halle, staying there as a *Privatdozent* until 1907. During this time he also undertook physiological studies at the institute of his father Ludwig, a professor of physiology. He returned to Göttingen in 1907 in order to take over the mathematics class at the Seminar of Insurance Studies. In 1911 he became an extraordinary professor for actuarial theory. In 1918 the Institute of Mathematical Statistics was founded under his direction, and in 1921, supported by Hilbert, he obtained there a full professorship of actuarial theory and mathematical statistics ([Frew81], 87–88). In 1919 he participated in drafting the concept of the German Reich's savings premium loans. In 1933, when the National Socialists had seized power, Bernstein, a Jew, did not return to Germany from a lecture tour in the USA. From 1933 to 1953 he held positions at Columbia University, NY, New York University, and Syracuse University,

---

[162] For example, Kurt Grelling as well as Paul Bernays had close contacts to Nelson.
[163] Cf. [Ble60], 22.
[164] Nelson to Hessenberg, dated Göttingen, 30 May 1904; GSA, N2210, Nelson Papers, No. 389, fols. 34–35.
[165] [Bor98]; cf. [Frew81], 84.

NY, and thereafter a Fulbright professorship in Rome. He died in Zurich in 1956.[166]

The correspondence between Zermelo and Nelson started when Nelson sent Zermelo his review of the second edition of Ernst Mach's *Erkenntnis und Irrtum*.[167] Zermelo sent the proof sheets of his new proof paper "as a revenge," supposing that its polemic part might interest Nelson.[168] In his postcard Zermelo does not discuss foundational questions, but comments on the thesis of the philosophizing lawyer Ernst Marcus stating ([Marc07], VIII) that "Kant's ethics is a true science, that ethical laws, even the highest legal laws, can be logically developed and determined with the precision of the mathematician." He considers the attempt to found ethics logically a "relapse into the times of the wildest rationalism," adding later[169] "that this 'panlogistic' metaphysics has absolutely nothing in common with Kant's doctrine."

The discussion about Bernstein was sparked by Bernstein's attempt to avoid the Burali-Forti paradox while permitting that the ordinals form a *set* W ([Bern05a]): For any element e not in W he allowed the formation of the set $W \cup \{e\}$, but he forbade stipulating that the ordering of W could be extended to $W \cup \{e\}$ by letting e follow all the elements of W (ibid., 189). He thereby attacked the validity of Zermelo's 1904 well-ordering proof which rests on the possibility of forming one point extensions for any well-ordered set, among them, according to Bernstein, possibly also the set W. Zermelo targets this "W-theory" as an "abolishment of set theory,"[170] calling it,[171]

> whichever way you look at it, *absolute nonsense*, according to the judgement of *all* thinking mathematicians. Whether I am already able to silence him [Bernstein] by now is of course a different matter; or rather absolutely impossible, considering the character of this "colleague."

Referring to his arguments against Bernstein which he had given in the new proof paper,[172] he concludes that Bernstein's principles were modifications made only arbitrarily and *ad hoc* (ibid.),

---

[166] For reference to Bernstein's biography cf. [Frew81] and [Scha87], especially 347–348. Cf. also [StrR83], 98, and [Gin57].

[167] [Nel07], enlarged version [Nel08]; new print in [Nel59], 119–178, and [Nel74], 233–281.

[168] Postcard of 5 December 1907.

[169] Letter of 14 December 1907.

[170] Postcard of 11 December 1907.

[171] Letter of 14 December 1907.

[172] [Zer08a], 123; [Zer67b], 194–195: "In fact, [Bernstein's] procedure followed in the justification of W amounts to this: the contradiction inherent in its definition is not resolved but *ignored*. If according to universal principles an assumption A yields two contrary consequences B and B', then A must be rejected as untenable. But in the present case, we are told, it is permissible to opt for one of these consequences, B, while prohibiting the other, B', by means of some special decree (Ausnahmegesetz) or veiling it by a change of name, lest it yield a contradiction with B. Since this procedure would obviously be applicable to every arbitrary hypothesis A, there

about the consequences of which the author remained completely unclear. Should B[ernstein] therefore oppose in the way insinuated by you, I will be able to serve him; I also have further weapons in stock. I was well aware of not shooting all cartridges at one go.

Zermelo considers his controversy with Bernstein symptomatic of the science politics in the field of mathematics, whose victim he felt he was after being a *Privatdozent* for so long:[173]

> I am convinced, by the way, that Mr. B[ernstein] himself no longer believes in his theory, but he wants, under all circumstances, to have the last word, by speculating on the pitiful scientific standard and the cliquey spirit which has, thanks to the "Klein School," become prominent among the generation of full professors of mathematics of today.[174]

In later years, exacerbated perhaps by the overall failure of his academic career, his resentment of "cliquey spirit" will become even stronger.

As regards Bernstein's dubious sketch of a proof of the continuum hypothesis which we mentioned already in 2.6.2 (cf. fn. 106),[175] his judgement becomes even more scathing:[176]

> In my opinion, Mr. B[ernstein] *can no longer be taken seriously, neither ethically nor scientifically,* after his peculiar behaviour in the continuum-question (where he [...] *against his better judgement,* as I can prove, publicly *claimed* to have a proof of Cantor's well-known theorem [...]).

Only a week later he reports[177] that Hilbert had talked about the new proof paper in a session of the Göttingen Mathematical Society,[178] Bernstein having been "very meek" if only for the time being, and that he had received "a very polite expression of thanks" for the proof sheets from Dedekind. Having also

---

would never be any contradiction at all; we could assert everything but prove nothing, since along with the possibility of a contradiction that of a proof would also be eliminated, and no science of mathematics could exist."

[173] Letter of 14 December 1907.

[174] Zermelo here addresses Felix Klein's influence on all matters concerning the appointment of candidates for mathematics chairs in Prussia. He believes that Klein's students are "now given preferential treatment everywhere" (letter to Hilbert, dated Arosa, 25 August 1907; SUB, Cod. Ms. D. Hilbert 447). Zermelo surely did not do Klein justice when it came to his own situation. Klein's dedication in supporting Zermelo's career is on record. For more information on Klein's politics of appointment and the fact that Klein's recommendations were *not* limited to his own students, cf. [Tob87].

[175] [Bern05c]; Bernstein there also announces the publishing of a detailed version of the proof in *Mathematische Annalen*, but never did so.

[176] Letter of 14 December 1907.

[177] Postcard of 22 December 1907.

[178] The report about the session of 17 December 1907 notes that Hilbert gave a talk about progress concerning his Paris problems (*Jahresbericht der Deutschen Mathematiker-Vereinigung* **17** (1908), 2nd section, 25–26).

received a "very friendly" letter from Julius König assuring him that "as far as the well-ordering is concerned, we continue to be irreconcilable – friends,"[179] and still further signs of support and reconciliation, his fighting attitude finally gives way to a triumphant mood:[180]

> With regard to the question of well-ordering the Göttingen "Count" [Bernstein] has suffered a spectacular defeat. König and Peano have written me obliging letters, and Jourdain is also retreating. Only Schoenflies gabbles on. Borel keeps silent.

2.9 The Axiomatization of Set Theory

In order to secure the proof of the well-ordering principle, to further its comprehensibility, and to clearly exhibit the role of the axiom of choice, Zermelo argued about his new 1908 proof that, apart from the axiom of choice, it does not use intricate set-theoretic results, but can be based on some "elementary and indispensable set-theoretic principles"[181] – "principles to form initial sets and to derive new sets from given ones."[182] By exhibiting these principles more systematically in his axiomatization paper ([Zer08b]), he formulated the first axiom system of set theory. Besides this methodological aspect, the "twin papers," the axiomatization paper and the new proof paper, are closely connected by mutual references as well as by temporal coincidence. Both were completed in July 1907 while Zermelo was staying in the village of Chesières above the Rhône valley in the western Swiss alps in order to recover from his tuberculosis.

2.9.1 Motivations

According to traditional assumption the driving force behind Zermelo's initial motivation in setting up an axiomatic system of set theory was an answer to the set-theoretic paradoxes, in other words, the attempt to create a consistent foundation of set theory.[183]

The preceding remarks offer another possibility: A strong motivation may also stem from the endeavour to "buttress his well-ordering theorem" ([Kan96] 11). In particular, this opinion has been advanced by Gregory H. Moore.[184] Moore believes that Zermelo "considered the paradoxes of limited concern."

---

[179]Letter from Zermelo to Hilbert of 19 December 1907; SUB, Cod. Ms. D. Hilbert 447, fol. 7.

[180]Postcard to Nelson of 20 January 1908.

[181][Zer08a], 110; [Zer67b], 186.

[182][Zer08a], 124; [Zer67b], 195.

[183]For the standard view cf., for example, [Bet59], 494; [Bou39], 47–48; [Kli72], 1185; [Qui66], 17; [vanH67], 199.

[184][Moo78], 326; [Moo82], 159–160.

Zermelo in 1907

The decisive point would have been to fortify the position of the axiom of choice by imbedding it in a rigorous system of axioms and letting it share an essential part in the well-ordering proof. Moore sums up ([Moo82], 159):

> Thus his axiomatization was primarily motivated by a desire to secure his demonstration of the Well-Ordering Theorem and, in particular, to save his axiom of choice.

Michael Hallett ([Hal84], 253) holds a similar opinion, going even further by trying to show "that the selection of the axioms themselves was guided by the demands of Zermelo's reconstructed proof" ([Hal84], xvi).

Hallett and Moore are right when realizing that the reactions against the 1904 well-ordering proof played a central role for Zermelo's future work and finally let him focus his research mainly on the foundations of set theory. Above all, however, one has to take into consideration how deeply Zermelo's axiomatic work was entwined with Hilbert's programme of axiomatization[185] and the influence of the programme's "philosophical turn" which was triggered by the publication of the paradoxes in 1903. It was after this turn that the axiomatization of set theory was recognized as a crucial step for the programme for establishing the concept of number.

Rather than being an argument for the standard view, this development shows that the causal influence of the paradoxes on the axiomatization was less important than the fact that the paradoxes revealed weaknesses in the foundational programme as a whole. It had become evident that it would not be possible to put the programme, and especially the consistency proof for the axioms of the real numbers, into practice, unless the sub-disciplines of mathematics and logic that were affected by the paradoxes had been well-founded themselves.

It is Zermelo himself, who shows that his work may be understood as the attempt to serve this higher purpose, i. e. to found the concept of number consistently, based on axiomatic set theory and carried out according to Hilbertian criteria. Apparently initiated by Frege's confession in the epilogue of the second volume of the *Grundgesetze der Arithmetik* ([Fre03]) that the basis of his foundation of number had been broken by the Zermelo-Russell paradox, he gave a talk before the Göttingen Mathematical Society on 12 May 1903 which shows the awakening Göttingen interest. The report says:

> E. *Zermelo* delivers a lecture about G. *Frege*: Grundgesetze der Arithmetik, Jena 1902 [!]. In particular, the lecturer offers, following G. *Frege*, Die Grundlagen der Arithmetik, Breslau 1884,[186] an exposition of *Frege*'s theory of the number concept, by parallelizing it with *Dedekind*'s and *Cantor*'s.[187]

And in the introduction to the axiomatization paper Zermelo evokes the wider Göttingen perspective:[188]

> Set theory is that branch of mathematics whose task is to investigate mathematically the fundamental notions of number, order, and function in their pristine simplicity, and to develop thereby the logical foundations of all of arithmetic and analysis.

Hilbert's insight that Zermelo needed publications to promote his career, but also the impact of the axiomatization on the Göttingen programme of a foundation of analysis might explain why Zermelo was urged by Hilbert to complete his paper without providing a consistency proof. In his letter to Hilbert of 25 March 1907[189] he says:

> According to your wish I will finish my set theory as soon as possible, although actually I wanted to include a proof of consistency. (OV 2.17)

---

[185] So also Herbert Mehrtens ([Meh90], 157): "Hilbert's programme and the Göttingen atmosphere need to be added as background of Zermelo's motives."

[186] [Fre1884], critical edition by Christian Thiel ([Fre86]).

[187] *Jahresbericht der Deutschen Mathematiker-Vereinigung* **12** (1903), 344–345.

[188] [Zer08b], 261; [Zer67c], 200 (modified).

[189] SUB, Cod. Ms. D. Hilbert 447, fols. 79seq.

The "certainly very essential" question of consistency[190] is deferred to a later investigation which – if started at all – did not result in a publication.

The axiomatization paper is the keystone of Zermelo's set-theoretic work during his first period of foundational research. As argued above, it is written in the spirit of Hilbert's foundational programme. Hilbert himself belonged to its admirers. In his 1920 course he says ([Hil20], 21–22):[191]

> The person who in recent years has newly founded [set] theory and who has done so, in my view, in the most precise way which is at the same time appropriate to the spirit of the theory, that person is Zermelo.

And he characterizes the axiomatization (ibid., 33) as

> the most brilliant example of a perfected elaboration of the axiomatic method [through which] the old problem of reducing all of mathematics to logic gains a strong stimulus.

Zermelo's next set-theoretic paper was to appear more than 20 years later ([Zer29a]). It opened his second period of foundational research which, unlike the first, was to manifest a growing estrangement from Hilbert, in parts directly contradicting Hilbert's views.

2.9.2 The Axiom System

When writing down his axiomatization of set theory, Zermelo paid due attention to the difficulties arising from the set-theoretic paradoxes, among them, in particular, the Zermelo-Russell paradox:[192]

> At present, however, the very existence of [set theory] seems to be threatened by certain contradictions, or "antinomies," that can be derived from its principles – principles necessarily governing our thinking, it seems – and to which no entirely satisfactory solution has yet been found. In particular, in view of the "Russell antinomy" [...] of the set of all sets that do not contain themselves as elements, it no longer seems admissible today to assign to an arbitrary logically definable notion a set, or class, as its extension. [...] Under these circumstances there is at this point nothing left for us to do but to proceed in the opposite direction and, starting from set theory as it is historically given, to seek out the principles required for establishing the foundations of this mathematical discipline.

This "method of exclusion" was proposed also in the new proof paper in an even more concise way. There[193] Zermelo writes with reference to his coming axiomatization that

---

[190][Zer08b], 262; [Zer67c], 201.
[191]Translation from [Moo02a], 57.
[192][Zer08b], 261; [Zer67c], 200.
[193][Zer08a], 115–116; [Zer67b], 189.

those who champion set theory as a purely mathematical discipline that is not confined to the basic notions of traditional logic are certainly in a position to avoid, by suitably restricting their axioms, all antinomies discovered until now.

When formulating his axioms, Zermelo follows a corresponding pragmatic directive:[194]

> In solving the problem we must, on the one hand, restrict these principles far enough to exclude all contradictions and, on the other hand, leave them sufficiently wide to retain all that is valuable in this discipline.

This indicates that his attitude lies in the middle of two extremes: On the sceptical side we find Russell for whom the Zermelo-Russell paradox leads to definite contradictions, and renders all reasoning about classes and relations, *prima facie*, suspect ([Rus06], 29). Having examined several ways out of this dilemma, Russell will finally build on the theory of types and, together with Alfred North Whitehead, create the impressive *Principia mathematica* ([WhiR10]).

The other extreme, passing over the problem of consistency, was widespread among those who considered set theory rather a mathematical discipline than a foundational one. It can be found with Grace Chisholm and William Henry Young's *Theory of Sets of Points* which "avoids controversial points" ([YouY06], 145) or in Hausdorff's influential *Grundzüge der Mengenlehre* ([Haus14]). In the preface of the second edition (1927) Hausdorff writes that, like in the first edition, he had not been willing to make up his mind about discussing foundational matters. Even more distinctly, in the introduction to his paper on ordered sets ([Haus08]), he declares that set-theoretic investigations such as the present one that tried to modestly augment the positive results of the young discipline in the sense of its creator, should not stop *prae limine* to discuss the principles – "an attitude which might cause offence to those wasting an unwarranted measure of astuteness for such a discussion."[195]

Zermelo gives his axiom system in the style of Hilbert's *Grundlagen der Geometrie* ([Hil1899]). Following this pattern, the axioms do not refer to a specific universe, but to an unspecified domain (Bereich) 𝔅 of unspecified objects (Dinge) together with a binary relation, the element relation over 𝔅. Some of the objects in 𝔅 are called sets. Later ([Zer30a], 29), the remaining objects will be called urelements. A set is either the empty set according to Axiom II below or an object that has an element. In other terms: Up to the empty set, the urelements are just the objects which do not contain elements.

---

[194][Zer08b], 261; [Zer67c], 200 (modified).

[195]For details about Hausdorff's foundational views and about characteristic features of early textbooks on set theory, cf. [Pur02].

We now list the axioms.[196] They constitute the *Zermelo axiom system*
Z *of set theory*. We use the original numbering and give both the English
translation and the original German formulation. For the term "definite" in
the axiom of separation we refer to its discussion below. We write "⊆" for the
inclusion of sets.

I. *Axiom of Extensionality*: If every element of a set M is also an element
of N and vice versa, if, therefore, both $M \subseteq N$ and $N \subseteq M$, then always
$M = N$; or, more briefly: Every set is determined by its elements.

Axiom der Bestimmtheit: Ist jedes Element einer Menge M gleichzeitig
Element von N und umgekehrt, ist also gleichzeitig $M \subseteq N$ und $N \subseteq M$,
so ist immer $M = N$. Oder kürzer: jede Menge ist durch ihre Elemente
bestimmt.

II. *Axiom of Elementary Sets*: There exists a (fictitious)[197] set, the *null set*,
0, that contains no element at all. If a is any object of the domain, there
exists a set $\{a\}$ containing a and only a as element; if a and b are any
two objects of the domain, there always exists a set $\{a, b\}$ containing as
elements a and b but no object x distinct from both.

Axiom der Elementarmengen: Es gibt eine (uneigentliche) Menge, die
"*Nullmenge*" 0, welche gar keine Elemente enthält. Ist a irgend ein Ding
des Bereiches, so existiert eine Menge $\{a\}$, welche a und nur a als Element
enthält; sind a, b irgend zwei Dinge des Bereiches, so existiert immer eine
Menge $\{a, b\}$, welche sowohl a als b, aber kein von beiden verschiedenes
Ding x als Elemente enthält.

III. *Axiom of Separation*: Whenever the propositional function $\mathfrak{E}(x)$ is definite for all elements of a set M, M possesses a subset $M_{\mathfrak{E}}$ containing as
elements precisely those elements x of M for which $\mathfrak{E}(x)$ is true.

Axiom der Aussonderung: Ist die Klassenaussage $\mathfrak{E}(x)$ definit für alle Elemente einer Menge M, so besitzt M immer eine Untermenge $M_{\mathfrak{E}}$, welche
alle diejenigen Elemente x von M, für welche $\mathfrak{E}(x)$ wahr ist, und nur solche
als Elemente enthält.

IV. *Axiom of the Power Set*: To every set T there corresponds another set
$\mathfrak{U}T$, the *power set* of T, that contains as elements precisely all subsets
of T.

Axiom der Potenzmenge: Jeder Menge T entspricht eine zweite Menge
$\mathfrak{U}T$ (die *Potenzmenge* von T), welche alle Untermengen von T und nur
solche als Elemente enthält.

V. *Axiom of Union*: To every set T there corresponds a set $\mathfrak{S}T$, the *union*
of T, that contains as elements precisely all elements of the elements of T.

---

[196] [Zer08b], 263–268; [Zer67c], 201–204.
[197] For Zermelo's scepticism about the empty set, cf. 3.5.

Axiom der Vereinigung: Jeder Menge T entspricht eine Menge $\mathfrak{S}\,T$ (die "*Vereinigungsmenge*" von T), welche alle Elemente der Elemente von T und nur solche als Elemente enthält.

VI. *Axiom of Choice*: If T is a set whose elements all are sets that are different from 0 and mutually disjoint, its union $\mathfrak{S}\,T$ includes at least one subset S_1 having one and only one element in common with each element of T.

Axiom der Auswahl: Ist T eine Menge, deren sämtliche Elemente von 0 verschiedene Mengen und untereinander elementenfremd sind, so enthält ihre Vereinigung $\mathfrak{S}\,T$ mindestens eine Untermenge S_1, welche mit jedem Element von T ein und nur ein Element gemein hat.

VII. *Axiom of Infinity*: There exists in the domain at least one set Z that contains the null set as an element and is so constituted that to each of its elements a there corresponds a further element of the form $\{a\}$, in other words, that with each of its elements a it also contains the corresponding set $\{a\}$ as an element.

Axiom des Unendlichen: Der Bereich enthält mindestens eine Menge Z, welche die Nullmenge als Element enthält und so beschaffen ist, daß jedem ihrer Elemente a ein weiteres Element der Form $\{a\}$ entspricht, oder welche mit jedem ihrer Elemente a auch die entsprechende Menge $\{a\}$ als Element enthält.

This axiomatization has a prehistory. In his new proof paper Zermelo emphasized[198] that his 1904 proof of the well-ordering theorem had aimed at avoiding all somehow questionable concepts, and that the paradoxes of set theory could be overcome by restricting the concept of a set in a suitable manner. His earlier approaches to an axiomatization already show that weakening comprehension to separation is an essential feature of such restrictions.

The earliest versions of axioms can be found as later supplements in a notebook for his course on set theory in the winter semester 1900/01 in Göttingen ([Zer00b]) and may stem from the time between 1904 and 1906.[199] There are two sketches of systems.[200] The second sketch consists in a system of four "axioms of the definition of set" ("Axiome der Mengen-Definition"). Its first axiom says that no well-defined set contains itself as an element. Obviously it is intended to block the Zermelo-Russell paradox.[201] The second axiom postulates that for any element s there exists the set $\{s\}$ and for any set S such that $s \notin S$ there exists the set $S \cup \{s\}$. It comes close to the axiom of

---

[198] [Zer08a], 119; [Zer67b], 192.

[199] For details cf. [Moo82], 155–157.

[200] A preceding list ends already with the first axiom, Axiom I, a rudimentary form of the axiom of separation.

[201] A similar solution of the paradox is proposed by Schoenflies ([Scho06], 22) on an informal level: "The set represents a *new* notion which is added to the separate notions which constitute the set. Hence, there cannot exist sets which contain themselves as elements."

elementary sets and gives a restricted form of the axiom of union, motivated maybe by its essential and criticized use in the 1904 well-ordering proof. The third axiom corresponds to the axiom of separation[202] and the fourth one to the axiom of the power set.[203]

A more elaborated axiom system can be found in the course "Mengenlehre und Zahlbegriff" from the summer semester 1906.[204] The first section of the manuscript, "The Axioms of Set Theory," gives nine axioms. Axioms III to IX correspond to Axioms I to VII of the axiomatization paper, partly with similar names. The only major difference here concerns the axiom of infinity, because the 1906 version follows Dedekind's definition of infinite sets in the *Zahlen* treatise, requiring that there is a set that is equivalent to a proper subset. Axiom I[205] and Axiom II[206] do not explicitly appear in the axiomatization paper, but are absorbed in the notions of definiteness and of domain, respectively. The axiom of the 1904/06 systems prohibiting sets that are elements of themselves, is missing. As we will discuss later, its obvious task – blocking the Zermelo-Russell paradox – was taken over by the axiom of separation. Only in 1930 will Zermelo come back to it and exclude sets S with $S \notin S$ by the axiom of foundation.

The 1906 lecture notes include a "Theory of Equivalence" which became part of the axiomatization paper, a "Theory of Well-Ordering," and a section "The Number Series." Obviously the plans of the axiomatization and elaboration of set theory had already been developed to a considerable extent in late spring of 1906. A letter of Poincaré ([Poi06b]) of 6 June 1906 indicates that Zermelo considered a quick publication. On 19 June 1906 he informed the Göttingen Mathematical Society about his results. The report[207] reads as follows:

> Herr Zermelo gives account of his attempt to found set theory axiomatically. As a basis he takes a domain of distinct "objects," some of which must be sets and some, indivisible objects. The property that an object a is

---

[202]For the exact formulation cf. 2.9.4, fn. 221.

[203]The first sketch consists in a system of three axioms, namely the first and the third axiom of the second sketch and an axiom stating that no set contains all elements.

[204]The course was not announced in the calendar of lectures of the University of Göttingen. As usually practised by Zermelo, his notes ([Zer06b]) are written, mostly in shorthand, in a notebook containing information on the semester, the day and the time of the lectures. Nelson's papers (ASD, 1/LNAA000005) contain notes in Nelson's hand of a course of Zermelo's entitled "Mengenlehre." They obviously concern the course of the summer semester 1906.

[205]"If a and b are objects, then it is determined ('definite') whether $a = b$ (identity) or $a \neq b$."

[206]"Some objects are 'sets.' All sets are objects. An object can be an element of a set. (Axiom of set)"

[207]*Jahresbericht der Deutschen Mathematiker-Vereinigung* **15** (1906), 407. The translation given here follows [Moo82], 156.

an "element" of a set M is treated as a primitive fundamental relation. Concerning sets and their elements, he proposes eight postulates, from which all the principal theorems of set theory, including those about equipollence and well-ordering, can be rigorously deduced. However, in order to obtain the order-type of the natural numbers, we must postulate the existence of some "infinite" set. Arithmetic and the theory of functions require this postulate, while "general set theory," which treats finite and infinite sets in the same fashion, can dispense with it.

The text indicates that Axiom II, the axiom of set, had already been incorporated in the notion of domain.

2.9.3 Dedekind's Influence

As stated in the introduction of the axiomatization paper, Zermelo aimed at axioms that were strong enough to yield "the whole theory created by G. Cantor and R. Dedekind." He thus pays due credit not only to the work of Cantor, but also to the contributions of Dedekind.[208] In order to exemplify them, we may think of the influence Dedekind may have had on Cantor by the intensive exchange of ideas both maintained over a long time,[209] and, of course, we may think of his *Zahlen* treatise ([Ded88]): When founding the concept of number on that of set, he not only provided an introduction to basic set theory, but also exemplified its power to model fundamental notions of mathematics. This endeavour already goes back to his treatise on continuity and irrational numbers ([Ded72]), where he gave a set-theoretic definition of the reals as cuts in the rationals. It started a development that finally led to the set-theoretic representation of the whole of mathematics. His procedure in the *Zahlen* treatise is not an axiomatic one in the strict sense; he gives an intuitive definition of sets accompanied by so-called explanations (Erklärungen) that partly re-occur as axioms with Zermelo, among them, for example, the axiom of extensionality and the axiom of union. As commented on below, the axiom of infinity also bears Dedekind's mark; according to Zermelo[210] it is "essentially due to Dedekind." When commenting on the *Zahlen* treatise, Emmy Noether stressed Dedekind's influence on modern set theory ([Ded1932], 390seq.):

> How strongly axiomatic set theory is influenced by Dedekind, can be seen by a comparison with Zermelo's axioms [...] which have been partly taken over directly from Dedekind's "explanations."

Besides the coincidence of axioms, similarities also include the character of the representation, namely a likewise strict argumentation, using only the axioms or explanations, respectively. Kanamori ([Kan96], 12) comments:

---

[208] Zermelo follows Hilbert with this view; cf. [Ferr99], 253seq.

[209] Cf. [NoeC37], [Gra74], [Gue02].

[210] [Zer08b], 266; [Zer67c], 204.

By Dedekind's time proof had become basic for mathematics, and indeed his work did a great deal to enshrine proof as the vehicle to algebraic abstraction and generalization. Like the algebraic constructs sets were new to mathematics and would be incorporated by setting down the rules for their proofs. Just as calculations are part of the sense of numbers, so proofs would be part of the sense of sets.

The technique of chains which Dedekind developed when modelling the natural numbers ([Ded88], §4) re-occurs in Zermelo's 1908 proof of the well-ordering theorem. A formal similarity together with a conceptual difference becomes evident with Zermelo's axiom of infinity. Dedekind gives an intuitive argument for the existence of an infinite set:[211] Starting with an object t of one's thoughts, a "thought-thing," one successively gets new thought-things by the thought of t, the thought of the thought of t, ...; hence, the set of all thought-things of a human being is infinite. However, as the set of all thought-things is a thought-thing itself, one arrives at an inconsistent concept, a fact Dedekind discusses in the third edition (1911) of his treatise.[212] Zermelo replaces the "thought of" operation by the formation of singleton sets, i.e. by the transition from an object a to the set $\{a\}$, and lets his axiom of infinity ensure the existence of a set that contains the empty set and is closed under this operation. The smallest set Z_0 satisfying the axiom of infinity – the intersection of all sets satisfying the axiom – can intuitively be given as

$$Z_0 = \{\emptyset, \{\emptyset\}, \{\{\emptyset\}\}, \{\{\{\emptyset\}\}\} \ldots\}.$$

Zermelo calls it the "number sequence," as "its elements can take the place of the numerals."[213]

The new proof paper contains a motivation for choosing the axiom of infinity in this way:[214]

Rather, Cantor (*1897*, p. 221)[215] defines the numbers of the second number class as the *order types* that can be associated with well-ordered denumerable sets, and he *proves* everything else from this definition. Only the existence of denumerable sets, or the type ω by which they are defined, is assumed; there is no further need for a further axiom. [...] In an analogous way every higher number class[216] is defined by means of the preceding one, *without* "any need of a new creation, or of a new axiom and the proof of its justification."

---

[211] [Ded88]; [Ded1932], 357.

[212] Dedekind speaks of "important and partially justified" doubts about the security of his views, justifying the new edition of his treatise "merely by the interest in and continuous demand for it."

[213] [Zer08b], 267; [Zer67c], 205.

[214] [Zer08a], 125; [Zer67b], 196.

[215] [Can97], 221; [Can32], 325.

[216] I. e., the set of ordinals of some fixed uncountable cardinality $\aleph_1, \aleph_2, \ldots$.

Zermelo's high estimation of the *Zahlen* treatise became apparent again ten years later. Edmund Landau, when preparing a memorial speech on the late Dedekind ([Lan17]), asked him to contribute an analysis of it, and, as reported there, he appreciated doing so.

In a certain sense Zermelo's proximity to Dedekind results in a greater distance to Cantor. Whereas Cantor's work centres around the notion of transfinite number, Zermelo – like Dedekind – shifted the emphasis to an abstract view of sets. In particular, the set-theoretic operations such as the operation of power set or union are open for an iterative application, thus allowing going beyond what was needed by Cantor or in mathematics. By admitting arbitrary choice functions or choice sets in his axiom of choice he went beyond the ideas of the time. Hence, so Kanamori ([Kan96], 12), Zermelo rather than Cantor should be regarded as the creator of abstract set theory.

2.9.4 The Axiom of Separation

In order to avoid the paradoxes, Zermelo followed the directive to restrict the set-theoretic principles "far enough to exclude all contradictions."[217] The key restriction that takes care of the paradoxes is embodied in his axiom of separation. Instead of allowing – in Cantorian terms – "the collection of well-distinguished objects of our perception or our thought" into a whole without any limitation, it confines comprehensions (1) to elements of a given set in order to take care, among others, of the Zermelo-Russell paradox and (2) interprets the notion of well-distinguishedness in a narrow sense in order to take care of Richard's paradox. In Zermelo's terms:[218]

> In the first place, sets may never be *independently defined* by means of [the axiom of separation] but must always be *separated* as subsets from sets already given; thus contradictory notions such as "the set of all sets" or "the set of all ordinal numbers" [...] are excluded. In the second place, moreover, the defining criterion must always be definite in the sense of our definition [...] with the result that, from our point of view, all criteria such as "definable by means of a finite number of words," hence the "Richard antinomy" and the "paradox of finite denotation" vanish.

Before considering Point (2), i.e. Zermelo's definition of definiteness, we will comment on his treatment of the Zermelo-Russell paradox. The paradox arises from an unrestricted comprehension with the property of not being an element of itself, leading to the "Zermelo-Russell set" $R = \{x \mid x \notin x\}$. By restricting the comprehension to a given set S, separation yields the set $R_S = \{x \in S \mid x \notin x\}$. Instead of arriving at the contradiction

$$R \notin R \text{ if and only if } R \in R$$

---

[217] [Zer08b], 261; [Zer67c], 200 (modified).
[218] [Zer08b], 263–264; [Zer67c], 202.

one now gets

$$R_S \notin R_S \text{ if and only if } R_S \in R_S \text{ or } R_S \notin S$$

which leads to $R_S \notin S$. Thus, instead of a paradox one gets a theorem, Theorem 10 with Zermelo: There is no universal set. In Zermelo's specific terms: For any set S there is a subset of S which is not an element of S. Zermelo comments (ibid., 265; 203):

> It follows from the theorem that not all objects x of the domain $\mathfrak{B}$ can be elements of one and the same set; that is, *the domain $\mathfrak{B}$ is not itself a set,* and this disposes of the Russell antinomy so far as we are concerned.[219]

Of course, he knew that blocking the usual road to the Zermelo-Russell paradox did not guarantee consistency in the strict sense (ibid., 262; 200–201):

> I have not yet even been able to prove rigorously that my axioms are consistent, though this is certainly very essential; instead I have had to confine myself to pointing out now and then that the antinomies discovered so far vanish one and all if the principles here proposed are adopted as a basis.

Coming now to Point (2), there was another paradox bothering Zermelo, Richard's paradox ([Ric05]). It can be derived by applying separation to the set of real numbers and the propositional function $\mathfrak{E}(x)$ of being definable by finitely many words. The resulting countable set S of real numbers can be enumerated (with repetitions) by alphabetically listing the definitions. Referring to this enumeration, let r be the real number $0.p_0 p_1 p_2 \ldots$ where for all natural numbers n we have $p_n = p + 1$ or $p_n = 1$ if the n-th decimal digit p of the n-th number of S is different from 8 and 9 or equals 8 or 9, respectively. As r is definable by finitely many words, namely by the definition just given, r belongs to S; however, by definition, r is different from any number in S.

Richard himself tried to eliminate the paradox by arguing that the definition of r is different from the definitions that were used to define the elements of S; for it will get its meaning only after the elements of S are defined. He thus anticipated the now common solution that is based on distinguishing between different levels of definition. Zermelo goes another way. In his axiom of separation he admits only properties which can be defined in purely set-theoretic terms, thus closing the road to the paradox. He calls such definitions *definite* (*definit*):[220]

> A question or assertion $\mathfrak{E}$ is said to be *definite* if the basic relations of the domain, by means of the axioms and the universally valid laws of logic, determine without arbitrariness whether it holds or not. Likewise a "propositional function" ("Klassenaussage") $\mathfrak{E}(x)$, in which the variable term x

---

[219]Schoenflies ([Scho11], 242) criticizes this passage because it treated the domain $\mathfrak{B}$ as a possible set. Obviously and against the Hilbertian conception of his axiom system, Zermelo has fallen back here into the standard interpretation of the element relation.

[220][Zer08b], 263; [Zer67c], 201 (modified).

ranges over all individuals of a class $\mathfrak{K}$, is said to be definite if it is definite
for *each single* individual x of the class $\mathfrak{K}$. Thus the question whether $a \in b$
or not is always definite, as is the question whether $M \subseteq N$ or not.

It is the vagueness of this definition which will lead to harsh criticism. A
principal weakness can be seen very quickly, an impredicativity observed by
Fraenkel ([Fra23]) who speaks of an inadmissible relation between the notion of
definiteness and the axioms that could even lead to concerns as for Richard's
paradox ([Fra27], 104): Definiteness refers to the axioms, but at the same
time it is an essential part of their formulation.[221] In 1922 Thoralf Skolem[222]
will identify the propositional functions with first-order ones, using only the
relation symbols $=$ and $\in$ for equality and the element relation, respectively.
In 1908 Zermelo is still far away from such an identification (and, in fact,
will never accept it). His concept of definability "has precious little to do
with language" ([Tay93], 546). Only twenty years later he will release a more
precise version, thereby performing a certain shift to a more linguistic attitude
([Zer29a]; cf. 4.6.1). The debate arising from this paper will deeply affect his
foundational views and dominate his set-theoretic efforts during the 1930s,
letting him finally end in isolation from the foundational mainstream. In this
sense the axiom of separation will be the dominant factor of his later scientific
destiny. Its significance for set theory itself – in the form Skolem had given it
– will then be fully acknowledged. Probably with respect to its effect on the
paradoxes, not only Zermelo ([Zer29b], talk 6), but also Fraenkel ([Fra24], 94)
considered it the most important axiom. For Gödel[223] it is "really the only
essential axiom of set theory."[224]

As we know, there was a third paradox bothering the Göttingen group,
Hilbert's paradox (2.4.3). It rests on the formation of unions of arbitrarily
many sets. In Zermelo's axiom of union the formation of unions is explicitly
restricted to *sets* of sets. So altogether the axioms of Z may owe typical features
to discussions in the Hilbert circle.

---

[221]In the system of four axioms from the time between 1904 and 1906 mentioned
above, there is a predecessor of the notion of definiteness which does not refer to the
axioms; the axiom of separation is formulated as follows: "If M is a well-defined set
and E is any property which an element m of M can possess or not possess, without
any arbitrariness being possible, then the elements m' which have the property E
form a well-defined set, a subset M' of M, as does the complementary set M''." ("Ist
M eine wohldefinierte Menge und E irgend eine Eigenschaft, die einem Element m
von M zukommen oder nicht zukommen kann, ohne daß noch eine Willkür möglich
ist, so bilden die Elemente m', welche die Eigenschaft E haben, eine Teilmenge M'
von M, sowie die komplementäre Menge M'' von M'.")

[222][Sko23]; [Sko70], 139.

[223][Goe40]; [Goe86], Vol. III, 178.

[224]Literally: "The axiom of reducibility or its equivalents, e. g., Zermelo's Ausson-
derungsaxiom, is really the only essential axiom of set theory." For a discussion of
this point cf., e. g., [Kan06], in particular §5.

2.9.5 The Range of the Axioms

One of the goals Zermelo had in mind was to formulate his axiom system of set theory "in a sufficiently wide way in order to retain all that is valuable in this discipline." So he aimed at "practical completeness." In his talk before the Göttingen Mathematical Society in June 1906 (cf. 2.9.2, fn. 207) he had given a criterion: the system should be strong enough to yield "all the principal theorems of set theory, including those about equipollence and well-ordering." The new proof paper had demonstrated that his system was adequate with respect to well-orderings. The axiomatization paper tests the system with respect to cardinalities, starting with the basic theory of equivalence (equipollence). Lacking a set-theoretic definition of ordered pairs, Zermelo defines functions between sets S and T only in case S and T are disjoint; they are conceived as subsets of the product $ST = \{\{x, y\} \mid x \in S \text{ and } y \in T\}$. The equivalence of arbitrary sets S, T is then defined by using a third set which is disjoint from both S and T.[225]

The elaboration continues with a proof of the Schröder-Bernstein equivalence theorem which he had already reported to Hilbert on 28 June 1905 in the following form:[226]

"If $S = (P, Q, S') = (P, S_1)$, with $S' \simeq S$ and P, Q, S' having no elements in common, then also $S_1 = (Q, S') \simeq S$."[227]

Namely, let (Q, V) be the common part of all subsystems of S which contain Q and, given the respective mapping of S onto S', their own images (i. e., which are mapped onto a part of themselves).[228] Then also (Q, V) has the same feature. Moreover, V contains only images of elements from (Q, V) because one could leave out each further element and would obtain a *smaller* system (Q, V_1) which would also be mapped onto itself. Hence V is a part of S', say $S' = (V, R)$. Now the image of (Q, V) is V itself, because Q does not contain any image and V no other elements. Thus, $(Q, V) \simeq V$ and $S_1 = (Q, S') = (Q, V, R) \simeq (V, R) = S'$. [Therefore] $S_1 \simeq S'$, q.e.d. Now the "equivalence theorem" follows in the known way.[229] (OV 2.18)

---

[225] A general set-theoretic definition of the notion of function based on a set-theoretic definition of ordered pairs was provided by Hausdorff in 1914 ([Haus14], 33).

[226] Postcard to Hilbert of 29 June 1905; SUB, Cod. Ms. D. Hilbert 447, 3. – The proof corresponds to that of Theorem 25 in the axiomatization paper.

[227] The symbol $\simeq$ denotes equivalence (equipollence) of sets and, for example, (P, Q) the disjoint union of the sets P and Q. The latter denotation follows Cantor ([Can95], 481; [Can32], 282).

[228] Let φ be a bijection from S onto S'. Then (Q, V) is the disjoint union of Q, $\varphi[Q]$ (i. e., the image of Q under φ), $(\varphi\varphi)[Q]$, $(\varphi\varphi\varphi)[Q], \ldots$.

[229] In cardinal notation, Zermelo has just shown that for all cardinals $\mathfrak{m}$, $\mathfrak{p}$, and $\mathfrak{q}$, if $\mathfrak{m} = \mathfrak{m} + \mathfrak{p} + \mathfrak{q}$ then $\mathfrak{m} = \mathfrak{m} + \mathfrak{q}$. Now the Schröder-Bernstein equivalence theorem follows as described in cardinal notation at the end of Zermelo's earlier proof given in 2.5.2.

Russell, when thanking Zermelo for providing him with preprints of the 1908 papers, writes about the published proof:[230]

> In particular I admire your new proof of the Schröder-Bernstein theorem because of its simplicity and its elegance. I always read carefully all your writings, always benefitting greatly from them.

Dedekind had found a similar proof already in July 1887, but had not published it. It is printed only in his collected works ([Ded1932], 447–448).[231]

The Schröder-Bernstein equivalence theorem was first stated without proof by Cantor in his letter to Dedekind of 5 November 1882[232] and in print in 1883[233] and reduced by him to the trichotomy of cardinalities in 1895.[234] On 24 September 1894 Ernst Schröder gave a flawed proof in his lecture "Ueber G. Cantorsche Sätze" at the Frankfurt meeting of the Gesellschaft Deutscher Naturforscher und Ärzte. In the proceedings the lecture is only reported by title ([WangT97], 43). It was not published before 1898 ([Schr98c]). Schröder's error was noticed by Alwin Reinhold Korselt who communicated it to Schröder in 1902 ([Kor11]). As mentioned already, Felix Bernstein found a proof in the winter 1896/97 which was published by Émile Borel in 1898 ([Bor98], 103–107).[235]

The axiomatization paper culminates with a generalization of König's inequality ([Koej05a]) which is known as the Zermelo-König inequality. Apparently Zermelo had been led to this investigation by the events during the 1904 Heidelberg congress of mathematicians (2.5.3). He presented his generalization to the Göttingen Mathematical Society on 25 October 1904. In the paper he characterizes it as "the most general theorem now known concerning the comparison of cardinalities, one from which all the others can be derived."[236] In modern terms it reads: If $(\kappa_i)_{i \in I}$ and $(\mu_i)_{i \in I}$ are families of cardinals such that $\kappa_i < \mu_i$ for $i \in I$, then the sum of the κ_i is less than the product of the μ_i. In Zermelo's own words it says:

> Let T and T' be two equivalent sets containing as elements the mutually disjoint sets M, N, $R, \ldots$ and the mutually disjoint sets M', N', $R', \ldots$, respectively. If T and T' are mapped onto each other in such a way that each element M of T is of lower cardinality than the corresponding element M' of T', the sum $S = \mathfrak{S}T$ of all elements of T is also of lower cardinality than the product $P' = \mathfrak{P}T'$ of all elements of T'.

The power of the Zermelo axiom system Z was thus demonstrated in a convincing manner.

---

[230] Letter to Zermelo of 23 March 1908; UAF, C 129/99.

[231] For further details see [Ferr99], 239seq.

[232] [Can1991], 86; English translation in [Ewa96], 874–878.

[233] [Can83], 583; [Can32], 201.

[234] [Can95], 484; [Can32], 285.

[235] For further information cf. [Gra00a], 132–134, [Moo82], 42–48, [Pec90b], 93–95.

[236] [Zer08b], 279; [Zer67c], 213.

2.9.6 Reception of the Axiom System

Zermelo's system was immediately used by Gerhard Hessenberg ([Hes10]) who based an extended theory of well-ordering on it. The tenor of the paper mirrors his high estimation for Zermelo's work. In fact, the relationship between Hessenberg and Zermelo was one of mutual appreciation. Zermelo read the galley proofs of Hessenberg's treatise on set theory ([Hes06]) and contributed Theorem XX on page 539 which excludes isomorphisms of a well-ordered set onto a part of it where some element is mapped to a smaller one. He used to recommend the treatise in his courses on set theory and in 1932 considered a new edition of it.

Explicit use of the axiom system was rare during the next years. Even more, there was criticism, for example, from Jourdain, Poincaré, and Russell.[237] Part of this criticism was aimed at the notion of definiteness; it was to get even stronger in the 1920s and, as we already know, was to be linked to Zermelo's scientific fortune. Russell doubted that the axioms really banned the paradoxes, viewing his theory of types the more appropriate way. The following passage from his letter to Jourdain of 15 March 1908[238] may be considered as being typical, also in hinting at concerns about impredicativity put forward by Poincaré:

> I thought [Zermelo's] axiom [of separation] for avoiding illegitimate classes so vague as to be useless; also, since he does not recognise the theory of types, I suspect that his axioms will not really avoid contradictions, *i. e.*, I suspect new contradictions could be manufactured specially designed to be consistent with his axioms. For I feel more and more certain that the solution [to the paradoxes] lies in types as generated by the vicious circle principle, *i. e.*, the principle "No totality can contain members defined in terms of itself." [. . .]
>
> But I think Zermelo a very able man, and on the whole I admire his work greatly. I am much obliged to you for writing to him about me and referring him to my articles.

Only in 1915 did Zermelo's axioms serve as a base for a major mathematical discovery, namely in Friedrich Hartogs' proof that the comparability of cardinalities implies the well-ordering theorem ([Har15]). Hausdorff, not well-disposed to foundational questions, did not share the axiomatic point of view; in particular, he did not do so in his influential *Grundzüge* ([Haus14]).

The 1920s brought the rise of axiomatic set theory with Z as its point of departure. The development started with Fraenkel's *Einleitung in die Mengenlehre* ([Fra19]), the first text book basing set theory on an axiom system, namely Z. The system also stimulated the von Neumann-Bernays-Gödel system NBG. For on 15 August 1923 von Neumann writes to Zermelo[239] that it

---

[237]See [Moo82], Ch. 3.3 for details.

[238]Cf. [Gra77], 109.

[239]Letter in UAF, C 129/85.

had been the axiomatization paper which "entirely" initiated his axiomatization.[240]

However, the growing influence of Z came with a weakness: In the early 1920s both Fraenkel ([Fra22b], 230seq.) and Skolem ([Sko23], 225seq.) discovered a gap: Let $Z_0 = \{\emptyset, \{\emptyset\}, \{\{\emptyset\}\} \ldots\}$ be as defined in 2.9.3 and $\mathfrak{U}$ the power set operation from the power set axiom. Fraenkel and Skolem argued that the system Z (presupposed its consistency) does not allow the existence of the set

$$\{Z_0, \mathfrak{U}(Z_0), \mathfrak{U}(\mathfrak{U}(Z_0)), \ldots\}$$

to be deduced. Both of them proposed a further axiom that would do the job, the axiom of replacement:

For any set S and any (definite) operation F mapping sets to sets there exists the set $\{F(x) \mid x \in S\}$.

If one takes the set Z_0 for S and a (definite) operation F mapping $\emptyset$ to Z_0, $\{\emptyset\}$ to $\mathfrak{U}(Z_0), \ldots$, the axiom yields the desired set. We will come back to the axiom of replacement in 3.5.

In the *Grenzzahlen* paper ([Zer30a]) Zermelo will enlarge his system Z by the axiom of replacement and a further axiom, the axiom of foundation. In order to honour Fraenkel's contribution concerning replacement, he will call the resulting system the *Zermelo-Fraenkel axiom system* ZF (4.7.2). Today ZF stands for the system without the axiom of choice and with definiteness understood as first-order definability, i. e., in the sense of Skolem, ZFC now meaning ZF together with the axiom of choice.

ZF and ZFC, respectively, became the standard axiom systems for pure set theory. However, the axiom of replacement as also the axiom of foundation are not needed if set theory is considered as "that branch of mathematics whose task is to investigate mathematically the fundamental concepts of number, order, and function in their original simplicity and thereby to develop the logical foundations of the whole of arithmetic and analysis."[241] As a matter of fact, both axioms get involved only in questions of a more intrinsic set-theoretic character. In view of the methodological novelties it provided, we are thus fully justified if we consider the appearance of the system Z in 1908 a landmark that impressively initiates the beginning of modern axiomatic set theory.

2.10 The Institutionalization of Mathematical Logic

2.10.1 Zermelo's Lectureship for Mathematical Logic

In 1907 Zermelo's *Privatdozenten* grant which he had received for six years (2.3.1) came to a definite end. Except for interruptions due to his illnesses,

---

[240] "Die Anregung [...] verdanke ich ganz Ihrer Arbeit über die 'Grundlagen der Mengenlehre'."

[241] [Zer08b], 261; [Zer67c], 200.

he had been teaching as a *Privatdozent* in Göttingen for 8 years. Now he was forced to find a remunerative position. In March 1907 he learned from a newspaper item that Philipp Furtwängler, professor of mathematics at the Landwirtschaftliche Akademie (Academy of Agriculture) in Poppelsdorf[242] and later a colleague of Hans Hahn and Moritz Schlick in Vienna, had accepted a position at the Technische Hochschule in Aachen. Zermelo applied for the now vacant position in Poppelsdorf. In a letter of 24 March 1907[243] he asked Hilbert for support, explaining that

> I cannot deny that my prospects regarding this position are also extremely meagre, because I have obviously not been recommended by any party. But I wanted to try everything because I will be forced to give up all academic teaching if I do not find a further position or grant.

An additional reason for going there may have been that his friend Erhard Schmidt was teaching at the University of Bonn at that time.

Hilbert obviously had other plans. Already on 7 March 1907 he had turned to the Ministry of Cultural Affairs in Berlin,[244] proposing that Zermelo

> receive, instead of the grant, a slightly higher remuneration during the following summer semester, but then, after his return [from a cure in Switzerland] in autumn, should be supported to continue an academic career by providing him with a permanent income.

The result of this initiative was that by the end of March Zermelo received a single payment equalling the grant to which he was no longer entitled, and another half of this sum at the beginning of May.[245] For him, however, this could only be the first step, and he made this clear to Hilbert:[246]

> I would of course always prefer a permanent position as soon as I had the possibility, even in a small academy or something similar, like the one in Bonn.

But the regular remuneration that Zermelo strove for was only possible if his academic status changed. On 25 July 1907 the directors of the Seminar of Mathematics and Physics therefore sent an application to the Prussian Minister of Cultural Affairs in Berlin, asking that

---

[242]Now a part of Bonn.

[243]From Arosa, Switzerland; SUB, Cod. Ms. D. Hilbert 447. For the original versions of quotations in this section, cf. [Pec90a], 36–40.

[244]Draft of the letter in SUB, Cod. Ms. D. Hilbert 494.

[245]Notifications of the Royal Curator No. 1283 from 18 March 1907 und No. 1881 from 2 May 1907; copies in UAG, 4/Vb 267a.

[246]Letter to Hilbert, dated Arosa, 25 August 1907; SUB, Cod. Ms. D. Hilbert 447, fol. 5, 1–2.

> Your Excellency may appoint the local Prof. Dr. Ernst Zermelo to a lectureship for mathematical logic and related matters and for that grant him an annual, fixed remuneration.[247]

The text of the application was outlined by Hilbert[248] and skilfully adjusted to the university politics of Friedrich Althoff, the mighty director of the First Educational Department in the ministry who was responsible in this case and who had made it his goal to promote the international standing of German science.

The application began with a few historical remarks reporting that subsequent to Augustus De Morgan and George Boole "an intermediate field positioned between logic and mathematics" had emerged, which was becoming more and more important. At the core of this intermediate field "called logical calculus or algebra of logic" lay a method which combined "the old questions of classical logic as well as certain epistemological problems of philosophy" with "questions arising from a purely mathematical ground about the nature of number." The algebra of logic, they added, had spread from England (Venn, Russell, Whitehead) first to America (Peirce), then to Italy (Peano, Burali-Forti, Veronese) and finally to France (Couturat, Liard) and Germany (Schröder, Frege).

> Although the science of mathematical logic in Germany has indirectly experienced very effective support through the purely mathematical investigations of Dedekind and Cantor, it still seems that, all in all, it is the foreign countries that have maintained the leading role in this scientific discipline. Especially among the younger generation in Germany only Ernst Zermelo can be considered a fully valid representative of this discipline.

By no means uncommon at that time, this quite original historical overview does not distinguish between the algebra of logic and Frege style mathematical logic.[249] The terms "mathematical logic," "logical calculus," "algebra of logic," and, since 1904, "logistics" were used synonymously. It is surprising that Louis Liard was mentioned as the main French representative of the algebra of logic, because Liard had not made any systematically relevant contribution to this discipline. But through his much-read, summarizing accounts of the newer English logic in the 1870s and 1880s he helped the algebra of logic to become known outside Great Britain.[250] For his work on the founda-

---

[247]GSA, Rep. 76 Va Sect. 6 Tit. IV No. 4 Vol. 4, fol. 269–270, edited in [Pec90a], 51–52.

[248]SUB, Cod. Ms. D. Hilbert 492.

[249]For more information on the emergence and reception of the algebra of logic in Great Britain and Germany cf. [Pec88], 179–203, and the pioneering, but partly out-dated study [Buh66]; cf. also [Pec97], Chs. 5 and 6, and [Vil02].

[250]Louis Liard, *Les logiciens anglais contemporains* ([Lia78], further editions 1883, 1890, 1901, 1907; German translation [Lia80] (2nd edition 1883)). Before that Liard had already published two essays about the systems of William Stanley Jevons and George Boole ([Lia77a], [Lia77b]).

tion of geometry, the Italian mathematician Giuseppe Veronese also cannot be credited with more than indirect support of the mathematical logic.[251]

The applicants emphasize that, as far as they knew, mathematical logic had not yet officially been included in a curriculum of a German university.[252] They point to the great variety of Zermelo's working fields and call him a "master" in set theory. Obviously alluding to the axiomatization paper ([Zer08b]) he was working on at that time, they mention a bigger treatise "which belongs to the field of mathematical logic and from which we expect essential support and clarification of the most difficult and most disputed questions in that field." Having characterized Zermelo's lectures as "quite excellent" for advanced students, they finally describe his financial situation. Even though he had held a position as a *Privatdozent* for nine years, he had never been offered a remunerative position. He had had a small private income that had by then almost completely vanished, especially due to the pneumonia that he had suffered from the year before and which had forced him to spend the last two semesters for convalescence in Switzerland. On top of that, he had only received very few students' fees for his lectures dealing with "purely theoretical" and "mostly special" objects. They therefore ask that he should be granted the teaching position together with proper payment in order to enable him "to continue his work, which is so valuable for our university and for science."

On 20 August 1907, only a few weeks after the application had been filed, the following directive, signed by Friedrich Althoff, was issued to Zermelo:[253]

> In compliance with the application filed by the directors of the Seminar of Mathematics and Physics of the university located there [i.e. the Göttingen University] I commission you to represent in your lectures, from the following semester on, also mathematical logic and related topics. (OV 2.19)

Once the scholarship ended, the new teaching position was connected with the payment of 1800 Marks annually "which means, after all, an augmentation

---

[251][Ver91], German [Ver94].

[252]Ernst Schröder had already given lecture courses about mathematical logic in Darmstadt and Karlsruhe, and Gottlob Frege in Jena. However, these courses had not been dependent on ministerial commissions. Schröder, for example, had given a course of lectures about "Logik auf mathematischer Grundlage" ("Logic Based on a Mathematical Foundation") in Darmstadt during the summer semester 1876, even before Frege's first lectures on *Begriffsschrift* in the winter semester 1879/80. During the summer semester 1878 he had held a course of lectures on "Logik als mathematische Disziplin" ("Logic as a Mathematical Discipline") in Karlsruhe, and between the winter semester 1883/84 and the summer semester 1888/89 he had taught, for six whole semesters, the "Algebra of Logic" there. Cf. the catalogues of the Polytechnical Schools (the later Technical Universities) in Darmstadt and Karlsruhe, and [Pec92]. With regard to Frege's teaching cf. [Krei83], [Krei01].

[253]Directive UI No. 17759, in a duplicate of the Royal Curator; UAG, 4/Vc 229.

by 50%," as Zermelo wrote to Hilbert after having received the directive.[254] About two years later, this subsidy was raised to an annual 3000 Marks,[255] then amounting to about half the payment of a full professor.

Zermelo's appointment marks the birth of mathematical logic as an institutionalized sub-discipline of mathematics in Germany. One might wonder why his teaching position was dedicated to "mathematical logic" and not, as one could conclude from Zermelo's fields of interest, to set theory. There are clear reasons. After the discovery of the paradoxes, Hilbert's revised axiomatic programme fostered the mutual development of logic and set theory as linked together. In one of his lectures on "Die logischen Principien des mathematischen Denkens" during the summer semester of 1905, Hilbert says:[256]

> The paradoxes show well enough that an examination and reconstruction of the foundations of mathematics and logic is absolutely necessary. It becomes immediately evident that the contradictions are based on the comprehension of certain totalities to a set, which seems not permissible; but nothing is won with that because every act of thinking is based on such comprehensions, and the problem remains to separate what is allowed from what is not allowed. (OV 2.20)

In that same course of lectures Hilbert had taken a first, provisional step and developed a propositional calculus that was based on the concept of logical equivalence. Zermelo seemed to be the person suited for enlarging logical competence in Göttingen.[257] Furthermore, after Ernst Schröder had died in 1902, there was nobody in Germany except Gottlob Frege who could be taken seriously as a representative of mathematical logic and who had an academic teaching position. This circumstance must have been favourable for the Minister's approval of the application, which, of course, was decisive for Zermelo's career as an academic researcher and teacher.

2.10.2 The 1908 Course "Mathematische Logik"

Zermelo's hope that he could begin lecturing during the winter semester 1907/08[258] was not fulfilled. Because of health problems in October 1907 he had to take leave "and put off the course on mathematical logic that is

---

[254]Zermelo to Hilbert, dated Saas-Fee, 27 August 1907; SUB, Cod. Ms. D. Hilbert 447, fol. 6, p. 2.

[255]Directive UI No. 15662 issued on 2 April 1909; GSA, Rep. 76 Va Sect. 6 Tit. IV No. 4 Vol. 5, fol. 41. Also letter from Royal Curator to Zermelo, dated 20 April 1907; MLF.

[256][Hil05b],139–140.

[257]Cf. also [Pec89], [Pec90a], [Pec92].

[258]Zermelo to Hilbert, dated Arosa, 25 August 1907; SUB, Cod. Ms. D. Hilbert 447, fol. 5, p. 1.

listed in the calendar of lectures."[259] On 25 October he was released from the obligation to teach.[260] He spent the winter in Glion near Montreux in the mild climate of the Lac Leman area. Thus, the first course on mathematical logic ever held at a German university and commissioned by a ministry took place in the summer semester of 1908. The Zermelo *Nachlass* contains lecture notes ([Zer08c]) written by Zermelo, mostly in shorthand, and also an elaboration ([Zer08d]) by Kurt Grelling who had attended the course.[261]

Zermelo starts his lectures with some general remarks on logic ([Zer08d], 1-5). He states that the term "mathematical logic" can be used in two ways: On the one hand, it can refer to the mathematical treatment of logic as in the work of Ernst Schröder. But it can also stand for scientific research on the logical elements of mathematics. It is the latter which he plans to work on in his course (ibid., 2).[262]

He continues with the then current debate on the nature of mathematical judgements, stating that in contrast to geometry where there is "a clear opinion" that its judgements are of a synthetic character, i.e., "rising from *Anschauung*," the question whether arithmetic is analytic[263] or synthetic was still under dispute. Whereas Frege, Peano and Russell stand for the analytic quality of mathematical judgements,[264] Poincaré supports the synthetic one.

---

[259]Zermelo to the Dean of the Philosophical Faculty, dated Glion, 21 October 1907; UAG, 4/Vb 267a. The course had been announced under the title "Die mathematischen Grundlagen der Logik" ("The Mathematical Foundations of Logic"). Zermelo had already announced a course entitled "Mathematische Behandlung der Logik" ("Mathematical Treatment of Logic") for the winter semester 1906/07. However, because of his illness, this course had had to be cancelled as well.

[260]The Royal Curator to the Philosophical Faculty of 25 October 1907, ibid.

[261]The course consisted of two one hour lectures a week. Zermelo's notes cover only the first three sections of Grelling's five sections. The notes are not paginated. Page numbers given here begin with "1" for the first page of the main text. Grelling's second part starts with page number 39. Besides Grelling, 20 students attended the course, among them Ludwig Bieberbach, Richard Courant, Ernst Hellinger, Leonard Nelson, and Andreas Speiser.

[262]Zermelo's distinction comes close to the present one between algebra of logic and Frege style mathematical logic. However, it does not take into account Schröder's efforts to employ, in his late work, the calculus of relatives in accordance with a logistic treatment of mathematics; cf. e.g. [Schr95a], [Schr95b], and the programmatic paper [Schr98a], published in English in a slightly modified version ([Schr98b]), as well as [Schr98c], [Schr98d]. For problems concerning the common interpretation of the difference between logistics and the algebra of logic cf. [Thi87], [Pec96], [Pec04a].

[263]"Where those judgements are to be called analytic which can be concluded solely by the basic principles of logic" ([Zer08d], 3).

[264]As regards Russell, this is not unproblematic: In his *Principles of Mathematics* ([Rus03], 457) Russell deduces, from Kant's conviction that mathematics is synthetic, the synthetic character of logic: "In the first place, Kant never doubted for a moment that the propositions of logic are analytic, whereas he rightly perceived that those of mathematics are synthetic. It has since appeared that logic is just as

At present Zermelo does not want to take sides in this argument ([Zer08d], 2):

> We assume, for the time being, that in arithmetic both synthetic and analytic judgements occur. We now make it our goal to isolate the analytic component.

The method applied is attributed to Euclid with a late completion by Hilbert and described as follows (ibid., 2):

> The method consists in completely resolving the proof of a theorem into syllogisms and placing the premises used for the proof in front as completely as possible. One now can incorporate the premises as hypotheses into the theorem instead of categorically asserting them. We can therefore say: Generally speaking, mathematical theorems are no analytic judgements yet, but we can *reduce them to analytic ones* through the hypothetical addition of synthetic premises. The logically reduced mathematical theorems emerging in this way are analytically hypothetical judgements which constitute the logical skeleton of a mathematical theory.

In order to give an example of such a reduction, Zermelo considers the theorem "$3 + 2 = 5$." He defines

$$(1) \qquad 2 = 1 + 1;$$
$$(2) \qquad 3 = 2 + 1;$$
$$(3) \qquad 4 = 3 + 1;$$
$$(4) \qquad 5 = 4 + 1.$$

From (1) he concludes

$$(5) \qquad 3 + 2 = 3 + (1 + 1).$$

Using the theorem

$$(9) \qquad a + (x + 1) = (a + x) + 1$$

to be treated later, he obtains

$$(6) \qquad 3 + 2 = (3 + 1) + 1.$$

Now (3) yields

$$(7) \qquad 3 + 2 = 4 + 1,$$

and, finally, (4) yields

$$(8) \qquad 3 + 2 = 5.$$

Hence, the analytically reduced theorem for (8) goes like this: If (1) to (4) and (9) are valid, then (8) must be valid as well (ibid., 4).[265]

---

synthetic as all other kinds of truth." With this he reverses Frege's line of argumentation, according to which it can be inferred from Kant's conviction that logic is analytic, that mathematics is analytic as well. Cf. [Wan57], 156–157.

[265] Possibly Zermelo adopted this argument from Frege ([Fre1884], 7–8).

The introductory remarks conclude with a possible objection to the analytic nature of mathematics ([Zer08d], 3):

> It has been argued that mathematics is not or, at least, not exclusively an
> end in itself; after all it should also be applied to reality. But how can this be
> done if mathematics consisted of definitions and analytic theorems deduced
> from them and we did not know whether these are valid in reality or not.
> One can argue here that of course one first has to convince oneself whether
> the axioms of a theory are valid in the area of reality to which the theory
> should be applied. In any case, such a statement requires a procedure which
> is outside logic. (OV 2.21)

By his position, Zermelo keeps the philosophical impact on mathematics to a
minimum. When doing mathematics, it is irrelevant whether the mathematical
objects have any real analogue, whether the signs used have a real world
reference, or whether the statements obtained are true in a referential sense.
In particular, the axioms are stipulated as hypotheses, forming the base of a
hypothetico-deductive system.[266]

Zermelo takes it as "self-evident" that his course could be structured as a
three-part programme (ibid., 5):

I   The elementary laws of logic.
II The logical form of the mathematical theory (definition, axiom, proof).
III Does arithmetic contain synthetic elements and which are they?

According to Grelling's notes only Part I was developed by giving a very detailed presentation of the propositional calculus and the calculus of classes. In
Zermelo's words it deals with what has "lately" been called "logistics, i. e., the
principles of logic, presented in a script of signs (Zeichenschrift) which is mod-
elled after that of mathematics,"[267] and uses the "modern division" of logic
into the theory of classes, propositions, and relations ([Zer08d], 6). In doing
this, Zermelo follows the standard works in mathematical logic of his time,
written by Guiseppe Peano[268] and Ernst Schröder.[269] His symbolism follows
Peano, but he also adopts elements of Schröder's pasigraphy (functional symbols and the symbol for subsumption) and of Frege's concept script, which he
does not understand correctly, though.[270]

---

[266]This position was introduced by Mario Pieri; cf. [Mar93], 292.

[267]The use of the term "logistics" for describing symbolic or mathematical logic was
introduced by Gregorius Itelson (1852–1926) in 1904 during the Second International
Congress for Philosophy in Geneva, and independently also by Louis Couturat and
André Lalande. Cf. the congress report by Couturat ([Cou04], esp. 1062–1063).

[268]Zermelo expressly mentions ([Zer08c], 18) volume 2.1 of Peano's *Formulaire de
mathématique* ([Pea95]), *Logique mathématique* ([Pea97]).

[269]Zermelo refers mainly to Schröder's unfinished, monumental *Vorlesungen über
die Algebra der Logik* ([Schr90], [Schr91], [Schr95a], [Schr05]).

[270]For details cf. [Pec90a], 40, or [Pec90b], 113.

The propositional calculus is not set up axiomatically. Zermelo rather gives a list of foundational theorems "for the time being without systematically reducing the number of principles and basic operations to a minimum" ([Zer08d], 6). He starts with two principles:

I. *Tertium non datur*: "All logically determined propositions are either right or wrong, fall apart into two classes, the true ones and the false ones" ([Zer08c], 19); $a = \bigvee$ means "the proposition a is true" and $a = \bigwedge$ means "the proposition a is wrong."

II. *Negation*: "To each proposition a corresponds a specific second proposition $\bar{a}$ which contradicts and is the opposite of a and is called the 'negation of a.' Of two 'opposite' propositions $a, \bar{a}$, one is always true and the other is false" (ibid., 23).

Equipollence ($a = b$), *conjunction* (ab or $a \cap b$), *disjunction* ($a \cup b$), and *implication* ($a \supset b$) are defined referring to the possible truth values of their constituents as given – only in Grelling's notes – by the following table ([Zer08d], 9):[271]

	$a =$	$b =$
1)	$\bigvee$	$\bigvee$
2)	$\bigvee$	$\bigwedge$
3)	$\bigwedge$	$\bigvee$
4)	$\bigwedge$	$\bigwedge$

For example, the definitions of equipollence and implication are given as ([Zer08d], 9)

> *Equipollence.* If only 1) or 2) take place and, hence, 2) and 3) are excluded, we write $a = b$.

and (ibid., 12)

> *Implication.* [...] $a \supset b$ [...] means that case 2) of our table is excluded.

Following Schröder's example (already in [Schr77]), Zermelo lists the laws concerning negation, conjunction, and disjunction via confrontation of the duals (ibid., 11). After having given a detailed treatment of implication (ibid., 13–26), he turns to effective procedures for reducing random compound propositions to simpler ones and solving logical equations by using disjunctive normal forms (ibid., 26–44).[272]

Grelling's notes contain two further sections. The first one, "General and Particular Propositions"[273] deals with the foundations of a theory of quantifi-

---

[271]This is an early anticipation of the truth table method which is most prominent in Ludwig Wittgenstein's *Tractatus logico-philosophicus* ([Witt21], bilingual German-English edition [Witt22]). For its history cf. [KneK62], reprint 1984, 532–533.

[272]These methods can already be found with Schröder; cf. [Schr90], 396–433.

[273] "Allgemeine und partikuläre Aussagen;" [Zer08d], 44–51.

cation, the second one, "Classes and Relatives"[274] sets up a calculus of classes dealing with "class propositions" a_ξ, ξ a variable; they determine "the class of all things for which this proposition is true."

With its eclecticism Zermelo's logic course is a remarkable document of a period of transition in the history of symbolical logic (cf. [Pec92]). Boole and his successors, but also Schröder, were primarily interested in reformulating the philosophical discipline of logic by invoking mathematical methods. With respect to mathematics, logic was merely regarded as an instrument, a reform aimed at only in a second step. Zermelo also considers mathematical logic this way, but he goes further. Mathematical logic is also designed as a basic discipline within the realm of mathematics. This seems to be in accord with Frege's logicistic programme, but differs in its lack of epistemological objectives.

It would be wrong, though, to conclude that Zermelo's logic course initiated or continued a "reform of logic." After all, Zermelo ignored Hilbert's alternative ideas for the avoidance of quantification and for the mutual development of mathematics and logic. He probably disregarded these ideas because he did not agree with them.[275]

In particular, Zermelo did not pursue the problem of the paradoxes which Hilbert had emphasized so strongly. This is astonishing, as some months before he had announced to Leonard Nelson[276] his intention of treating "the logical question of the paradoxes" in this course. Nelson had sent him the galley proofs of a paper which he had written together with Kurt Grelling ([GrelN08]). It treats the paradoxes of Russell and Burali-Forti, analyzing their structure and critizising the solutions attempted so far. In his answer Zermelo mainly argued that the word "Paradoxie" should not be used here, and that the word "Antinomie" was much more precise.[277] For the word "Paradoxie" denoted a statement "which conflicts with the general opinion, it contains no interior contradiction," whereas antinomies were due to "incompletely limited concepts," and "the task here would be to find criteria for this."[278]

Zermelo's next course on mathematical logic, "Über die logischen Grundlagen der Mathematik" ("On the Logical Foundations of Mathematics")[279] took place in the winter semester 1909/1910. It consisted of one lecture a week and did not go beyond the framework of his 1908 course. Paul Bernays, Erich Kamke, and Hugo Steinhaus were among the 102 students attending it.

---

[274] "Klassen und Relative;" ibid., 52–61.

[275] In connection with criticism of Russell's logicism, Hessenberg reports in 1906 that "Zermelo is of the opinion that Hilbert's logicism is also not practicable" (letter from Hessenberg to Nelson of 7 February 1906; ASD, 1/LNAA000271).

[276] Postcard of 1 January 1908; ASD, 1/LNAA000399.

[277] Postcard of 22 December 1907; ASD, 1/LNAA000399. Cf. also fn. 62.

[278] Postcard of 1 January 1908; ibid. For a related discussion cf. also [MooG81].

[279] Lecture notes in UAF, C 129/153.

Apparently his occupation with mathematical logic led Zermelo to considering writing a book on this subject. The *Nachlass* contains an undated and still unsigned contract to be concluded between "Ernst Zermelo in Göttingen" and the Göschen publishing house.[280] It concerns a book entitled *Einführung in die mathematische Logik* which was to appear in the Schubert collection, a series of small and inexpensive text books. Apparently work on it never started.

There were still other projects which Zermelo considered in 1908. Having heard of difficulties which the *Abhandlungen der Fries'schen Schule* were facing because of low sales, he turned to Nelson on 14 March 1908.[281] Expressing fears that the journal's "school-like" title and its "casual" publishing might deter too many people, and referring to the narrow scope of the *Abhandlungen* because of its ties to the school, he suggests:

> A *periodical* journal directly for the *"philosophy of mathematics"* or something similar could in my opinion gain a much greater acceptance among those who are involved. What does Hessenberg[282] think about this? I myself would give such an up-to-date project my *full* support, while I would have my doubts if I committed myself to a special philosophical system. (OV 2.22)

Nelson rejected this proposal.[283] Apparently in April, during the International Congress of Mathematicians in Rome, Zermelo, Hessenberg, and Hugo Dingler, then a *Privatdozent* in Munich, conceived a new project, now aiming at establishing a quarterly journal for the foundations of the whole of mathematics, *Vierteljahrsschrift für die Grundlagen der gesamten Mathematik*. In May Hessenberg negotiated the terms with the Teubner publishing house in Leipzig. However, the plans finally failed.[284]

2.10.3 Hilbert and Zermelo

More than thirty years after he had left Göttingen, then in his seventies and living in seclusion in Freiburg, Zermelo wrote to Paul Bernays ([Zer41]):

> Even though I am still scientifically interested and active, I very much miss the scientific exchange of ideas which I enjoyed so much in former times, particularly during my time in Göttingen. (OV 4.53)

---

[280] UAF; C 129/277.

[281] ASD, 1/LNAA000399. – See also 2.8.4.

[282] Then a co-editor of the *Abhandlungen der Fries'schen Schule*.

[283] Letter from Nelson to Hessenberg of 29 March 1908; GSA, 90 Ne 1, Nelson Papers, No. 389, fols. 170–171.

[284] Besides economical concerns from Teubner's side, a further reason may be seen in Teubner's opinion that such a journal had to be a forum not only for the "positivistic direction," thereby surely meaning the Hilbertians, but also for the "epistemological" one represented, e. g., by the neo-Kantian Paul Natorp and the phenomenologist Edmund Husserl (letter from Hessenberg to Nelson of 15 May 1908; ASD, 1/LNAA000273). For further details, cf. [Pec04d].

Complaining that he had not received due acknowledgement and due respect for his scientific work, he continued that apparently this was "the fate of everybody who is not backed by a 'school' or a clique," thereby addressing his own independence from and aversion to any school as in the letter to Nelson of 14 March 1908 quoted above.

During his time in Göttingen Zermelo was backed by Hilbert. He was Hilbert's student in the sense that he oriented his own foundational research according to Hilbert's suggestions, taking up Hilbert's initial ideas and developing them further.

In his letter to Bernays he states that he owed many of his achievements to the people around him and to the scientific atmosphere they created. This is true, for example, for the proof of the well-ordering theorem that was promoted by conversations with Erhard Schmidt as well as for the axiomatization of set theory whose development was shaped by discussions and criticism in Hilbert's group. In this sense, Zermelo's set-theoretic achievements are not ingenious strokes through which new fields of research were introduced for the first time – they grew out of a stimulating "scientific exchange of ideas" strongly influenced by Hilbert.

But Hilbert's ideas were not meant as dogmatic instructions; rather, they were programmatically formulated hints at possible fields of research, whose imagined end products were hardly outlined. Moreover, his relation to Zermelo was characterized by a fruitful collaboration. In his biographical sketch, Otto Blumenthal reports on Hilbert's foundational work in the period between 1904 and 1914 ([Blu35], 422):

> For a long time Hilbert's work on these questions paused, although he touched on them several times in lecture courses on the foundation of mathematics. In the meantime he followed E. Zermelo's set-theoretic ideas, his postulate of choice and his axiomatics of set theory with a most lively interest. At the same time, he also familiarized himself with the logical calculus in its different versions, because he had recognized immediately after 1904 that he could not proceed on the desired path without a clear and complete formalization of the logical modes of inference.

Blumenthal thus indicates that Hilbert relied on Zermelo's pioneering work in set theory, pioneering in the direction he had given, but that he also recognized the indispensability of research in mathematical logic.

After having pointed the way in foundational studies, Hilbert seems to have seen his main task in supporting young researchers going this way. His help was not restricted to programmatic recommendations concerning promising fields of research, it also covered institutional support. Indisputably, Felix Klein was the decisive academic leader of university politics in Göttingen,[285] but Hilbert felt himself entitled to say ([Hil71], 79):

---

[285] For Klein's university politics see [Mane70], [Tob81], and especially [Tob87], [Row89].

In all questions concerning organization *Klein* had the undisputed and absolute leadership; I have never cared about matters of administration. However, I have always actively taken part in matters regarding essential decisions, especially appointments, creating new positions, etc.

After the turn of the century Hilbert supported, above all, Ernst Zermelo to whom he attributed an essential role in his foundational programme and who was linking his work closely and successfully to his mentor's ideas. Despite all the difficulties due to Zermelo's prolonged illness, Hilbert's commitment never weakened.

However, there were limits. When Hilbert's colleague and friend Hermann Minkowski died unexpectedly on 12 January 1909,[286] it became clear that Zermelo's lobby was unable to help him get the vacant chair. Instead of Zermelo, Edmund Landau, then a *Privatdozent* in Berlin, became Minkowski's successor. Gerhard Hessenberg had foreseen this. Two weeks after Minkowski's death he reported to Nelson that the succession was already being openly discussed:[287]

It would be best if they finally gave Zermelo a position. This neglect of one of our most brilliant scientists is already a public scandal and a serious disadvantage of the regiment of Felix the One and Only [meaning Felix Klein] which is otherwise fairly successful. Every foreigner to whom this is still a novelty is outraged when he learns that Zermelo is still a *Privatdozent* – I am afraid that, *ahead of* Zermelo, Landau and Blumenthal are considered first. I plainly begrudge the latter this; Landau would deserve such a success very much, but Berlin would again lose one of its rare good teachers, without a hope of replacing him. (OV 2.23)

In a letter quoted below (cf. 2.11.3) Hilbert explained that "Landau was the suitable person for us – for many reasons." There also was a formal hurdle. As a rule, university regulations in Germany prohibited a scholar from getting his first professorship at the university where he had made his *Habilitation.*

As proved by his commitment for Zermelo's lectureship and later for the extraordinary professorship (2.11.4), Hilbert clearly aimed at enabling Zermelo to make a life for himself as researcher and academic teacher. However, as emphasized already, apart from personal and social motivations, he was committed to following a long term research strategy. Through the revision of his foundational programme, problems of set theory were considered problems of mathematical logic, and research in set theory became relevant for mathematical logic. When planning Zermelo's lectureship for mathematical logic, he made sure that Zermelo's valuable work was integrated into the pursuance of his programme. In addition, the official institutionalization and legitimization of the new academic subject was a matter of research politics. It affected the

---

[286] For details about Minkowski's death which shocked the Göttingen mathematicians, cf. [Rei70], 114–116.

[287] Letter of 27 January 1909, ASD, 1/LNAA000273.

education of future young scientists and strengthened his own position in the institute.

2.11 Waiting for a Professorship

As already described, Zermelo's academic career progressed only slowly. After he had been under discussion for an extraordinary professorship at the University of Breslau in 1903, years were to pass until he was considered again for a permanent university position. Strongly supported by Hilbert and scientifically recognized, he should have been successful much earlier. In a letter of the directors of the Göttingen Mathematical Seminar of 24 January 1910 quoted below (2.11.4) it is written that the criticism of his 1904 well-ordering paper, despite being "not justified, caused harm to him." However, altogether, it may have been his state of health that formed the dominant reason.

2.11.1 Deterioration of Health

Zermelo's school certificates had already given evidence of his weak constitution. At the beginning of 1905 he fell seriously ill. In a letter to his sister Margarete of 24 February 1905[288] he speaks of "phenomena of inflammation" ("Entzündungserscheinungen") which caused utmost fatigue and forced him to stay in bed. The Hilbert family and, in particular, Carathéodory payed regular visits to him. In order to recover, he was advised to spend the spring and early summer in the mild climate of Italy. Probably, he was suffering from an inflammation of the respiratory system. At the beginning of 1906 he fell ill again, now from pleurisy, and had to take a longer break.[289] Finally, in June 1906, his doctors diagnosed tuberculosis of the lungs and after clinical treatment recommended a longer stay in the mountains.[290] For the time being, Zermelo spent some time at the seaside. He returned to Göttingen "in a refreshed state."[291] However, as the recovery did not last, he had to cancel his course of lectures in the winter semester 1906/07 and took several longer cures in the Swiss mountains.

Obviously, it was his absence in Göttingen between the beginning of 1905 and the spring of 1908 together with the interruption in his scientific publications that was a contributing factor in making him a victim of what was called "*Privatdozentenmisere*" at that time, i. e., the problematic situation for those who were qualified for a professorship and who had the obligation to

---

[288]Copy in MLF.

[289]Letter from Hermann Minkowski to Wilhelm Wien of 12 October 1906; DMA, prel. sign. NL 056-0585. The letter is quoted in the next subsection.

[290]Cf. letters from Leonard Nelson to Gerhard Hessenberg, dated Göttingen, 4 June 1906 and 8 June 1906; GSA, N2110, Nachlaß Nelson, fols. 79–80.

[291]Letter from Minkowski to Wien; as in fn. 289.

teach, but were nevertheless unable to obtain a paid professorship.[292] In this sense Hilbert allegedly said about Zermelo in 1907:[293]

> Althoff [Friedrich Althoff, the Prussian under-secretary responsible for university politics] does everything for the *Privatdozenten*. The faculties do nothing. Do you know any faculty or do you believe that there is any which is going to offer Zermelo a position? From Göttingen, yes, but not from Arosa.[294]

With regard to the time up to 1905, and perhaps thereafter, there probably was another reason for Zermelo's failure to obtain a permanent position, a reason put forward by Minkowski as a crucial one, described by him as "a nervous haste which shows in his speaking and conduct."[295] This haste will again and again return in later judgements and remarks about Zermelo in varying terms and surely formed a basic feature of his personality and constitution.

In 1909 his state of health seemed to have stabilized. Moreover, his axiomatization of set theory had been published and impressively underlined his scientific qualities. Opportunities promptly opened up, first at the University of Würzburg in 1909. In the following we will give a description of the procedure for filling the position which has some kind of preludium in 1906. The assessments of Zermelo that were written by Göttingen mathematicians and physicists in this connection are given almost completely as they provide valuable information of Zermelo near the end of his time in Göttingen.

2.11.2 Würzburg 1906

In the autumn of 1906 the University of the Frankonian episcopal town of Würzburg on the River Main had to fill a professorship in mathematics. Hermann Minkowski had been asked for suggestions for possible candidates. On 12 October 1906 he replied,[296] naming Zermelo and Herglotz from Göttingen, Erhard Schmidt from Bonn, Hahn from Vienna, and Carathéodory from Göttingen in this order. Whereas the assessments of Herglotz, Schmidt, Hahn, and Carathéodory are rather short, the assessment of Zermelo fills several pages, giving a detailed and sympathetic picture of Zermelo's achievements, personality, and academic fate:

> Zermelo is really a mathematician of the highest qualities, of broadest knowledge, of quick and penetrating grasp, of rare critical gift. Recently he has been very occupied with so-called set theory [...]. In this field which touches philosophy Zermelo has become an authority [...]. Finally he is

---

[292] Details about the situation of the young generation of scientists can be found in Christian von Ferber's statistics, cf. [Fer56], especially 75–138.

[293] Letter from Hessenberg to Nelson of 31 July 1907; ASD, 1/LNAA000273.

[294] A resort in the Swiss alps particularly suited for illnesses of the lungs.

[295] Letter from Minkowski to Wien; as in fn. 289.

[296] Letter from Minkowski to Wilhelm Wien; DMA, prel. sign. NL 056–0585.

very experienced and interested in mathematical physics [...]. He has several times succeeded in very skilfully mastering problems which he learnt of in an embryonic state from mathematical pains (mathematischen Wehen) of applied colleagues here. [...] Despite his importance he hasn't yet received external acknowledgement. This is an injustice which is a constant topic in my conversations with Hilbert. Once, Descoudres[297] called Zermelo the prototype of a tragic hero. Above all, his conspicuous lack of good luck stems from his outer appearance, his nervous haste which shows in his speaking and conduct. Only very recently is it giving way to a more quiet, serene nature. Because of the clarity of his intellect he is a first class teacher for more mature students for whom it matters to penetrate into the depth of science. They, like also all younger lecturers here to whom he is a closer friend, appreciate him extraordinarily. However, he is not a teacher for beginners. [...] Unfortunately, Zermelo's state of health has not been satisfactory recently. Last winter he was so incautious as to contract pleurisy; he had to take a longer break. Recently he returned from the sea in a refreshed state; however, the doctor advised him to spend the winter in the south. Up to now I have not heard what he has decided. (OV 2.24)

Minkowski seems to be still unaware of the seriousness of the illness he mentions. In the end it cost Zermelo his academic career by leading to his early retirement in 1916 from the professorship which he had obtained at Zurich University in 1910. When contrasting these features with his impressive scientific achievements, Des Coudres' picture of the prototype of a tragic hero may not be called inadequate.

Obviously Zermelo was not seriously considered. The full professorship was given to the extraordinary Würzburg professor Georg Rost (1870–1958).

2.11.3 Würzburg 1909

On 3 July 1909, extraordinary Würzburg professor Eduard von Weber was promoted to a full professorship. The Senate of the University was ordered by the Bavarian Minister of Education to submit a suggestion for a possible successor for the extraordinary professorship to be filled on 1 October1909.[298]

With a note of 11 July 1909 the Senate asked the Philosophical Faculty to make suggestions. Only three days later the Faculty sent back a list of three mathematicians with Zermelo as first choice. The report mentions that Zermelo had been a non-budgetary extraordinary professor in Göttingen since 1905 with a well-paid lectureship for mathematical logic. Obviously referring to Minkowski's assessment from 1906 and an assessment by Hilbert quoted below, it continues:

---

[297]The physicist Theodor Des Coudres (1862–1926), since 1903 at the University of Leipzig as the successor of Ludwig Boltzmann.

[298]The record of this procedure can be found among the personal files of Emil Hilb, University Archives, University of Würzburg, UWü ZV PA Emil Hilb (no. 88).

Zermelo with colleagues in Göttingen[299]

He is a mathematician of highest qualities, of versatile knowledge, a quick keen intellectual grasp, and a rare critical gift. In the fields of set theory, axiomatics, logic[,] in questions of the principles of mathematics on the whole, he is the first authority; in the field of the calculus of variations he has had a really epoch-making effect. At the University of Göttingen he developed varied and successful teaching activities; he is regarded as an excellent teacher especially among mature students who are interested in penetrating into the depths of science. (OV 2.25)

The other two mathematicians on the list were Emil Hilb and Zermelo's friend Gerhard Hessenberg. Hilb was just 27 years old, a student of Hilbert and Felix Klein when he finished his university studies in Göttingen in the summer semester 1904, and a *Privatdozent* of mathematics at the University of Erlangen, working mainly on the theory of integral equations. Hessenberg was a professor at the Agricultural Academy in Poppelsdorf and a *Privatdozent* of mathematics at the University of Bonn.

On 17 July 1909 the Senate followed the suggestion of the Philosophical Faculty with full consensus, naming Zermelo in position one, Hilb in position two, and Hessenberg in position three and stressing that all of them were not only distinguished by their scientific activities, but also by their outstanding teaching faculties. They finally emphasized that Zermelo and Hilb outclassed Hessenberg in their scientific achievements.

---

[299]Front row from left to right: Hermann Minkowski, Max Abraham, and Zermelo.

The Minister in Munich decided not for Zermelo, but for Hilb, shortlisted in the second position. Hilb stayed in Würzburg until his death in 1929.[300]

A letter from Arnold Sommerfeld to Wilhelm Wien, then professor of physics in Würzburg, shows that already in an early state of the discussion Zermelo's illness may have played a role. At the beginning Sommerfeld writes:[301]

> With regard to the matter of appointment, I consulted Schwarzschild in order to get information about Zermelo's state of health. But so far I have not received an answer. Couldn't you perhaps appoint Einstein? The professorship is said to have been an astronomical one formerly. Hence, the space-time theorist deserves it at least as much as the number or set theorist.

Apparently Zermelo's teaching qualities had also been questioned. For in Schwarzschild's assessment it says:[302]

> The question whether Zermelo's lectures are fairly understandable, can be answered very easily, as sometimes his lectures are in fact quite excellent. He looks the picture of health, has kept up his course, and, as far as I know, has not complained about his state of health. I suppose that he has been so happy to get through the critical age for tuberculosis candidates. (OV 2.26)

When sending this letter to Wien, Sommerfeld adds (ibid.; undated) that also Minkowski praised Zermelo's talks in the Göttingen Mathematical Society as "particularly ingenious, clear, and perfectly structured." Furthermore he offered to ask Hilbert, too, for an assessment. Hilbert did so on 5 May 1909,[303] at the same time not hiding displeasure at his vain endeavours to help Zermelo in obtaining a position:

> I no longer recommend mathematicians from Göttingen without being asked. According to my experience this is no help to them. In particular, I no longer believe that a faculty could be as wise as to offer Zermelo a position. I therefore provided him with a teaching post for math[ematical] logic already quite a time ago, and I used the offer of a chair to Landau to have the Ministry make a substantial increase in his remuneration. We could not make him a full professor to succeed Minkowski, as Landau was the suitable person for us – for many reasons. Fortunately, Zermelo has become thick and fat, and his lungs are totally cured. He feels very well in all respects. His lecture courses are always very successful: in this semester he is giving a course in differential and integral calculus with 90 participants and exercises together with Toeplitz where these two colleagues have also about

---

[300] On Hilb's biography, cf. [Hau33], [Vol95].

[301] Letter of 21 April 1909; DMA, prel. sign. NL 056-0703.

[302] Letter from Schwarzschild to Sommerfeld of 19 April 1909; DMA, prel. sign. NL 056-0702.

[303] Letter to Wilhelm Wien; DMA, prel. sign. NL 056-0077.

Zermelo with his sister Elisabeth

100 participants. In the fields of set theory, axiomatics, logic, in questions of the principles of mathematics, he is the first authority who has proved himself only recently against Poincaré in a very striking manner. (OV 2.27)

The last remark concerns Poincaré's stay in Göttingen in April 1909. In his last will the Darmstadt professor of mathematics Paul Friedrich Wolfskehl (1856–1906) had donated a prize of 100 000 Marks[304] for a proof of Fermat's last theorem. Until the prize was claimed, the interest accruing from it was at the disposal of the Gesellschaft der Wissenschaften zu Göttingen. The academy had formed a committee to decide about the awards. The first award, 2500 Marks, was used to invite Poincaré to Göttingen. Poincaré gave a series of six lectures ([Poi10a]), the majority dealing with integral equations. The fifth one on transfinite numbers emphasized his predicative point of view, arguing against Zermelo's objection[305] that a request for predicativity made

---

[304]Equivalent to more than 1 Million Euros in 2005; in 1997, when given to Andrew Wiles, it had melted down to 30 000 £ by inflation and monetary reforms.

[305][Zer08a], 117; [Zer67b], 190–191.

large parts of mathematics invalid (among them the fundamental theorem of algebra) and concluding that according to his point of view and despite Zermelo's proof the well-ordering principle was either still open or senseless ([Poi10b]).

Altogether, the assessments of Hilbert and Schwarzschild lead to the impression that Zermelo seemed to be no longer seriously ill. They emphasize his ability to teach advanced students, and they praise his scientific qualities. Why did the Minister nonetheless refuse to follow the recommendation of the Faculty? Remembering Minkowski's letter from 1906, one cannot exclude that certain features of Zermelo's constitution such as his "nervous haste" together with a possible weakness in teaching beginners may have resulted in a negative momentum. After all it is "wisdom" which Hilbert asks for with a faculty offering him a position. The Würzburg Faculty as the station of competence in the procedure had shown this "wisdom." So we are left without any reason which is more than an item of speculation.

As regards Hilbert, his assessment bears witness to an unconditional belief in Zermelo's qualities, and its tenor may bear paternal features. This impression is confirmed when in 1911 the Wolfskehl committee, with Hilbert as its chairman, donated an award of 5000 Marks to Zermelo whose tuberculosis had flared up again. The award was given "for his achievements in set theory and as an aid in the full recovery of his health" ([Rei70], 135). Hilbert's attitude thus corresponds in an impressive way to Zermelo's conviction quoted at the beginning of this chapter that Hilbert was his "first and sole teacher."

2.11.4 An Extraordinary Professorship

Zermelo's personal controversy with Bernstein (2.8.4) was linked to arguments within the Göttingen Philosophical Faculty. These arguments arose partially in connection with Hilbert's demand for an increased influence on the appointment of candidates for philosophy chairs. They brought to the fore the events of January 1910 which were triggered off when the Faculty decided to appoint Bernstein as an extraordinary professor of insurance mathematics,[306] two years after he had become a *Privatdozent*.

On 24 January 1910 the board of directors of the Seminar of Mathematics and Physics applied to the minister[307]

> to make the local *Privatdozent*, Prof. Dr. Ernst Zermelo, an extraordinary professor of mathematics at our university while letting him keep his present lectureship for mathematical logic and related objects and retaining his current remuneration.

---

[306]The application is dated 26 January 1910; GSA, Rep. 76 Va Sect. 6 Tit. IV No. 1 Vol. 22, fols. 282–283. It was meant to renew the post of an extraordinary professorship of insurance mathematics which had, until 1907, been an extraordinary professorship of theoretical astronomy.

[307]GSA, Rep. 76 Va Sect. 6 Tit. IV No. 4 Vol. 5, fols. 176–178; outline in UAG, 4/Vc 229, edited in [Pec90a], 53–54.

They emphasized that they only aimed at giving Zermelo personal recognition and "in no way" intended to counter the application of the Philosophical Faculty concerning Bernstein. Even so, they said, the fact that they announced their request at the same time as the Philosophical Faculty made their application, was "not *coincidental.*" Their goal was to make sure that Zermelo was not passed over by Bernstein. Zermelo was not elegible for the extraordinary professorship for which Bernstein had been chosen, because he was specialized in a different field of research. So one could argue that Zermelo might remain solely a *Privatdozent.* However, the whole matter became "curiously" complicated as

> Bernstein's scientific achievements until now were primarily, and his lectures partly, of a purely mathematical kind. And not only that! The works of both scientists meet in the field in which each of them has achieved the most, in *set theory.* According to the judgement of the mathematical experts among us Zermelo is – as far as mathematics is concerned – not only much superior to his younger colleague Bernstein, but we regard Zermelo as one of the most important of all younger German mathematicians.

The fact that Zermelo had, up to that point, not been offered a chair, could in their opinion partly be explained by the reactions to his well-ordering paper of 1904 which by the well-ordering principle "gave set theory the foundation that was, until then, missing:"

> It was just this paper of Zermelo's that came under some *criticism* which is, according to the experts among us, *not justified.* Among those critics who unwittingly harmed him is just colleague *Bernstein,* and *we therefore ask most respectfully, so as to avoid a wrong impression, that you use the occasion of Bernstein's promotion to pay Zermelo the respect that he has been deserving for a long time, and to promote also him to an extraordinary professor.*

After having briefly described Zermelo's academic situation, emphasizing that "a considerable number of today's professors were educated" ("haben Belehrung geschöpft") by his clear and gripping lectures, they summed up:

> We would regard it as a well-deserved acknowledgement of Zermelo's pioneering papers in science and of the help he has given us *for eleven years* by his successful teaching if Your Excellency would not deny him promotion into the next step of the academic career.

The application was signed by all professors of the Seminar, among them Hilbert, Klein, Landau, and Prandtl. In the end it became irrelevant because two months later Zermelo was offered a full professorship at the University of Zurich. Already on 16 January Hilbert had sent an assessment there which may have been decisive for the offer (3.1.2). Both the application and the assessment coincide in many details of the description of Zermelo's achievements. They thus represent the concluding picture which Hilbert had of Zermelo at the end of their time together in Göttingen.

Zurich *1910–1921*

3.1 A Full Professorship

3.1.1 Legends

Zermelo's time in Zurich from 1910 to 1916 can be seen as an outstanding period in his life: It was the only time that he held a paid university professorship. Moreover, it provided the setting for some widely circulated anecdotes of doubtful truth. One of their sources is a long footnote in Abraham A. Fraenkel's autobiography *Lebenskreise* which is worth quoting completely ([Fra67], 149, fn. 55):

> Although it does not belong here, some less known characteristics may be reported of this ingenious and strange mathematician whose name has held until today an almost magical sound. Being a bad teacher, he could not get ahead in Göttingen despite having published in 1904 a paper of three pages – proof of the "well-ordering theorem" – which got the whole mathematical world excited – positively and negatively; he grappled with his opponents in a treatise which has in respect to its sarcasm nothing like it in mathematical literature. In 1910 he was finally offered a full professorship at the University of the Canton Zurich. Shortly before the World War he spent a night in the Bavarian Alps. He filled the column "Nationality" in the hotel's registration form with the words: "Not Swiss, thank goodness." Misfortune would have it that shortly after that the head of the Education Department of the Canton Zurich stayed at the same hotel and saw the entry. It was clear that Zermelo could not stay much longer at the University of Zurich. He retired in 1916 and moved to Germany, where I was in frequent scientific contact with him, even from Jerusalem. When I asked Zermelo's friend Erhard Schmidt on one occasion, why Zermelo had almost stopped publishing, he replied that he could not expect to annoy anyone anymore with his publications.

It may be that Zermelo was not a gifted teacher as Fraenkel maintains. He was, however, an inspiring one for mature students as described in numerous assessments by the Göttingen mathematicians and physicists. It should have become clear from the considerations in the preceding chapter that in any case his teaching qualities had nothing to do with his problems of making a career in Göttingen. The hotel story will be discussed after the circumstances of Zermelo's retirement have been clarified as far as the available documents allow.

3.1.2 The Chair in Zurich

In 1910 Zermelo was offered a full professorship of mathematics by the Philosophical Faculty II[1] of the University of Zurich, the university of the Swiss Canton Zurich.[2] He became the successor of Erhard Schmidt whom he knew from Göttingen, where Schmidt had taken his doctorate under Hilbert's supervision in 1905. After his *Habilitation* in Bonn in 1906, Schmidt had become a full professor at the University of Zurich in 1908. In 1910 he left Zurich for Erlangen. He then went to Breslau. Finally, in 1917, he became the successor of Zermelo's teacher Hermann Amandus Schwarz in Berlin.

The search committee for the professorship had considered a number of candidates, among them also younger Swiss mathematicians, but finally recommended two mathematicians working in Germany: Ernst Zermelo from Göttingen and Issai Schur, a *Privatdozent* in Berlin. They supported their choice by assessments from Frobenius for Schur and from Hilbert for Zermelo. The proposal was discussed at the meeting of the Faculty on 21 January 1910.[3] Schmidt, who was a member of the committee, strongly confirmed Hilbert's opinion about Zermelo and supplemented it by his own impressions. The Faculty agreed on presenting only one candidate to the higher authorities, but to mention also the *aequo loco* recommendation of the committee. Obviously impressed by Hilbert's very positive assessment and Schmidt's support, they voted 15:1 *pro* Zermelo.

Details of the discussion can be found in the report which the Dean of the Faculty sent to the Council of Education of the Canton Zurich on 22 January 1910.[4] In the beginning the report lists the requirements which the candidates should fulfil (ibid., 1–2). They should be representatives of "mathematical analysis" and stimulating teachers, and they should have an independent command of the mathematical disciplines and an extensive knowledge of the literature in order to be able to assign unsolved scientific problems as topics

---

[1] I. e., the Section for Science and Mathematics of the Philosophical Faculty.

[2] Besides the university, founded in 1832, Zurich was also home to the Eidgenössisches Polytechnikum, founded in 1855 and renamed in Eidgenössische Technische Hochschule (ETH) in 1911.

[3] Minutes of the meeting, given in Protokolle der Philosophischen Fakultät II; SAZ, Z 70.2906 AA 10 3, p. 298.

[4] In Zermelo's personal files; SAZ, Einzelne Professoren, U 110b.2, No. 45. If not stated otherwise, items quoted in this section belong to these files. – Details of the whole procedure can also be found in the minutes of the Zurich *Regierungsrat* of 24 February 1910 (ibid.) and in [FreiS94], 17–19.

for dissertations and to evaluate the solutions presented by the candidates. At the same time the Faculty aimed at a "certain stability of teaching," because "a great fluctuation of teachers would have to be regarded as a bad thing for the students." The choice of Zermelo was justified by quoting Hilbert's assessment[5] extensively (ibid., 5–6):

> In all his publications Zermelo presents himself as *a modern mathematician who combines versatility with depth in a rare way. A thorough expert in mathematical physics,* as unanimously confirmed by men like [Woldemar] Voigt, Carl Runge, [Arnold] Sommerfeld, [Walther] Nernst, he is at the same time *the* authority in mathematical logic. *Together with this he combines in his person the understanding of the most far-flung parts in the enormous field of mathematical knowledge.* For 11 years Zermelo has conducted here [in Göttingen] *a most versatile lectureship.* He has repeatedly given even elementary lecture courses such as *analytical geometry, differential and integral calculus* with good success for a larger audience (70–80 participants), who stayed with him for several semesters in undiminished numbers. When he lectures *on chapters of higher mathematics to more mature students,* his lectures become original achievements. He presents in them the modern standpoint of the respective branch of science; they are among the best to be heard here. A considerable number of today's *Privatdozenten* and professors have profited from them. My colleague Landau, who is now regularly attending Zermelo's course of lectures on the logical foundations of mathematics, has just praised the clarity and fascination of his talk. Years ago he was ill, but he must now be considered completely cured. [...] *He would undoubtedly follow the offer of a chair in Zurich with the greatest pleasure, and I am convinced that he would acquit himself in his position with utmost devotion to duty.* (OV 3.01)

The report continues (ibid., 7):

> Colleague Schmidt declares that he fully shares Prof. Hilbert's assessment of the scientific significance of Zermelo's [mathematical] papers. [...] He further mentions that Zermelo is a German although his name does not sound German. Based on longstanding personal relations in Göttingen, he gives special praise *not only of his superb character, but also of his positive traits of strict objectivity, honesty and integrity of judgement.* (OV 3.02)

Schur was said to be one of the most important living mathematicians. While working in a narrower field, he was a conscientious, practical teacher and very popular with students. The following argument pro Schur is most noteworthy:

> It was reported to the Faculty that, as a Jew and a Russian, he has only a small chance of getting a position at a Prussian university. Therefore there is a possibility of binding him to Zurich for a greater number of years.

However, this argument proved to be wrong. Already in 1913, Schur was called to Bonn University as Felix Hausdorff's successor.[6]

---

[5]Outline in SUB, Cod. Ms. D. Hilbert 492, fols. 4/1-2.
[6]For Schur's biography cf. [Sta04], ix–xii.

Zermelo during his time in Zurich
Courtesy of Universitätsarchiv Zürich

On 4 February the Council of Education informed the Faculty that they had stopped the procedure for filling the position. Reviving previous considerations of discontinuing the professorship they demanded proof of its necessity by providing a convincing curriculum. On the same day the Faculty decided to take appropriate measures.[7] On 23 February they sent a report to the Council proposing a curriculum and arguing:

> The impression should not be allowed to arise that the Canton Zurich is attempting to sponge improperly on the Federal Government [...] by using the teachers of the Polytechnical University who are paid by the Federal Government to an incorrect degree for the education of its students. Such an impression would, however, definitely arise if Zurich University were

---

[7]SAZ, Z 70.2906 AA 10 3, p. 300–301.

to discontinue its very important chair of mathematics and encourage its students to attend in growing numbers lecture courses at the Polytechnical University.

These arguments may have convinced.[8] On 24 February the *Regierungsrat* gave its approval. Zermelo was offered the chair initially for a period of six years as was then usual, with the stipulation to represent the field of pure mathematics according to the curriculum which the Council of Education had set. The teaching load was fixed to 10–12 hours. The assumption of office was on 15 April 1910.[9]

3.1.3 Under Consideration at the Technical University of Breslau

Two years later Zermelo was close to being offered another chair. On 6 February 1913 the Prussian Minister for Spiritual and Educational Matters (geistliche und Unterrichtsangelegenheiten) informed Rector and Senate of the Technical University of Breslau[10] that he intended to appoint Constantin Carathéodory, then still budgetary professor in Breslau, to a full professorship in Göttingen effective from 1 April. He asked them to require the Department of General Sciences to make provisions for a replacement.[11] On 21 February the Senate informed the Minister[12] that they had voted 6 to 2 for the request of the Department to name Zermelo in the first place, Issai Schur, *Titularprofessor* in Berlin, and Max Dehn, extraordinary professor in Kiel, *aequo loco* in the second place, and Hermann Weyl, *Privatdozent* in Göttingen, in the third place. The achievements of Zermelo are described as follows:[13]

> Zermelo is one of the German mathematicians who are best-known here as well as abroad; his scientific papers are excellent [...] and have partly even caused a real sensation, for example a paper on the kinetic theory of gas and his investigations in set theory, in particular the proof of the famous well-ordering theorem. Although he is working here in the perhaps most abstract areas of mathematical logic, his time as an assistant to Max Planck and the papers in theoretical physics stemming from then show that Zermelo totally masters those applications of mathematics which are most important for a technical university. At any rate, the liveliness of his mind enables him to adapt himself quickly and easily to the most varied demands. (OV 3.03)

---

[8]Probably also Zermelo himself was involved in the discussion, because on 17 February the Council of Education asked him by cable to turn up with them on 19 February (MLF).

[9]Certified copy of protocol in MLF.

[10]Whereas the University of Breslau was founded already in 1702, the Technical University was founded only in 1910.

[11]The presentation follows documents in UAW, TH 156, and in GSA, I. HA Rep. 76 Kultusministerium, V b Sekt. 8 Tit. III Nr. 6.

[12]GSA, ibid., fol. 75.

[13]UAW, ibid., 13–14; GSA, ibid., fol. 75.

The files show that there was resistance to the list because of doubts which were raised about Zermelo's state of health and about his ability to teach at a technical university. On 18 March Ernst Reichel, a professor at the Technical University of Berlin, sent information about Zermelo to the ministry which he had received from Aurel Stodola, a professor of mechanical engineering at the ETH Zurich who was recognized as the pioneer of thermal turbo engines.[14] Stodola wrote:[15]

> Your fears concerning [Zermelo's] state of health are not without reason; in fact, he recently had to take a whole semester off and stayed in Davos. It was said, however, that it was not an illness of his lungs, but rather an illness of his nerves. Apart from this he is a kind, but somewhat strange personality; half an eccentric, rather restless, perhaps more suited for universities than for technical universities. It would surprise me very much if he were interested in *technical* problems.
>
> This is the little and vague information I can give you. (OV 3.04)

It is now obvious that Stodola's strong criticism of Zermelo's interests in technical matters was unjust. There are numerous documents showing his steady occupation with technical problems. They range from a talk about lifting devices for the Dortmund-Ems Canal which he gave to the Göttingen Mathematical Society as early as 6 June 1902[16] to various applications for technical patents worked out in Zurich and later also in Freiburg. On 21 February 1911, e. g., together with a partner, he applied for two patents[17] at a Zurich patent attorney. The first patent, "On an oscillation-free regulator to produce a constant number of revolutions which can be arbitrarily fixed while the machine is running,"[18] was for a kind of cruise control for engines, the second for a related means "to produce a constant torque at one of two coupled shafts."[19] On 7 March 1911, these applications were supplemented by an application for an "elastic and shock-free clutch." There is no document in the *Nachlass* which tells whether these applications were successful. In 1931 Zermelo would work out another patent for automatic gears which might be seen as a derivative of his 1911 ideas.

It is certain that in Breslau Zermelo had been supported by Ernst Steinitz and Gerhard Hessenberg, both professors in the Department of General Sciences, Steinitz even being its head. When they heard of the difficulties Zermelo's appointment was facing, Hessenberg turned to the president of the Province of Silesia for support. The president refused to take official mea-

---

[14]GSA, ibid., fol. 80.

[15]Ibid., fols. 79, 81.

[16]*Jahresbericht der Deutschen Mathematiker-Vereinigung* **11** (1902), 357.

[17]UAF, C129/275.

[18]"Oszillationsfreier Regulator zur Herstellung einer konstanten, auch während des Laufs willkürlich einstellbaren Tourenzahl einer Maschine."

[19]"Selbstätige Reguliervorrichtung zur Herstellung eines konstanten Drehmoments an der einen von zwei miteinander gekoppelten Wellen."

sures, but declared his willingness to inform the ministry about any statement which might reach him from mathematicians.[20] On 7 June he forwarded a statement by Adolf Kneser and Erhard Schmidt, both professors at the University of Breslau and strongly in favour of Zermelo. Kneser and Schmidt argued as follows:[21]

> 1) The professors of mathematics at the University and at the Technical University maintain a lively scientific and cooperative exchange; we would regard being joined by such an excellent researcher as Zermelo in our circle as a valuable gain for the scientific life of our city.
>
> 2) With our lively consent the students of our university make use of the teaching opportunities of the Technical University. Since one of us [Schmidt] knows from personal impressions which he got in Göttingen that Zermelo stimulates his students a good deal, we would therefore highly welcome his appointment also in the interest of our students.

On 10 July the minister offered the professorship to Max Dehn,[22] the co-founder of combinatorial topology. Dehn had also been a student of Hilbert and since then was a close friend of Zermelo's. He accepted the offer and stayed in Breslau until 1921 when he left for a professorship at the University of Frankfurt.

3.1.4 Zermelo's Medical History

In January 1910 Zermelo had undergone a check-up by the Zurich expert for pulmonary diseases Oskar Wild[23] who diagnosed tuberculosis of the upper parts of the lung in an apparently not alarming state. However, soon things changed. Zermelo's health problems became notorious in Zurich as Stodola's assessment shows, and Zermelo's personal files there are filled with his medical history and his attempts to get an adequate substitution for courses he had to cancel.

As early as on 28 January 1911 he applied for time off for the coming summer semester. As attested by a physician, his lung disease made a stay of several months at a health resort necessary. The Council of Education agreed on 3 February (MLF). For the time being Zermelo took the cure at the Schatzalp sanatorium for lung patients above Davos.

On 29 June 1911 he reported to the Dean that the cure had taken effect but that the physicians had advised him to extend his stay into the winter in order to prevent relapses. This was confirmed by the doctors in charge.[24] They attested that Zermelo's disease consisted in an older tubercular infection of the right part of the lung which was already mainly scarred. On 18 July

---

[20]GSA, ibid., fol. 85.
[21]Ibid., fol. 82.
[22]UAW, ibid., p. 22.
[23]Wild's report of 18 March 1916.
[24]Medical certificate of 10 July 1911.

Zermelo officially asked for leave for the winter semester. A second assessment, dated 24 July 1911, confirmed that the winter in Zurich would involve certain risks and that it appeared to be very doubtful whether he would be able to give any lectures at all. On 4 August the Faculty supported his application which was finally granted. In the meantime he had succeeded in arranging Alfred Haar, then a *Privatdozent* in Göttingen, to take over his lectures in the winter semester.

In January 1912 Wild diagnosed the disease of the right lung to have considerably progressed since January 1910 and already produced a cavern. In March 1914 Zermelo underwent surgery of the thorax during which the cavern was reduced with the help of a paraffin wax filling.[25] The operation was carried out by Ferdinand Sauerbruch, a pioneer of thorax surgery.[26]

The worsening of his state of health affected Zermelo's confidence in a final recovery to such a degree that he made definite plans for a publication of his collected works. Written on letter paper of a hotel in Davos, his concept provided for three groups of papers, those on the calculus of variations, on mechanics and the theory of heat, and on set theory and mathematical logic, respectively, comprising altogether 285 estimated pages. The title he had in mind was *Gesammelte Abhandlungen aus der reinen und angewandten Mathematik (Collected Papers From Pure and Applied Mathematics)*.[27]

In spring 1915 there was a second serious recurrence of the disease. At the end of April a cold made it impossible for Zermelo to speak. On 17 May he asked for time off for the complete summer semester because a medical assessment of 14 May had attested that he suffered from a chronic larynx catarrh. He suggested that the main lecture course on the differential and integral calculus should be transferred to Paul Bernays who had already assisted him in the preceding semester.

The larynx catarrh proved to be a tubercular infection of the vocal chords which had to be treated surgically several times in autumn 1915.[28] As it needed further care, Zermelo's leave had to be extended through the whole winter semester. It was only during the last weeks that he felt strong enough to hold his seminar, examinations, and a small special course.[29]

3.1.5 The End of Zermelo's Career

On 3 February 1916 the University Commission expressed its opinion that

---

[25] Wild's report of 18 March 1916.

[26] Letter from Sauerbruch to Zermelo of 25 February 1914; MLF. At the end of the letter it says: "On this occasion I would like to reassure you once more that the surgery is quite harmless and that you need not be concerned about it."

[27] UAF, C 129/242. The collection ends with his chess paper ([Zer13]) which is attributed the year 1912.

[28] Medical assessment of 14 October.

[29] Protocol of the Council of Education of 4 January 1916.

it was due time to give all orders [to ensure] an undisruptive, adequate course of classes in the mathematical disciplines through the summer semster 1916. (OV 3.05)

During the next weeks, two alternatives were discussed in the educational institutions. The first one allowed Zermelo to extend his 6-year period which was ending on 15 April, provisionally for a further semester; the second one aimed at an instant retirement. On 1 March the Faculty held the opinion that one should for the time being refrain from insisting on a definite decision because Zermelo himself hoped to be fully active again in the summer semester and as it was already too late for arranging a substitution. However, the University Commission remained sceptical, letting its recommendation depend on a medical examination which was carried out on 18 March by Wild. Wild's diagnosis ruled out the possibility that Zermelo would be able to start lecturing at the beginning of the summer semester. Three days later, the University Commission pleaded for an instantaneous retirement. In his letter of 5 April to the *Regierungsrat*, written in Arosa, Zermelo agreed:

[Earlier] I informed the Educational Secretary that I still had reservations about giving up my professional activity so early merely because of a probably only temporary inability to work. I had therefore requested postponing this question to the next semester. [...] As I infer from your letter, the higher authorities, following a medical assessment, did not accept the proposal of my Faculty to extend my term of office provisionally for a further semester. They would prefer an immediate decision in the interest of regular lessons. Therefore I am ready now to accept the arrangement which the government proposes and herewith request my transfer into retirement for reasons of health. I gratefully appreciate the size of the pension which I was promised as a friendly concession of the authorities. However, I intend to make use of this offer only until I shall have regained my ability to work and succeeded in finding a sphere of activity corresponding to my strength.

Hoping that the now imminent reappointment of the lectureship may attribute to the thriving of both science and our University, I sign with highest esteem. (OV 3.06)

He was retired as from 15 April. The official pension was fixed at 3800 Francs per year. Together with payments from the pension fund his total pension amounted to 5505 Francs, a little bit more than his salary before. In January 1918 the official pension was raised to 4800 Francs.

Zermelo's professorship had come to an end, due to an illness which had interfered with his scientific work for years and very probably had cost him several offers of other university positions. The imposed retirement was to leave its scars. In later years he will feel ill-treated by the "bigwigs," those more successful in their university careers.

In autumn 1916 he got a sign of acknowledgement. On 31 October he was awarded the annual Alfred Ackermann-Teubner Prize for the Promotion of the Mathematical Sciences of the University of Leipzig[30] for his 1904 and 1908 papers on the well-ordering theorem and on the axiom of choice ([Zer04], [Zer08a]), in particular for the latter one.[31]

About that time he had himself decided to pay a longer visit to Göttingen in order to spend the winter semester 1916/17 at his "home university."[32] On 7 November 1916 he gave a talk there on his new theory of ordinal numbers (3.4.3).

Altogether, documents concerning the time between Zermelo's retirement and his move to Freiburg (Germany) in October 1921 are sparse. From remarks he made during his time in Freiburg, one can conclude that he continued to work mathematically and started to have a critical look at the emerging intuitionism.

His registration files are not complete. They show that he kept his apartment at Schönberggasse 9 in Zurich, but did not stay there, choosing his places of residence according to his state of health and returning to Zurich only now and then for short periods. After his return from Göttingen he spent the rest of the year in the mountains. In April 1917 we find him in the Alpine resort of Arosa,[33] in October 1917 in Davos where he showed signs of "a remarkable recovery."[34] On 9 October 1919 he moved back to Zurich from Seewis, a village higher up in the mountains near Davos. Only some weeks later he left Zurich again for the town of Locarno on the north bank of Lago Maggiore in the mild climate of the Southern Alps where he stayed until early 1921.[35] During the spring of 1921 he stayed in Southern Tyrol and during the summer in Bavaria.[36] Later he will look back on this time as "a roving life in guest-houses."[37]

His generous pension evidently allowed him to live without financial concerns. This situation was to change when he moved to Germany. On 18 September 1922 the Council of Education stated that Zermelo, unmarried, living in Freiburg (Germany) and still unfit for work, meanwhile was receiving a total pension of 6505 Francs which amounted to about four times the

---

[30] Alfred Ackermann-Teubner Gedächtnispreis zur Förderung der Mathematischen Wissenschaften, donated in 1912 by Alfred Ackermann, then one of the owners of Teubner-Verlag and the grandson of its founder. Later prizewinners include, for example, Emil Artin and Emmy Noether.

[31] Letter from the Rector of Leipzig University of 21 November 1916; MLF. Cf. *Jahresbericht der Deutschen Mathematiker-Vereinigung* **25** (1917), 2nd sect., 84.

[32] So in his letter to Courant of 4 February 1932; UAF, C 129/22. – Zermelo was registered in Göttingen from 1 November 1916 until 15 February 1917.

[33] Postcard from Bernays to Zermelo of 31 March 1917; MLF.

[34] Letter from Zermelo to his sisters, dated Davos-Platz, 8 November 1917 (MLF).

[35] Unfinished *curriculum vitae*, undated; MLF. Zermelo was not officially registered in Locarno.

[36] According to letters which he wrote to Abraham A. Fraenkel during this time; cf. 3.5.

[37] Letter to Marvin Farber of 31 December 1924; UBA, Box 25, Folder 25.

Zermelo in 1921
Courtesy of Margret and Ronald B. Jensen

salary of a German full professor. Therefore the official pension was reduced from 4800 Francs to 1000 Francs. The act was based on a law from 1919 which allowed the adjustment of pensions after changes in the conditions under which they had been fixed.

When in 1923 inflation in Germany accelerated drastically,[38] Zermelo's financial situation became very difficult. Hence, on 26 May 1923 he asked for a revision of the reduction. On 6 December 1923 he repeated his request,

---

[38] From January to mid November 1923 wholesale prices climbed by a factor greater than 250 000 000.

adding a detailed list of his expenses and of food prices[39] in order to prove that the shortened pension

> amounts to about as much as I needed 30 years ago as a young student for an extremely plain way of life and is absolutely not in accordance with the present rise of prices. I am already forced to spend more than I receive. Before long, when the savings of the previous year have been used up, I will not be able anymore to ensure a nutrition sufficient for my weakened state of health. (OV 3.07)

The *Regierungsrat* now accepted Zermelo's arguments, re-raising the pension to the height of 1916:[40]

> Though Prof. Zermelo has lectured at the University of Zurich for only a few years, his pension should – from a purely human point of view – nevertheless be fixed in such a way that Prof. Zermelo can live on it in a modest way. (OV 3.08)

The partial abolition of the reduction did not lead to a lasting betterment. In later years, Zermelo again and again complained about the low level of his pension.[41] Financial matters also affected the relationship with Zurich University. Rudolf Fueter, Zermelo's successor, arranged a series of talks on set theory which Zermelo was to hold in April 1932. When Zermelo asked for a reimbursement of travel expenses, Fueter answered:

> We always thought that you could give the talks to do a favour to your home university, since we unfortunately have to do without your activity because of your health, whereas Freiburg enjoys this advantage.[42]

Not being ready to take over the financial burden himself,[43] Zermelo cancelled the talks,[44] giving as reasons that he was being pressed by the publisher to complete the edition of Cantor's collected works ([Can32]).

At the beginning of this chapter we have quoted a footnote in Fraenkel's autobiography telling us that Zermelo lost his Zurich professorship because he had given his nationality in a Bavarian hotel as "Not Swiss, thank goodness." In the Zurich files there is no document which supports Fraenkel's report. Of course, this does not prove that the hotel story is wrong. It can, however, be ruled out that a corresponding incident was the decisive reason for Zermelo's dismissal. Zermelo served in Zurich for 12 semesters, for 4 semesters he

---

[39]The prices are given in a gold-based currency unit which the German government had introduced in October 1924 in order to fight inflation.

[40]Letter of 15 December 1923.

[41]Because of a waiting time of 80 years the Zurich files do not allow to give a documentation.

[42]Letter of 5 January 1932; UAF, C 129/35. Fueter refers to courses of lectures which Zermelo was giving at Freiburg University as an honorary professor; cf. 4.1.

[43]Letter from Zermelo to Courant of 4 February 1932; UAF, C 129/22

[44]Letter to Fueter of 8 January 1932; ibid.

was on sick leave, namely in the summer semester 1911, the winter semester 1911/12, the summer semester 1915, and the winter semester 1915/16. When he resigned it was not foreseeable when he could take up his work again.

No doubt, Zermelo was capable of acting in the described way. The originality of his scientific work had a counterpart in his character. One reason for believing that the hotel story did really take place is that it was well-known during Zermelo's lifetime. This is confirmed by the letter of denouncement which was written by Eugen Schlotter, an assistant at the Freiburg Mathematical Institute, on 18 January 1935[45] and led to the loss of the honorary professorship Zermelo had got there. The parts in which Schlotter refers to Zermelo's time in Zurich run as follows:

> A generation ago, by giving a foundation of set theory, Professor Zermelo made himself an immortal name in science. [...] Because of his famous name, a benevolent patron provided him with a professorship at the University of Zurich. When he was sure of this position, Zermelo started travelling without bothering about his duties at the university. He lost his professorship and lived on the generous pension granted to him. He repaid his benefactors' hospitality with unsurpassable malice. Once he registered in a Swiss hotel with the following words: "E. Zermelo, Swiss professor, but *not* Helvetian!" This resulted in a considerable reduction of his pension. (OV 4.36)

The exposition in this section shows that Schlotter's defamation is untenable and distorted. On the other hand, documents and personal impressions from his time in Freiburg bear evidence that Zermelo had certain reservations about Switzerland, which may be rooted in the unfortunate events he experienced in Zurich. Together with his inclination to jocular anecdotes (such as that about the origin of his name) they may have led him to remarks which finally became a story by decoration and sharpening, and, with this, part of the Zermelo folklore.

3.2 Colleagues and a Friend

Zermelo attempted to integrate himself into the intellectual life of Zurich. For example, in March 1911 he became a member of the "Antiquarische Gesellschaft in Zürich," an association for history and classical antiquity, founded in 1832 and still active today.

Among his friends, Paul Bernays has to be mentioned first. Zermelo had already had a close relationship with Bernays during his time in Göttingen. A Swiss citizen from the Canton Zurich, Bernays was born on 17 October 1888 in London.[46] He was first educated in Berlin and studied mathematics, philosophy, and physics, starting at the Technical University Charlottenburg in 1907, changing to the University of Berlin half a year later, and

---

[45]UAF, B 24/4259. The letter is quoted in 4.11.2.
[46]On Bernays' biography cf. [Moo90].

Paul Bernays
Courtesy of Gerald L. Alexanderson and Birkhäuser Verlag

continuing at the University of Göttingen in 1909. In Göttingen he became a member of Hilbert's working group, but also of Leonard Nelson's school of Neo-Frieseanism. He got his doctorate in 1912 with a number-theoretic thesis supervised by Edmund Landau. In the enclosed curriculum vitae he mentions Zermelo as one of his Göttingen teachers ([Ber12]). Shortly afterwards he took the opportunity to follow Zermelo to Zurich in order to make his *Habilitation* there. His thesis on modular elliptic functions ([Ber13]) was accepted in 1913. He then served as an assistant to Zermelo and as a *Privatdozent* until spring 1919.

In the autumn of 1917 Hilbert visited Zurich and gave his landmark talk "Axiomatisches Denken" ([Hil18]). He invited Bernays to move to Göttingen to assist him in his investigations on the foundations of arithmetic. Bernays accepted and quickly wrote a second *Habilitationsschrift* on the completeness of propositional logic ([Ber26]). He became Hilbert's closest collaborator in foundational issues, co-author of the "Hilbert-Bernays," the two-volume textbook set on the foundations of mathematics ([HilB34]) which gave the first codification of metamathematics and proof theory, Bernays being regarded as its real architect. After the war he became a professor at the ETH Zurich. Highly honoured, he died in Zurich on 18 September 1977.

The files document the fruitful collaboration between Zermelo and Bernays. Their personal relationship was marked by mutual esteem. Both continued to exchange letters at least on special occasions, thereby showing a particular intimacy. The condolence letter which Bernays wrote to Zermelo's wife Gertrud after Zermelo's death contains a warm-hearted characterization of Zermelo which is among the most personal ones we know of.

Besides Bernays, Ludwig Bieberbach also made his *Habilitation* under Zermelo's patronage. Bieberbach had accompanied Zermelo from Göttingen to Zurich. In 1910 he already became a *Privatdozent* there; Zermelo's positive evaluation was shared by Albert Einstein, then an extraordinary professor at Zurich University. Four months later, Bieberbach took up a lectureship in Königsberg (cf. [FreiS94], 19).

Apparently, Zermelo's relationship to Einstein got closer when Einstein, who had left Zurich in 1911 for the University of Prague, returned to the ETH in 1912. On a postcard which Einstein wrote to him on 4 October 1912 (MLF), it says:

> I am always in my apartment in the evenings [. . .] and will be delighted if you pay a visit to me. (OV 3.09)

Entries in Zermelo's diary of 1914 (MLF) show that in February, two months before Einstein left Zurich for Berlin, both met several times a week for hours. Gertrud Zermelo remembered remarks of her husband that Einstein appreciated these conversations because of Zermelo's competence in both, mathematics and physics. In 1919 Zermelo attended and took notes of a series of lectures on special relativity theory which Einstein gave in Zurich in January and February.[47]

There are no documents which give evidence that Zermelo came into closer contact with Hermann Weyl.[48] Both had got to know each other when Weyl had studied mathematics in Göttingen between 1903 and 1908, then also attending lectures of Zermelo's. Having made his doctorate under Hilbert's supervision in 1908 and his *Habilitation* in 1910 and having lectured then as a *Privatdozent* in Göttingen, Weyl obtained a full professorship at the ETH Zurich in 1913. As early as in his *Habilitation* address ([Wey10]), Weyl had criticized Zermelo's notion of definiteness. He intensified this criticism in *Das Kontinuum* ([Wey18]) which he finished in Zurich in 1917 (cf. 4.6.1). Around that time he also started to favour features of intuitionism.[49] In the preface of *Das Kontinuum* he states that "essential parts of the house [of analysis] are built on sand,"[50] and three years later he created the term "foundational

---

[47]UAF, C 129/146.

[48]There is only one item: Zermelo's diary of 1914 shows that both had an appointment in February.

[49]For Weyl's views cf., e. g., [Ferr99], 338seq. and 357seq.

[50]He thereby takes up a metaphor of Hilbert's which we have encountered already in 2.4.3 as "the house of knowledge" ([Hil05c], 122).

crisis" for the state of scepticism and uncertainty arising from the paradoxes ([Wey21]), giving it the following vivid description (ibid., 30):

> Usually the antinomies of set theory are considered as boundary disputes which concern only the most remote provinces of the mathematical empire and by no means can endanger the inner sturdiness and security of the empire itself, of its true heartlands. [...] Any serious and honest reflection has to lead to the conclusion that those unhealthy events in the border zones of mathematics must be rated symptoms; they bring to light what is hidden by the superficially shining and frictionless work in the centre: the inner insecurity of the foundations upon which the construction of the empire is based.

Twelve years later Zermelo will speak of "a somewhat noisy appearance of the 'intuitionists' who in vivacious polemics announced a 'foundational crisis' of mathematics [...] without being able to replace it by something better" ([Zer30d]). On a postcard to Fraenkel of 8 July 1921 (cf. 3.5) he writes that Weyl's 1921 paper can be taken more seriously than *Das Kontinuum*; however, "instead of all criticism and anticriticism [I prefer] positive achievements" ("Doch statt aller Kritik und Antikritik lieber positive Leistung!"). It may have been the deep discrepancy in foundational views which blocked a fruitful scientific exchange.

3.3 Teaching in Zurich

Accounts for fees (MLF) show that in the winter semester 1910/11 Zermelo had 39 paying listeners in his beginners' course on the differential and integral calculus and 20 listeners in his course on differential equations. So he could certainly have gained some gifted students for more intensive mathematical studies, thus establishing his own working group. In fact, he instantly started to improve the situation for his discipline. On 27 March 1911 he complained that a single professorship[51] was totally insufficient to represent the wide field of higher mathematics, and continued:

> I may remind you that since my appointment I have eagerly tried to win excellently gifted younger mathematicians for our University. However, I already failed with two of them who were offered more favourable prospects elsewhere. (OV 3.10)

Apparently his eagerness had let him leave the official channels, thus causing irritation in the Faculty. His reaction of 10 June 1911 marks one of the rare instances where his letters to the Faculty go beyond their polite and friendly style:

---

[51] Besides Zermelo's full professorship there was an extraordinary professorship for geometrical subjects.

As it is most important for me to regain my ability to work, I would not cling to a position which my own colleagues try to hinder when possible. (OV 3.11)

His endeavours were finally successful, and Bernays became his assistant.

Altogether, Zermelo failed in establishing a school. Obviously, the long interruptions of his lecturing activities and his absence from Zurich forced by his illness prevented him from constantly taking care of the mathematical development of his students. On the other hand, he tried to fulfil his educational duties as well as possible. The Zurich files show that he felt quite responsible for the lecture courses he had to cancel and exercised great care in organizing replacements.

Only one dissertation was done in Zurich under his supervision, Waldemar Alexandrow's elementary foundations of measure theory dealing with the measurability of point sets ([Ale15]). In his assessment of 9 July 1915[52] Zermelo characterizes it as "of a more reviewing and collecting nature," but nevertheless "a valuable scientific achievement." He intimates his own essential contributions which he "put at the author's disposal for a first publication." Later he would publish them in the *Journal für die reine und angewandte Mathematik* ([Zer27a]; cf. 4.2, p. 148).

3.4 Scientific Work in Zurich

After having taken up his professorship in Zurich, Zermelo continued his academic activities and his research as far as his illness would allow. As in Göttingen, his interests ranged from physics, even engineering techniques, via applied mathematics to algebra and set theory. The dominant feature of his research in Zurich is diversity; no topics are pursued in a systematic manner.

We recall that in the axiom paper ([Zer08b]) he had promised to come back to the question of the consistency of his set-theoretic axiom system – a question of essential importance for Hilbert's enterprise to secure arithmetic. However, there is no evidence that he really pursued this problem in Zurich. As he had not yet separated himself from the core objectives of the Göttingen group, this fact may indicate the burden which his illness together with the new duties of his professorship imposed on him.

3.4.1 Game Theory

Zermelo was an enthusiastic player of chess. His passion motivated two papers. The later one, conceived in 1919, but published only in 1928 ([Zer28]), uses a maximum likelihood method to calculate the outcome of a chess tournament (4.2.1). The earlier one ([Zer13]) concerns an application of set theory to the game of chess. It represents one of the two addresses that he gave during

---

[52]SAZ, U 110e.19.

the Fifth International Congress of Mathematicians at Cambridge in 1912.[53] Bertrand Russell, who had invited him to the congress, seems to have been disappointed when Zermelo proposed a talk on chess. In his letter of 8 April 1912[54] he stressed that a lecture on the foundations of set theory would be of the highest interest. The theory of chess would also be interesting, but he had no idea in what section the lecture could be placed. His obvious reservations proved to be unwarranted, faced with the undoubted significance attributed to the paper today. In fact, the "ingenious" ([Koed27], 126) chess paper, "very full of content" ([Kal28], 66), marks the beginning of what is now called the "theory of games." In a certain sense made more precise below, its main theorem may be regarded as the first essential result in the field and finally became known as "Zermelo's theorem." The literature contains varying versions which may more or less deviate from what Zermelo really proved. Moreover, the theorem is often connected with the so-called method of backward induction, i. e. an induction in a game tree which leads from the leaves to the root, and the evaluation method using this tool is known as "Zermelo's algorithm" – despite the fact that nothing about such inductions appears in his paper.[55]

Again, Zermelo's work constituted a novelty for the mathematical community, like the axiom of choice and later his concept of infinitary languages and infinitary logic. However, in contrast to the serious controversy connected with the introduction of the axiom of choice where Zermelo's opinion finally gained acceptance, and in contrast to his committed fight against finitary mathematics as represented by Kurt Gödel and Thoralf Skolem where he lost, the game-theoretic excursion may be regarded as the first step of an uncontroversial mathematical development. This development did not start immediately, but only fifteen years later in the work of Dénes König ([Koed27]) and László Kalmár ([Kal28]) and found its continuation with new perspectives by John von Neumann ([Neu28b]).

Zermelo considers chess-like games played by two opponents, White and Black, who move alternately. He emphasizes that his considerations are independent of the special rules and are "valid in principle just as well for all similar games of reason (Verstandesspiele) in which two opponents play against each other with the exclusion of chance events." Motivated by the game of chess, he assumes that there are only finitely many positions. A move of a player means a transition from a given position p to some possible position q. For simplicity he assumes that a game which starts from a position p_0, either ends after finitely many moves with a win of White or with a win of Black, or it goes on forever; it then ends with a draw.

Zermelo aims at answering the following questions:

---

[53] For the other talk see 4.3.1.

[54] UAF, C129/99.

[55] For information about this situation we refer to [SchwW01]. Quotations below follow the English translation of Zermelo's paper given there.

(1) Is it possible to characterize in an objective mathematical manner those positions p_0 which are winning positions for White, in the sense that White can force a checkmate?

> It seems to me worth considering whether such an evaluation of a position is at least theoretically conceivable and makes any sense at all in other cases as well, where the exact execution of the analysis finds a practically unsurmountable obstacle in the enormous complication of possible continuations, and only this validation would give the secure basis for the practical theory of the "end games" and the "openings" as we find them in textbooks on chess.[56]

(2) If p_0 is a winning position for White, how many moves will White need to force a win? If p_0 is a winning position for Black, how many moves can White enforce on Black until Black has won?

In pursuing these aims, Zermelo uses set-theoretic means. However, there are no set-theoretic arguments involved as the title may suggest – Zermelo demonstrates that set theory provides a comfortable framework to give an answer to Question (1) which is familiar today, but was a novelty in 1912. To illustrate his procedure, we describe his definition of what it means that in position p_0 where it is White's turn, White can enforce a win in at most r moves. This is true if and only if there is a non-empty set $U_r(p_0)$ which consists of sequences of allowed moves of the form $p_0 p_1 \ldots p_s$ with $s \leq r$ such that p_s means a win of White and which has the following property: If $p_0 p_1 \ldots p_s \in U_r(p_0)$, $s' < s$, Black is in turn in position $p_{s'}$, and $p_{s'}q$ is a possible move for Black, then there is a sequence in $U_r(p_0)$ which starts with $p_0 \ldots p_{s'}q$. There is no exact definition of strategy. This will be given only 15 years later, partially by Dénes König ([Koed27]) and fully by László Kalmár ([Kal28]).

The answer Zermelo obtains, for example, in regard to the first part of question (2) is the essential content of Zermelo's theorem and can be summarized as follows: Starting from any position with any opponent's turn, either White can enforce a win in less than N moves (where N is the number of positions) or Black can enforce a win in less than N moves, or both can enforce a draw.

In order to prove that in case p_0 is a winning position for White, the number N of positions yields an upper bound for the number of moves necessary to checkmate, Zermelo uses a simple non-repetition argument: If White, starting in position p_0, enforces a win in a game corresponding to the sequence $p_0, p_1 \ldots p_s$ of moves with $s \geq N$, then (ibid., 503)[57]

> the end game $p_0, p_1 \ldots p_s$ would have to contain at least one position $p_\alpha = p_\beta$ a second time and White could have played at the first appearance of it in the same way as at the second and thus could have won earlier.

---

[56] Ibid., 501.
[57] Zermelo uses "q" instead of "p".

Dénes König ([Koed27]) realised that this argument contains two gaps: The instruction given to White does not apply if it is Black's turn in position p_α. Moreover, the argument is only valid if the number of moves of the different end games that White may play according to Black's behaviour, is bounded. Faced with these gaps, Zermelo provides proofs for both points that are given in the appendix of König's paper ([Zer27b]). The proof which closes the first gap does not use a repetition argument. The closure of the second gap amounts to König's own proof in the paper which is based on what is now called König's lemma ([Koed27], 121). In the case considered here it says that an infinite rooted tree where each node has only finitely many successors has an infinite branch. If the lengths of the end games starting in p_0 by which White forces a checkmate, were unbounded, the game tree with root p_0 arising in a natural way from the possible end games would be of the kind in question and, hence, would contain an infinite branch, i. e. Black could enforce a draw.

As already mentioned above, Kalmár ([Kal28]) provides exact definitions of strategies and winning positions. In an even more general framework including games of infinite length and using genuine methods of set theory such as transfinite induction, he answers Zermelo's questions in a conclusive form. Modern game theory starts with von Neumann's "Zur Theorie der Gesellschaftsspiele" ([Neu28b]). A new feature characteristic for it is the strategic interaction between the players. Subsequently, without knowing the anticipations of Zermelo and König, Max Euwe, one of the most influential chess players of his time and world champion from 1935 to 1937, independently published parts of the results in an intuitionistic framework ([Euw29]).

Did Zermelo's paper have any influence on the development of game theory? There is no clear answer. König remarks ([Koed27], 125) that it was von Neumann who proposed to him closing the second gap in Zermelo's proof by an application of his lemma. In a letter to Zermelo of 13 February 1927[58] he writes that von Neumann did not know Zermelo's paper when informing him about this idea,

> I was the one who drew his attention to it. Now he is working on a paper about games which will appear in the *Mathematische Annalen*. (OV 3.12)

This passage leaves open several possibilities. Von Neumann may have started to think about games either by himself or may have been inspired by König who knew Zermelo's paper and was working on related questions. One may also assume that it was Zermelo's paper which prompted him to think about games more thoroughly. But von Neumann does not refer to it – neither in the *Gesellschaftsspiele* paper ([Neu28b]) announced by König nor in the comprehensive book on game theory which he wrote together with Oskar Morgenstern ([NeuM53]). To stay on stable grounds, we will therefore confine ourselves to giving Zermelo the credit for having written the first paper that mirrors the spirit of modern game theory.

---

[58]UAF, C129/64.

3.4.2 Algebra and Set Theory

In late 1913 Zermelo completed an algebraic paper ([Zer14]) that involves his well-ordering theorem in an essential way. Ten years earlier, Georg Hamel had proved the existence of bases for the vector space of the reals over the field of the rationals, later called "Hamel bases" ([Ham05]). He defined them along a well-ordering of the reals, adding a new element to the basis under construction if it was linearly independent of the elements added so far. Zermelo transfers the method to algebraic independence in order to show the existence of rings whose quotient field is the field of reals and the field of the complex numbers, respectively, and whose algebraic numbers are algebraic integers. Later, Emmy Noether continued these investigations by giving a classification of all such rings ([Noe16]).

3.4.3 Ordinal Numbers

In 1923, John von Neumann gave a set-theoretic definition of the ordinals ([Neu23]) which finally became generally accepted. The ordinals in his sense – the *von Neumann ordinals,* as they are termed today – start as

$$\emptyset, \{\emptyset\}, \{\emptyset, \{\emptyset\}\}, \ldots.$$

The well-ordering that corresponds to the sequential presentation coincides with the element relation. In particular, any ordinal is the set of ordinals smaller than it, and the less-than-or-equal relation coincides with the subset relation.

As stated, e. g., by Paul Bernays ([Ber41], 6) and documented by Michael Hallett,[59] Zermelo conceived and worked with the von Neumann ordinals perhaps as early as 1913. The main body of his considerations fills a small writing pad ([Zer15]) which he bought in a stationer's shop in Davos probably either during his stay there in 1911/1912 or after 1915. On the first 21 pages he anticipates von Neumann's definition of ordinal numbers and establishes their basic properties, attributing some of them to Paul Bernays.[60] The main fact

---

[59][Hal84], 277–280; [Hal96], 1218.

[60]In a letter to Zermelo of 12 March 1912 (UAF, C 129/48) Hessenberg, when explaining his use of "equal" and "identical," expounds that "a number, for instance the number 2, can be entirely '*individualized*' [...], for example, according to Grelling by the definition $2 = \{0, \{0\}\}$," thus attributing the basic idea leading to the von Neumann ordinals to Kurt Grelling. Hessenberg refuses to follow this definition in mathematics as there "'two' is not $\{0, \{0\}\}$, but a quality which happens to two legs, two eyes, two feather points, *etc.*" Grelling's Ph. D. thesis ([Grel10]) which would have been a suitable place to discuss the von Neumann definition of natural numbers, does not mention it. – For the early history of the von Neumann ordinals see also [Hal84], 276–277.

saying that any well-ordering is isomorphic to the well-ordering of an ordinal is missing; its proof would have required the axiom of replacement or a suitable substitute. Von Neumann was well-informed about this:[61]

> A treatment of ordinal numbers closely related to mine was known to Zermelo in 1916, as I learnt subsequently from a personal communication. Nevertheless, the fundamental theorem, according to which to each well-ordered set there is a similar ordinal, could not be rigorously proved because the replacement axiom was unknown.

The terminology of Zermelo's notes changes from "Ordnungszahl" ("ordinal number") in the beginning to "Grundfolge" ("basic sequence").

The second part of the notes, an elaboration of 10 pages entitled "Neue Theorie der Wohlordnung" ("A New Theory of Well-Ordering"), gives an axiomatic treatment of the von Neumann ordinals and may be the beginning of a more systematic elaboration of the preceding part. None of the two parts was ever rewritten or worked out for publication. However, Zermelo gave a talk on his results on 7 November 1916 in the Göttingen Mathematical Society.[62] Its title "Über einige neuere Ergebnisse in der Theorie der Wohlordnung" ("On Some Recent Results in the Theory of Well-Ordering") recalls that of the second part and may be read as confirming von Neumann's statement above. Hilbert mentioned Zermelo's theory in his course "Probleme der mathematischen Logik" in the summer semester of 1920. Bernays, obviously not informed about the talk which Zermelo had given in 1916, reported to him:[63]

> Your theory of well-ordering numbers was also mentioned in the course, namely in connection with Burali-Forti's paradox; it leads to a more succinct version of it. I suppose you agree that I have told Hilbert about your ideas. In any case they were mentioned in the course, giving your authorship explicitly; you were hence called the originator of the most comprehensive mathematical axiomatics. (OV 3.13)

In published writings Zermelo did not return to the von Neumann ordinals before 1930. In his *Grenzzahlen* paper ([Zer30a]) he used them under the name "basic sequences" without mentioning his or von Neumann's authorship. Later, when elaborating some sketches given there, he also considered axiomatizations of well-orderings in the spirit of the second part of the notes in the writing pad, but now with the specific intention of axiomatizing well-orderings of extraordinary length (4.7.4).

---

[61][Neu28b], 321; translation according to Hallett ([Hal84], 280).

[62]*Jahresbericht der Deutschen Mathematiker-Vereinigung* **25** (1917), 2nd section, 84.

[63]Postcard to Zermelo of 30 December 1920; UAF, C 129/10.

3.5 The Fraenkel Correspondence of 1921 and the Axiom of Replacement

In March 1921 Abraham A. Fraenkel turned to Zermelo with some questions about his plan for showing the independence of the Zermelo axioms of set theory. The letter was the beginning of a short, but intensive correspondence. The exchange suffered from Zermelo's moving from guest-house to guest-house, staying in Southern Tyrol between March and June, and then at different places in Bavaria, finally arranging his move to Freiburg in October. Whereas Fraenkel's letters are lost up to his letter of 6 May 1921, several of Zermelo's letters have been preserved at least in part.[64]

Zermelo appreciates Fraenkel's plans: "I am very pleased that you are working on the question of the independence of the axioms of set theory; I also consider it a promising task."[65] He shows a particular interest in the independence of the axiom of choice (ibid, 3) and gives good advice on this point. When Fraenkel published his results, he thanked Zermelo for providing helpful arguments ([Fra22a], fn. 3).

Apart from proposals concerning Fraenkel's project, Zermelo's letters reveal many a judgement about features of his axioms. For instance, he comments several times on the empty set. It is "not a genuine set and was introduced by me only for formal reasons."[66] In his letter of 9 May he is even more explicit:

> I increasingly doubt the justifiability of the "null set." Perhaps one can dispense with it by restricting the axiom of *separation* in a suitable way. Indeed, it serves only the purpose of formal simplification. (OV 3.14)

We recall Zermelo's comment about the character of the axiom of choice, which seems to counteract his aim of replacing the idea of successive or simultaneous choices by the abstract notion of choice set (cf. Ch. 2, fn. 151). The explanatory remark of Fraenkel ([Fra22b], 233) quoted there originates in the following passage of Zermelo's letter of 31 March (p. 3):

> I call [the axiom of choice] "axiom of choice" not in a true sense, but only for the sake of customary terminology. My theory does not speak of *real* "choices." The [comment] should be taken as an annotation that does not touch the theory. For me [the axiom of choice] is a pure axiom of existence, nothing more and nothing less, [...] and "simultaneous choices" of elements and their comprehension into a set [...] are not more than a mode of imagination which serves the purpose of *illustrating* the idea and the (psychological) necessity of my axiom. (OV 3.15)

---

[64]I thank Benjamin S. Fraenkel for the permission to quote from the letters of his father.

[65]Zermelo to Fraenkel, dated Meran-Obermais, 31 March 1921, p. 1.

[66]Letter of 31 March 1921, p. 2.

Abraham A. Fraenkel in 1939
Courtesy of the Fraenkel family

An essential theme of the correspondence is the axiom of replacement. Taken together, the letters provide valuable information about the process which led to Fraenkel's conception of this axiom.

In early May Fraenkel wrote to Zermelo[67] requesting a reply to two letters which he had written in mid-April from Amsterdam. He continues:

> Even if this repeated request for your answer is the main purpose of this letter, I use this opportunity to bring up two further points.
>
> 1) Let Z_0 be an infinite set (for example, the set which you name Z_0)[68] and $\mathfrak{U}(Z_0) = Z_1$, $\mathfrak{U}(Z_1) = Z_2$, *etc.*[69] How can you show in your theory [...] that $\{Z_0, Z_1, Z_2, \ldots\}$ is a set, and, hence, that, e. g., the union of this set exists? If your theory were insufficient to allow such a proof, then obviously the existence of, say, sets of cardinality $\aleph_\omega$ could not be proved.
>
> 2) [...][70]
>
> If your time does not allow more, I would appreciate very much at least a short comment on my example from Amsterdam and on the first question

---

[67]Letter dated Marburg, 6 Mai 1921; UAF, C 129/33.

[68]I. e., the set $\{\emptyset, \{\emptyset\}, \{\{\emptyset\}\}, \ldots\}$.

[69]$\mathfrak{U}$ denotes the power set operation.

[70]This point concerns a specific question about definiteness.

of the present letter. By the way, I believe that I need no longer rely on your kindness in the near future. (OV 3.16)

Zermelo received Fraenkel's letter in the evening of 9 May. Already on the next day, he answered:

Your remark concerning $Z^* = Z_0 + Z_1 + Z_2 + \cdots$ seems to be justified, and I missed this point when writing my [axiomatization paper [Zer08b]]. Indeed, a new axiom is necessary here, but which axiom? One could try to formulate it as follows: If the objects $A, B, C, \ldots$ are assigned to the objects $a, b, c, \ldots$ by a one-to-one relation, and the latter objects form a set, then $A, B, C, \ldots$ are also elements of a set M. Then one only needed to assign $Z_0, Z_1, Z_2, \ldots$ to the elements of the set Z_0, thus getting the set $\Theta = \{Z_0, Z_1, Z_2, \cdots\}$ and $Z^* = \mathfrak{S}\Theta$.[71] However, I do not like this solution. The abstract notion of assignment it employs seems to be not "definite" enough. Precisely this was the reason for trying to replace it by my "theory of equivalence."[72] As you see, this difficulty is still unsolved. Anyway, I appreciate you having brought it to my attention. By the way, with such an "axiom of assignment" ("Zuordnungs-Axiom") your intended proof of independence would fail. [...] But is there another axiom which works? As soon as I will encounter something promising, I will inform you. On the other hand, I am highly interested in any remarks of you referring to this question. (OV 3.17)

Zermelo thus formulates a second-order form of the axiom of replacement where the replacements admitted are restricted to injective ones. His instantaneous criticism, rooted in the deficiency of definiteness, would be settled only a decade later when he added the full second-order axiom of replacement to his 1908 axioms (cf. 4.7.2).

Zermelo's postcard of 8 July 1921, now written from Munich, answers a letter from Fraenkel of 19 May where Fraenkel had evidently reported on his intended proof of the axiom of choice by making use of the axiom of replacement, namely by replacing the sets of a system of non-empty and pairwise disjoint sets by one of their elements and applying replacement to obtain a choice set.[73] Zermelo is sceptical:

Your "proof" of the axiom of choice by use of the "axiom of replacement" is rather questionable to me [...]. "Replacement" asks for a one-to-one assignment of elements m to sets M, and just this is not possible here.

Well, the difficulties consist in finding a sufficiently precise version of "replacement" or "assignment." However, in order to master these difficulties, one has to proceed in a way different from that in my "theory of equivalence." (OV 3.18)

---

[71]$\mathfrak{S}$ denotes the union operation.

[72]In his letter to Dedekind of 3 August 1899 ([Can99]) Cantor writes that "two equivalent multiplicities either are both 'sets' or are both inconsistent."

[73]Cf. [Fra21], 97, or [Fra22b], 233.

This quotation shows that Zermelo still understands the axiom of replacement as referring to injections. It is here that the name "Ersetzungsaxiom" ("axiom of replacement") appears for the first time in his letters. The context indicates that it is used in Fraenkel's letter of 19 May.

Contrary to Zermelo's scepticism, Fraenkel retained the new axiom. Already on 10 July 1921 he finished a paper ([Fra22b]), where he first describes the gap which he had discovered in Zermelo's system, stating that it can easily be verified by going through the axioms, and then introduces the new axiom in an even more general form by admitting arbitrary replacements instead of only injective ones (ibid., 231):

> *Axiom of Replacement.* If M is a set and each element of M is replaced
> by [a set or an urelement] then M turns into a set again.[74]

Some weeks later, at the 1921 meeting of the Deutsche Mathematiker-Vereinigung at Jena, he announced his results in an address given in the afternoon of 22 September. The report ([Fra21]) states that Zermelo took the floor during the discussion following the talk. He accepted Fraenkel's axiom of replacement in general terms, but voiced reservations with respect to its extent.

On 6 July 1922 Thoralf Skolem delivered a talk to the 5th Congress of Scandinavian Mathematicians in Helsinki ([Sko23]), where he described the same gap in Zermelo's system as Fraenkel, giving a rigorous proof and proposing a similar solution. Skolem's axiom of replacement obeys Zermelo's requirement of definiteness by restricting the replacements allowed to those being definite in the sense of being first-order definable:

> Let U be a definite proposition that holds for certain pairs (a, b) in the
> domain B; assume further, that for every a there exists at most one b such
> that U is true. Then, as a ranges over the elements of a set M_a, b ranges
> over all elements of a set M_b.[75]

When reviewing Skolem's paper,[76] Fraenkel simply states that Skolem's considerations about the axiom of replacement correspond to his own ones.

Skolem's version of the axiom of replacement has now become the standard one. As a matter of fact, Zermelo would never accept it, a disapproval that would separate him from the future development of set theory and logic.[77]

---

[74] "*Ersetzungsaxiom.* Ist M eine Menge und wird jedes Element von M durch ein 'Ding des Bereichs $\mathfrak{B}$' [in the sense of Zermelo's [Zer08b]] ersetzt, so geht M wiederum in eine Menge über." – Fraenkel's form can be reduced to Zermelo's special case; instead of replacing x by $F(x)$, one replaces it by $(x, F(x))$.

[75] "Es sei U eine definite Aussage, welche für gewisse Paare (a, b) im Bereich B gültig ist und in der Weise, dass für jedes a höchstens ein b vorkommt, sodass U wahr ist. Durchläuft dann a die Elemente einer Menge M_a, so durchläuft b die sämtlichen Elemente einer Menge M_b." (English translation from [vanH67], 297.)

[76] *Jahrbuch über die Fortschritte der Mathematik* **49** (1922), 138–139.

[77] For further details about the history of the axiom of replacement and for a discussion of Fraenkel's, von Neumann's, and Skolem's part, cf. [Hal84], 280–286. See also [Wan70], 36–37.

4

Freiburg *1921–1953*

In October 1921, some months after his 50th birthday, Zermelo moved to Freiburg im Breisgau in southwestern Germany, where he then lived for more than three decades until his death in 1953. The mid-1920s saw a resumption of his scientific activity in both applied mathematics and set theory. Moreover, they also provided the opportunity to lecture at the University of Freiburg as an honorary professor. However, the mid-1930s brought a change. As his foundational work was directed against the mainstream shaped by inspired members of a new generation such as Kurt Gödel and John von Neumann, he ended in a kind of scientific isolation and also in opposition to his mentor David Hilbert. As his political views disagreed with the Nazi ideology, he lost his honorary professorship. Thus in 1935 the promising new beginning came to an end which left him in deep disappointment and led him to a life of seclusion. His last years were brightened by his wife Gertrud whom he had married in 1944.

4.1 A New Start

Zermelo's move to Freiburg meant a return to the city where he had studied mathematics and physics during the summer semester of 1891 with, among others, the mathematician Jakob Lüroth and the physicist Emil Warburg. He had of course experienced the advantages the town might offer to his illness: the mild climate of the upper Rhine valley together with its location at the foothills of the Black Forest promising fresh air from the nearby mountains and ample forests. In fact, he later made a habit of extended walks in order to improve his state of health. In the evenings he seems to have enjoyed the local wine. On the other hand, the University with its emphasis on arts and humanities also accomodated his strong interests in philosophy and the classics. In the winter semester 1923/24, he attended Edmund Husserl's course "Erste Philosophie" ("First Philosophy")[1] and Jonas Cohn's course "Logik und Erkenntnistheorie" ("Logic and Epistemology"), taking notes in shorthand.[2]

For the time being his still uncurtailed Swiss pension permitted a comparably carefree life in economically ruined post-war Germany. However, after the cut in 1922 mentioned above, the situation changed. Nevertheless he finally found an apartment in a spacious villa[3] built around the turn of the century and situated in one of the finest residential areas, requiring only short walks in park-like surroundings to both the adjacent mountains and the centre of the city with the University. His next apartments[4] also offered these amenities, but were simpler in order to suit his then more limited financial means.

Documents concerning Zermelo's first years in Freiburg are sparse. The main source consists of his correspondence with Marvin Farber,[5] with Marvin's brother Sidney, a medical doctor, and with Sidney's wife Norma, furthermore in notes which Marvin Farber took about his discussions with Zermelo in Freiburg.[6] The discussions led to plans for a co-authored book on mathematical logic (4.3.2).

Beyond the Farber letters, the correspondence in the *Nachlass* comprises one important piece: On 15 August 1923 John von Neumann sent Zermelo a carbon copy of his dissertation.[7] In the accompanying letter[8] he sketches the contents, underlining the novelties, and asks for Zermelo's comments. The new point he mentions first is his notion of ordinal number, and he explicitly writes down the first finite ordinals together with ω and $\omega + 1$. As described above (cf. 3.4.3), Zermelo had the same idea already as early as around 1915. There is no document concerning his reaction.[9] According to comments in his 1929 paper on definiteness ([Zer29a]) he appreciated the precise version of

---

[1]Cf. [Hus1956].

[2]UAF, C 129/147 (Husserl), C 129/247 (Cohn).

[3]Karlstraße 60. It was Zermelo's fourth apartment in Freiburg. In the first one (Kybfelsenstraße 52) and the next ones (Reichsgrafenstraße 4 and Reichsgrafenstraße 17) he stayed only for shorter periods. These apartments were situated in likewise good and quiet residential areas.

[4]Maximilianstraße 26 and Maximilianstraße 32. The registration files are incomplete. Apparently Zermelo sometimes neglected registration.

[5]Marvin Farber (1901–1980), professor of philosophy at the State University of New York in Buffalo (1927–1961, 1964–1974), and professor emeritus (1974); studies at Harvard (Ph. D. 1925) and at Berlin, Heidelberg, and Freiburg in Germany; founder and editor of the journal *Philosophy and Phenomenological Research*, and author of books and articles on philosophy.

[6]The Zermelo part of the correspondence and Marvin Farber's notes are contained in the Marvin Farber *Nachlass* (UBA, the letters in Box 25, Folder 25, and the notes in Box 33; the Farber letters are contained in the Zermelo *Nachlass*, Marvin's letters in UAF, C 129/29, Sidney's and Norma's letters in UAF, C129/30.

[7]UAF, C 129/229.

[8]UAF, C 129/85.

[9]Von Neumann's remark quoted on p. 134 indicates that Zermelo informed von Neumann about his anticipation.

Karlstraße 60 and Maximilianstraße 26 in Freiburg
Photos by Lothar Klimt

definiteness which implicitly comes with von Neumann's system, but did not feel very pleased with von Neumann's way of basing the theory on the notion of function instead on the notion of set, because it led to a "foundation being too involved and understandable only with major difficulties" (ibid., 341).

After years of retreat from the mathematical community, Zermelo began to re-enter the scene. He announced a talk for the annual meeting of the Deutsche Mathematiker-Vereinigung in Innsbruck (Austria), which was scheduled for 25 September 1924.[10] Although he had to cancel it,[11] it was the first clear sign of the new start.

Zermelo presumably came soon into contact with the two full professors of mathematics at the Mathematical Institute of the University, Lothar Heffter (1862–1962) and Alfred Loewy (1873–1935). Heffter was a stimulating teacher who had written influential textbooks, e. g. on differential equations and analytic geometry. Loewy likewise worked in the area of differential equations, but also in insurance mathematics, and later mainly in algebra, attracting good students and collaborators to Freiburg, among them Reinhold Baer, Wolfgang Krull, Bernhard H. Neumann, Friedrich Karl Schmidt, Arnold Scholz, and Ernst Witt. Mathematical life at the University was supported by the

---

[10] *Jahresbericht der deutschen Mathematiker-Vereinigung* **33** (1924), 2nd section, 59.

[11] In a letter to Sidney Farber of 21 September 1924 (UBA, Box 25, Folder 25) Zermelo refers to "well-known reasons" and an imminent change of his apartment.

Freiburg Mathematical Society. Here Zermelo found a group to report on and discuss "his new results."[12]

Both Baer, who was Loewy's teaching assistant until 1928, and Scholz, who followed Baer, were mainly interested in algebra – Baer in group theory[13] and Scholz in algebraic number theory.[14] However, they were also open to the foundations of mathematics and became Zermelo's closest scientific companions. During his Freiburg time, Baer published several papers in set theory[15] and worked on first-order axiomatizations of set theory.[16] Around 1930 – then at the University of Halle – he became Zermelo's most important partner during a decisive time in the latter's foundational research. In the thirties Scholz took over this role, also becoming a thoughtful friend.

Heffter appreciated the contacts with Zermelo so much that he proposed to the Faculty of Sciences and Mathematics that they should offer him an honorary professorship, that is, a professorship with the right but not a duty to lecture ([Hef52], 148). Zermelo agreed with pleasure. In a letter to his sisters Elisabeth and Margarete of 27 March 1926[17] he says:

> I would appreciate it very much if I could again take up an academic activity *at all* and could have cooperative contacts to the University. (OV 4.01)

On 2 February 1926 Heffter and Loewy, the two directors of the Mathematical Institute, applied to the Ministry to nominate Zermelo a "full honorary professor." There was no salary, but he was given the fees which the participants of his courses had to pay. In the application[18] which was passed by the Faculty on 19 February we read that during his stay in Freiburg his state of health had improved so far as to allow him a modest teaching activity, and that such an activity would attribute to the attractiveness of the University:

> Professor Zermelo is an excellent mathematician. He enjoys an international reputation, in particular through his papers on set theory. It would provide an extraordinarily valuable enrichment for the mathematical teaching programme and, hence, for the Faculty of Science and Mathematics if he could be incorporated in the teaching personnel of our university. However, we do not feel that it is correct for a man of his age and importance to join the Faculty as a *Privatdozent*. We therefore request his appointment as a full honorary professor. (OV 4.02)

Asked by the Ministry for a short appreciation of Zermelo's previous research and teaching, Heffter and Loewy[19] speak of a "happy connection of knowledge

---

[12]Letter from Heffter and Loewy to the Ministry of Cultural and School Affairs of 23 March 1926; UAF, B 24/4259.

[13]For Baer's life and work see [Gru1981].

[14]For Scholz's life and work see [Tau52].

[15]For instance, [Bae28] and [Bae29].

[16]Letter from Skolem to Baer of 13 December 1927; copy in MLF.

[17]Copy in MLF.

[18]UAF, B 24/4259.

[19]Response of March 3; UAF, B 24/4259.

in physics and analytical sharpness," praising the talks he had given in the Freiburg Mathematical Society, and emphasizing his multifarious erudition and varied interests. On 22 April the Ministry granted their request.[20] From a later letter of Heffter to Zermelo[21] we learn that parts of the Faculty had opposed the appointment "considerably." As the letter was written during a severe personal controversy which we will describe below, we cannot estimate the intensity of the resistance. In any case, it might have been partly responsible for the fact that in 1935, when Zermelo faced dismissal because of his anti-nationalist behaviour, none of his colleagues in the Faculty tried to help him. In particular, Heffter's letter seems to indicate that Loewy had not been very enthusiastic about Zermelo's professorship.[22] The scepticism may have been a mutual one. For we are astonished to observe the coolness Zermelo will show when Loewy – as a Jew – was dismissed by the Nazis in 1933. Maybe hinting at the fact that Loewy, an enthusiastic teacher, continued to lecture despite having become blind,[23] he frankly states[24] that the dismissal was not an accident because Loewy "would never have retired voluntarily." So all in all the personal contacts in the Faculty he really enjoyed were those with Baer and later with Scholz, and the influential person he really could rely on was Heffter. Heffter was deeply convinced of Zermelo's scientific abilities and not concerned that they were sometimes accompanied by nervous and irritable behaviour. In his recollections ([Hef52], 148), written at the age of about ninety, he characterizes Zermelo as a researcher of international fame who had personally been extremely strange and fidgety (höchst wunderlich und zappelig), adding that Oskar Bolza's[25] old mother had unintentionally, but aptly varied his name to "Tremolo."

As a rule, Zermelo gave one course in each semester, the subjects ranging from applied mathematics to foundations, sometimes also accompanied by exercises or a seminar. Among the courses we find "Foundations of Arithmetic" (winter semester 1926/27),[26] "Introduction to Complex Analysis" (summer semester 1927),[27] "General Set Theory and Point Sets" (winter

---

[20]Letter of 27 April 1926; UAF, B 24/4259.

[21]Of 14 March 1929; UAF, C 129/42.

[22]In a letter to Marvin Farber of 25 May 1926 (UBA, Box 25, Folder 25) Zermelo speaks of a delay in the nomination procedure because of Loewy's "unclear mode of expression" which had led to "misunderstandings" that were overcome by Heffter's "very friendly engagement."

[23]Cf. [Rem95].

[24]Letter to Heinrich Liebmann of 7 September 1933; UAF, C 129/70.

[25]Bolza had lived in Freiburg since 1910 and held an honorary professorship at the Mathematical Institute.

[26]Lecture notes from Marvin Farber (UBA, Box 33) give the following topics: Formal Foundations of Mathematics (Elements of Logistics, General Axiomatics), The Origin of the Notion of Number (in particular Cardinal Number), The Number Systems (Set, Group, Field).

[27]Lecture notes in UAF, C 129/158.

semester 1927/28),[28] "Calculus of Variations" (winter semester 1928/29),[29] "Set Theory" (winter semester 1929/30, summer semester 1932, and winter semester 1933/34),[30] "Real Functions" (winter semester 1932/33),[31] "Differential Equations with Exercises" (summer semester 1933),[32] a seminar on the nature of mathematics (summer semester 1934) and "Number Theory" (winter semester 1934/35).[33] The list shows that Zermelo was fully integrated in the teaching programme of the Mathematical Institute to an extent quite unusual for an unpaid honorary professor. When he lost his position in 1935, he touched this point with a sign of bitterness when he complained about the shameful treatment he had to suffer, adding that this happened "after I had worked at the University for seven years without any salary." As far as his style of lecturing is concerned, we know of only one report. As a student of mathematics and biology, the later logician Hans Hermes attended parts of the course on set theory which Zermelo gave in the summer semester 1932. He remembered[34] that Zermelo's style was strange, because he used to erase with his left hand what he had written on the blackboard with his right hand.

The appointment to the honorary professorship coincides with a remarkable increase in Zermelo's scientific activity. It launched his second period of intensive research, about twenty years after his time in Göttingen. One might therefore be tempted to view the professorship as a stimulant. On the other hand, his improved health might also be taken into consideration. The following years until the mid-1930s will see a broad spectrum of activities: Besides his participation in the teaching programme of the Mathematical Institute he first returned to applied mathematics, publishing papers on different topics. As regards the foundations of mathematics, he tried to gain influence on the new developments in mathematical logic and the developments in set theory coming with them. Furthermore, he edited the collected mathematical and philosophical works of Georg Cantor.

Before these activities are presented in greater detail, it should be emphasized that Zermelo's interests went far beyond mathematics and the sciences. Various assessments emphasize his broad cultural interests. We have already mentioned his encounter with Husserl's philosophy. He was familiar with ancient Greek and Latin. Around 1926 he started a translation of Homer's Odyssey.[35] He there aimed at "a *liveliness* as immediate as possible"

---

[28] Lecture notes in UAF, C 129/157.

[29] Lecture notes in UAF, C 129/159.

[30] Lecture notes of the third one in UAF, C 129/164.

[31] Lecture notes in UAF, C 129/160–161.

[32] Lecture notes in UAF, C 129/162–163.

[33] Lecture notes in UAF, C 129/264.

[34] Oral communication, November 2000.

[35] UAF, C 129/220 (in shorthand), C 129/221 (7th and 8th song), C 129/222 (5th and 6th song), the latter ones typewritten and also in UBA, Box 25, Folder 25. A sample from the 5th song is given in the Appendix (OV 4.03).

("möglichst unmittelbare *Lebendigkeit*").[36] In an annotation of a published sample the editor writes ([Zer30g], 93):

> Our readers will recognize from the sample offered here that the author possesses not only a secure knowledge of language, but also an unusual empathy and articulateness.

At the same time Zermelo participated in the educational programme which the Association of the Friends of the University organized in the towns, health resorts, and sanatoria of the more far-flung surroundings.[37] In the winter semester 1932/33 we find him with a talk "Selected Samples of a Modern Translation of Homer (Odyssey V–VI)" (MLF). In spring 1932 he took part, together with his friend Arnold Scholz, in the "Hellas Tour" for teachers and students of German secondary schools. The cruise with the steamship "Oceana" from 19 April to 4 May led to various ancient places in Greece, Italy, and Northern Africa. His pleasure at reflecting events and thoughts in little poems and other forms of poetry led him, for example, to an ironic dialogue between Telemachos and Nausicaa, "Tele" and "Nausi," to represent features of Nausicaa as described by Homer and in Goethe's respective fragment.[38]

Zermelo was a motorcar fan. In the early 1930s he owned a small three-wheeled car which could be driven without a driving licence. When he finally got his licence at the end of 1935, he was no longer able to realize the dream of an automobile of his own because of a shortage of means. However, he not only enjoyed driving, but was also interested in related mechanical problems. On 4 March 1931 he submitted an – in the end unsuccessful – application to the national German Patent Office[39] asking for a patent on a construction which used a gyroscope to stabilize bicycles and motorcycles during a stop, at the same time providing energy for a new start. From a letter of Richard von Mises[40] we learn about his interest in monorail trains and in steam and gas turbines. In 1932, probably remembering that his application for a patent in 1911 had dealt with constructions which could be used for automatic gears (3.1.3), he began to work on this problem. The *Nachlass* contains a two page sketch[41] entitled "Über eine elektrodynamische Kuppelung und Bremsung von Kraftwagen mit Explosionsmotoren" ("On an Electrodynamic Clutch and

---

[36] Letter to his sisters Elisabeth and Margarete of 27 March 1926; copy in MLF. Zermelo continues that he could complete a translation of the whole of the Odyssey within one to two years if he had a publisher.

[37] The *Verband der Freunde der Universität Freiburg* was founded in 1925 by Heffter in order to strengthen the relations between the University and its surroundings and in order to give support in academic and social need.

[38] "Homer's und Goethe's Nausikaa. Ein Gespräch im Elysium." Manuscript, 6 pages; UAF, C 129/223.

[39] UAF, C 129/274.

[40] Letter to Zermelo of 5 June 1931; UAF, C 129/78.

[41] Of June 1932; UAF, C 129/219.

Zermelo's three-wheeled car

Braking of Motor Cars with Internal Combustion Engines"). The introduction may mirror what he had experienced when driving his three-wheeled car on the winding roads of the Black Forest:

> In the following we describe a method to dispense with the shifting of different "gears" by cogwheels, namely by an elastic clutch between the shaft of the engine and the driveshaft of the car, thus facilitating a shockfree starting and braking of the car as well as an even and quiet run in a hilly terrain, in particular also on mountain roads. (OV 4.04)

Apparently he did not prepare an application for the Patent Office. A success would have been doubtful: In 1927 the Daimler company had already equipped a car with hydrodynamic automatic gears which had been developed two years earlier. In spite of the high degree of technical maturity, these gears were not pursued to production. Work on the problem was taken up again in 1932 by General Motors in the USA and brought to production maturity in 1939 by the Hydramatic system.[42]

Despite the diversity of projects which he started and performed and despite all the personal contacts they involved, Zermelo remained caught in a feeling of loneliness. In many of his letters to the Farber brothers (between 1924 and 1929) he mentions his longing for company ("Gesellschaft") and scientific conversations. Concerning the latter point, his contacts to the Mathematical Institute were not as fruitful and the openness for his logical work not as wide as he wished.[43] Moreover, after the winter semester 1923/24, during

---

[42]Details may be found in [Stu65].
[43]Letter to the Farber brothers of 31 July 1925; UBA, Box 25, Folder 25.

which he had still attended Husserl's lecture course, he had lost interest in
the philosophical trends represented in Freiburg, among them in particular
Husserl's phenomenology,[44] and hence also any contact with Husserl.[45] This
separation deepened when Heidegger's philosophy gained influence:[46]

> There is nobody here to whom I could talk [about logic]. Prof. Becker[47]
> is not interested either: The "intuitionists" despise logic! And even worse
> when Heidegger will come:[48] the "worry of the worrier" and the "throw of
> the throwedness" – such a philosophy of "life" and "death," respectively,
> has nothing in common with logic and mathematics. (OV 4.05)

As the Farber correspondence reveals, loneliness came together with the feeling of financial distress and lack of acknowledgement, rooted perhaps in the
loss of his Zurich professorship and the cuts in his pension. The latter was
combined with a high sensitivity against the "bigwigs," the full professors.
When Marvin Farber had brought up the idea of a professorship in the United
States, Zermelo was open to such a possibility, but remained pessimistic about
its realization:[49]

> I do not have illusions about the possibility of getting there *myself* [. . .].
> Such offers are always made only to those scientists who enjoy sufficient
> acknowledgement already in their homeland and who are well-positioned
> like Sommerfeld, Weyl, and others. But for those, who are already "put
> to chill," one does not have a soft spot abroad either. Well, that is the
> "solidarity" of the international bigwig-cracy! (OV 4.06)

Nevertheless he agreed when Marvin Farber proposed to provide and publish
an English translation of his collected works. On 9 July 1925 Farber wrote to
the Open Court Publishing Company[50] to learn about their attitude on the
matter, at the same time suggesting that "the Chicago or Harvard mathematicians (Professors Osgood, Birkhoff, Whitehead, Sheffer) would no doubt

---

[44]Letter to Sidney Farber of 31 May 1924; ibid.

[45]Postcard to Marvin Farber of 12 August 1924; ibid.

[46]Letter to Marvin Farber of 24 August 1928; ibid.

[47]Oskar Becker (1889–1964), a student of Husserl, who later became a professor
of philosophy at the University of Bonn. He attended one of Zermelo's courses in
set theory and became Zermelo's main partner in questions of the philosophy of
mathematics. For more information on Becker cf. [Pec05a].

[48]Martin Heidegger (1889–1976) became the successor of Edmund Husserl in 1928.

[49]Letter to Marvin Farber of 21 September 1924; UBA, Box 25, Folder 25.

[50]UBA, Box 25, Folder 25. The enclosed list of Zermelo's papers includes three
manuscripts in preparation: "Über das System der transfiniten Ordinalzahlen,"
"Über den Isomorphismus algebraisch abgeschlossener Körper," and "Essay on the
Principle of Substrates." Apparently, none of them was completed. The second
manuscript is also mentioned in a letter from Kurt Hensel to Zermelo of 1 August
1926 (UAF, C 129/44). For the third one cf. 4.3.2.

heartily welcome such a project." As a matter of fact, the project was not realized.[51]

4.2 Research in Applied Mathematics

I still clearly remember our conversation in Breslau where you urged me to take up again the calculus of variations instead of the fashionable foundational research. Well, you see that I'm not a hopeless case where "classical" mathematics are concerned, and that my old, even though hitherto mostly unhappy love for the "applications" has secretly kept glowing. (OV 4.07)

With these words Zermelo accompanies the announcement of a talk for the 1929 meeting of the Deutsche Mathematiker-Vereinigung in Prague.[52] The talk dealt with "navigation in the air as a problem of the calculus of variations" as given in the extended abstract ([Zer30c]) and was instantly recognized as an excellent piece of applied mathematics. However, work on problems of applied mathematics had already started earlier with a paper on the evaluation of chess tournaments ([Zer28]) the origins of which go back to his time in Zurich.[53] Also the preceding publication on the measure and the discrepancy of point sets ([Zer27a]), the first publication to appear after a break of nearly 15 years, had originally been drafted in the Zurich period. It gives a definition of the Lebesgue measure by replacing the standard role of the inner measure by its difference to the outer one. The investigations were first presented in Waldemar Alexandrow's Ph. D. thesis ([Ale15]) written under his supervision in Zurich (cf. 3.3), among them a "particularly simple" proof of the Borel covering theorem (Heine-Borel theorem).[54]

We shall now discuss the Freiburg papers in applied mathematics in more detail.

4.2.1 The Calculation of Tournament Results

Zermelo was very fond of playing chess and was also interested in its theory. We recall his paper on a set-theoretic treatment of chess ([Zer13]). The *Nachlass* contains a set of 22 postcards written in 1926/27 which are part of a chess game by mail,[55] a handbook of chess from 1921,[56] notes of chess games,[57]

---

[51]In December 1924, Farber had informed Zermelo (letter of 15 December 1924; UAF, C 129/29) that Sheffer had been enthusiastic about these plans and had tried to win Osgood as a supporter, but had not been successful.

[52]Letter to Adolf Kneser, then chairman of the Deutsche Mathematiker-Vereinigung, of 7 September 1929; UAF, C 129/63.

[53]First notes stem from July 1919; UAF, C 129/262.

[54][Ale15], 18–11; the then customary German term "Borelscher Überdeckungssatz" stems from here (cf. [ZoreR24], 882 and 885).

[55]With F. Gudmayer, Vienna; UAF, C 129/37.

[56]UAF, C 129/184.

[57]UAF, C 129/249.

and parts of a manuscript by him on end games.[58] It is reported that his relationship with Baer comprised not only a common interest in set theory, but also regular games of chess.[59]

The tournament paper deals with the problem of how to measure the result of a chess tournament, thereby introducing essential principles of modern tournament rating systems. Zermelo criticizes the practice of defining the relative strength of the participants of a tournament by the number of games they have won. He argues that in a chess round robin tournament performed by one excellent player and k weak players, the relative strength of the excellent player differs from that of the next one by a factor $\leq 2 + \frac{2}{(k-1)}$ which does not aptly reflect the strength of the excellent player. Moreover, the method would fail if the tournament is not completed. Proposals which had been made hitherto – such as that by Landau ([Lan14]) – would not considerably improve this situation.

In his paper Zermelo proposes "a new procedure for calculating the results of a tournament which seems to be free of the concerns raised so far and leads to a result that obeys [. . .] all sensible requests" ([Zer28], 437). Essentially, he uses the method of maximum likelihood that arose in statistical mechanics. His argument runs as follows: Each participant p is attributed a yet unknown *relative strength* u_p which, up to a constant factor, is interpreted as the probability for p to win a game of the tournament. Games which are drawn may count as two games where each of the opponents has won one game, and games won by a player may count as two games won by this player. Using this interpretation and given the result R of the tournament in terms of the results of the games performed at it, one can explicitly calculate the probability $w((u_p)_p, R)$ for R to happen. Now the relative strength v_p of each participant p is defined to be the choice for u_p which yields a maximum of $w((u_p)_p, R)$. As Zermelo shows, the v_p exist and are uniquely determined.[60] He describes an iterative procedure of how to calculate them and gives examples, among them the chess masters tournament of New York in 1924 with eleven participants. For the winner Emanuel Lasker,[61] for the participant ranking next,[62] for the participant ranking in the middle, and for the weakest one he obtains

---

[58]UAF, C 129/289.

[59]Oral communication by Otto H. Kegel.

[60]Up to certain exceptional tournaments, for example, a tournament without any game.

[61]Emanuel Lasker (1868–1941) was the second chess world champion, holding this position from 1894 until 1921. Being also a gifted mathematician, he studied mathematics at the universities of Berlin, Göttingen, Heidelberg, and Erlangen, finishing his Ph. D. thesis at the latter university in 1900. Lasker published several papers on mathematics and philosophy, among them an important paper on modules and ideals ([Las05]). His dream of obtaining a professorship in mathematics did not realize. Being a Jew, he left Germany early in 1933. He spent his last years in New York.

[62]José Raúl Capablanca, then the chess world champion.

(normalized by $\Sigma_p v_p = 100$):

$$v_1 = 26.4; \quad v_2 = 18.4; \quad v_6 = 7.12; \quad v_{11} = 2.43,$$

the respective points gained (1 for a win, 0.5 for a draw) being 16, 14.5, 10, and 5. The given numbers indicate that preferences for weak players as illustrated by the example above may vanish.

Lasker appreciates "the first approach to use probability theory for such problems" and believes that the method might have a promising future.[63] But he was only partially right: Like his game-theoretic paper on chess ([Zer13]) the tournament paper also had no immediate influence. His method came into effect only when it was re-invented by Ralph Allen Bradley and Milton E. Terry ([BraT52]). Herbert A. Davis comments:[64]

> There are many roots to the ubiquitous Bradley-Terry model. Indeed it was originally proposed by the noted mathematician E. Zermelo (1928) in a paper dealing with the estimation of the strength of chess players in an uncompleted round robin tournament. However, Zermelo's excellent paper was an isolated piece of work. His model was independently rediscovered by Bradley and Terry (1952).

The Bradley-Terry model or, as one might formulate more exactly, the Zermelo-Bradley-Terry model now provides the standard mathematical framework for rating systems. Extending Zermelo's original goal, these rating systems also provide ratings which take into consideration the achievements in former tournaments as popularly exemplified by tennis rating systems.

Modern chess rating systems, however, are based on variants developed in the 1950s by Airped Elo ([Elo78]). These systems aim at the expected score rather than at the probability of winning. They originate in the Ingo system which was introduced by Anton Hößlinger in 1947 and was the official rating system in Germany from 1974 to 1992.[65]

4.2.2 The Zermelo Navigation Problem

According to Richard von Mises ([Mis31], 373) Zermelo was the first to formulate and solve a problem that "could not be without interest for practical aeronautics." Carathéodory characterizes its solution as "extraordinary elaborate" ("außerordentlich kunstvoll") ([Car35], 378). The problem came to Zermelo's mind when the airship "Graf Zeppelin" circumnavigated the earth in August 1929.[66] He considers a vector field given in the Euclidean plane that describes the distribution of winds as depending on place and time and treats

---

[63]Letter to Zermelo of 12 January 1929; UAF, C 129/66.

[64][Dav88], 13. In the 1st edition of 1963 Zermelo is not mentioned.

[65]Details may be found in [Gli95] or in [GliJ99].

[66]Letter from Zermelo to the Notgemeinschaft der Deutschen Wissenschaft of 9 December 1929; UAF, C 129/140.

the question how an airship or plane, moving at a constant speed against the surrounding air, has to fly in order to reach a given point Q from a given point P in the shortest time possible. With

- $x = x(t)$ and $y = y(t)$ the Cartesian coordinates of the airship at time t,
- $u = u(x, y, t)$ and $v = v(x, y, t)$ the corresponding components of the vector field representing the speed of the wind, and
- $\varphi = \varphi(x, y, t)$ the angle between the momentary speed (u_0, v_0) of the airship against the surrounding air and the x-axis,

and normalizing to $|(u_0, v_0)| = 1$, one has

$$\frac{dx}{dt} = u + \cos\varphi \ \text{ and } \ \frac{dy}{dt} = v + \sin\varphi.$$

Using the calculus of variations, Zermelo obtains the following differential equation for the intended solution φ, his "navigation formula:"

$$\frac{d\varphi}{dt} = -u_y\cos^2\varphi + (u_x - v_y)\cos\varphi\sin\varphi + v_x\sin^2\varphi.$$

He thus provides a necessary condition for the optimal steering function φ. By considering a situation where $\varphi = 0$, i.e., where $\frac{d\varphi}{dt} = -u_y$, he illustrates that in changing winds "the helm has to be turned to that side where the (u, v)-component against the steering direction increases." Applying the Weierstraßian field construction, he then discusses sufficient conditions for a solution.

As already stated above, Zermelo presented his results in a talk given at the 1929 meeting of the Deutsche Mathematiker-Vereinigung in Prague and published an extended abstract in the proceedings ([Zer30c]). Having been confronted with an error detected by Tullio Levi-Civita and having found "surprising" new results,[67] he decided himself to publish a revised extended and more systematic version ([Zer31a]), where he added a similar treatment of navigation in 3-dimensional space.

The talk and the papers found a positive echo and initiated various alternative methods of addressing the problem. Levi-Civita provided a different reduction to the calculus of variations ([Levi31]). Following principles of geometrical optics, Richard von Mises succeeded in avoiding the calculus of variations at least in the two-dimensional case ([Mis31]). Philipp Frank built on the analogy with light in a streaming medium ([Fran33]).[68] On 25 April 1933 Constantin Carathéodory informed Zermelo[69] that he had found a simple solution in the case of a stationary distribution of winds.[70]

---

[67] Letters to von Mises of 19 and 23 September 1930; UAF, C 129/78.

[68] Cf. also [Mani37].

[69] Letter to Zermelo; UAF, C 129/19.

[70] Carathéodory presents such a solution and also an alternative general solution in his later monograph ([Car35], 234–242 and 378–382, respectively).

In a natural way, Zermelo's results ask for similar investigations in other spaces. The corresponding "Zermelo navigation problem" has even been treated still most recently.[71] Zermelo himself made a first step to extend the scope of the problem: When Johann Radon was in search of a problem for his Ph. D. student Kurt Zita, he suggested treating navigation on the sphere.[72]

4.2.3 On Splitting Lines of Ovals

Apparently due to the successful resumption of research in applied mathematics, Zermelo was invited to contribute a paper to a special issue of *Zeitschrift für angewandte Mathematik und Mechanik* in honour of its founder and editor Richard von Mises on the occasion of his 50th birthday. The invitation from the editorial board[73] addressed him as "a 'pure' mathematician who, as a friend and colleague, was personally and scientifically close to von Mises." Zermelo accepted and contributed an investigation on splitting lines of ovals with a centre ([Zer33a]). The subtitle "Wie zerbricht ein Stück Zucker?" ("How Does a Piece of Sugar Break?") indicates that his starting point is the question, how a piece of sugar breaks if one grips it at two opposite corners and then tries to divide it. In the introduction he justifies his choice by a double coincidence with the "wise" physicist and philosopher Gustav Theodor Fechner (1801–1887), who had treated a likewise everyday question ("Why are sausages not cut straight?"[74]) in a humorous way and published his treatise under the pseudonym "Dr. Mises" as part of his *Kleine Schriften* ([Fec75]).

The paper is of an elementary character. Zermelo treats the two-dimensional case. By a physically motivated minimization procedure[75] he shows that there is exactly one splitting line passing through the centre of the piece of sugar. He also treats the case of ovals with a centre (for sufficiently smooth parametrizations), finding two necessary conditions for diameters to be a splitting line. The first one is met by ellipses with the minor axis as a splitting line. The second one may be satisfied by a class of ovals that can be viewed as some kind of smooth generalizations of parallelograms.

---

[71] For example, in [BaoCS04] where the authors "study Zermelo's navigation problem on Riemannian manifolds and use that to solve a long standing problem in Finsler geometry."

[72] Letters from Radon to Zermelo of 22 July 1930 and 17 June 1931; UAF, C 129/92. The finished thesis, [Zit31], is also contained in Zermelo's *Nachlass* (UAF, C 129/186).

[73] Letter from Hans Reissner of 23 July 1932; UAF, C 129/93.

[74] "Warum wird die Wurst schief durchschnitten?"

[75] Let S be a rectangular piece of sugar, A and B opposite corners of S, and L a line through the centre of S of length l in S and distance d to A and B, respectively. According to Zermelo, L is a splitting line of S if the force p applied in A and B in order to break S is minimal such that - in suitable units - the (absolute value of the) torque dp equals the splitting resistance which is proportional to l, i.e., if $\frac{l}{d}$ is minimal.

4.3 The Return to the Foundations of Mathematics

But in the meantime the question of the "foundations" had also again got
underway with the somewhat noisy appearance of the "intuitionists" who in
vivacious polemics announced a "foundational crisis" of mathematics and
declared war on nearly all of modern science – without being able to replace
it by something better. "Set theory as a special mathematical discipline will
no longer exist," one of their most eager disciples decreed – while at the
same time new textbooks on set theory ran to seed. These facts caused me
to come back to investigations of foundational problems – due to lengthy
illness and isolation in a foreign country I had nearly become estranged
from scientific production. I did not take sides in this proclaimed quarrel
between "intuitionism" and "formalism" – actually I take this alternative
to be a logically inadmissible application of the "Tertium non datur." But
I thought I could help to clarify the questions under consideration, not
as a "philosopher" by proclaiming "apodictic" principles which, by adding
to already existing opinions, would merely augment confusion, but as a
mathematician by showing objective mathematical connections – only they
can provide a safe foundation for any philosophical theory.

With these words Zermelo describes the reasons which brought him back to the
foundations of mathematics when he left Switzerland and settled in Freiburg
([Zer30d], 1–2). Following the maxim expressed here, he will explicate a view
of mathematics that separates him from the main directions under discussion
such as intuitionism in the sense of Brouwer or Weyl and formalism as conceived in Hilbert's proof theory, and also from the important developments
initiated by Gödel, Skolem, and others.

4.3.1 The Return

Up to the early 1920s Zermelo does not comment explicitly on his epistemological position – with perhaps one exception: At the mathematical congress
in Cambridge in 1912 he gave a talk entitled "On Axiomatic and Genetic
Methods with the Foundation of Mathematical Disciplines." There is no abstract or elaboration available. However, from remarks made around 1930 and
a related discussion of Hilbert[76] one may guess what he had in mind: In principle, infinite domains can be defined in two ways: "genetically" (sometimes
also termed "constructively"), e. g. by some inductive definition, or "axiomatically, i. e. by a system of postulates which should be satisfied by the intended
domain."[77] Examples of the first kind include the natural numbers as arising
from the number zero by finitely often applying the successor operation. Examples of the latter kind include the natural numbers as given by the Peano
axioms and the real numbers as given by the axioms for completely ordered
fields. Infinite domains "can never be given empirically; they are set ideally

---

[76][Hil00a], 180–181. Here Hilbert introduces the term "genetic" ("genetisch").
[77]Cf., e. g., [Zer29a], 342; [Zer32c], 1.

and exist only in the sense of a Platonic idea" ([Zer32c], 1). Hence, so his conclusion ([Zer29a], 342), in general they can only be defined axiomatically; any inductive or "genetic" way is inadequate.[78] So the 1912 talk may have contained a plea for idealism.

Around 1930 idealism is clearly predominant in Zermelo's thinking. He *expressis verbis* identifies "mathematical" with "ideal" ([Zer30a], 43), and emphasizes ([Zer32a], 85) that "the true subject of mathematics consists in the *conceptual-ideal relations* between the elements of infinite varieties which are set in a conceptual way."

Even in old age he will not change his point of view. As late as 1942, then 71 years old, he will send his infinity theses (4.9.1) to Heinrich Scholz. The second thesis strongly advocates idealism: "The infinite is neither physically nor psychologically given to us in the real world, it has to be comprehended and 'set' as an idea in the Platonic sense."

The quotation from the letter to the *Notgemeinschaft* which opens this section shows that Zermelo's move to Freiburg came together with his decision to re-enter the field of mathematical foundations and to fight against intuitionism and formalism by providing mathematically convincing alternatives. His first foundational encounter was promoted by discussions with members of the Philosophical Seminar at Freiburg University and led to two major projects: to a logic project with Marvin Farber which was not realized (4.3.2), and to the edition of Cantor's collected works which was completed in 1932 (4.3.3).

There were specific points which brought Zermelo back to set-theoretic research: Starting in the early 1920s, his axiom system from 1908 became an object of growing criticism because of the vagueness of the notion of definiteness. Moreover, it also suffered from a certain insufficiency which had been discovered by Fraenkel and Skolem and healed by the axiom of replacement (cf. 3.5). The criticism of definiteness caused Zermelo to reconsider this notion and publish his results in the definiteness paper ([Zer29a]), the first paper in set theory to appear after a break of more than 15 years. His insight into the role of the axiom of replacement (together with the role of the axiom of foundation) led him to the idea of a far-reaching cumulative hierarchy as presented in his *Grenzzahlen* paper ([Zer30a]) only one year later.

4.3.2 The Logic Project

During the summer semester 1923 and the winter semester 1923/24 Marvin Farber stayed in Freiburg to study Husserl's phenomenology. Very probably Zermelo and Farber got to know each other in one of Husserl's courses. The

---

[78]With Hilbert it says (translation according to [Ewa96], 1093): "*Despite the high pedagogic and heuristic value of the genetic method, for the final presentation and the complete logical grounding of our knowledge the axiomatic method deserves the first rank.*" – For the early discussion of "construction" vs. "axiomatization" cf. [Ferr99], 119seq.

Zermelo with Marvin Farber around 1924

encounter was the beginning of a longer scientific cooperation and also of a friendship which included Marvin's brother Sidney and Sidney's wife Norma and is reflected in an extended correspondence which we addressed already several times. In his letters and postcards Zermelo reveals personal features and his opinion about the people around him. Above all they document the loneliness he felt in Freiburg and surprise with many a blunt judgement about his colleagues in the Philosophical Seminar and the Mathematical Institute.

Marvin Farber had a strong interest in logic. On 4 July 1923 he gave a talk in the Philosophical Seminar entitled "Vorlesung über die Symbolische Logik" ("Lecture on Symbolic Logic") which was intended "to outline the achievements of the Harvard philosopher Henry M. Sheffer in symbolic logic, starting from Peano, Frege, and Russell."[79] On 27 November 1923, obviously introduced by Zermelo, he gave a talk on the axiomatic method before the Freiburg Mathematical Society. In the time following, discussions on logic between Farber and Zermelo seem to have intensified. There are notes by Farber taken of discussions between 24 April and 10 June 1924 touching a variety of topics and open questions in logic and set theory and summarized

---

[79]The present section rests on documents in UBA, Box 33 (Math, Logic-Foundations, Zermelo etc.).

in an 8 page typescript "Notes on the Foundations of Mathematical Logic" with the footnote "Discussed with Zermelo in 1924." They may reflect essential features of Zermelo's views on the philosophy of logic. As central issues reappear in lectures which Zermelo gave in Warsaw in 1929 (cf. 4.4), we confine ourselves to some characteristic points.

The notes start with a consideration on different kinds of axioms. They distinguish (ordinary) *axioms,* for instance the axioms of set theory, from *postulates* such as Hilbert's axiom of completeness in the *Grundlagen der Geometrie.* Postulates are defined as "requirements imposed upon a system of axioms."[80] The task is (ibid., 2) to present clearly

> the defining characteristics of axioms and postulates [...]. A further question is, then, what kinds of combination or organization of postulates obtain. Can they, too, be systematized? [...] And is there a medium of intertranslatability with the axioms? Or, what, apart from their regulative function, is their essential difference from axioms?

Besides the question for a medium of intertranslatability there is no further comment on the possibility of replacing postulates by equivalent axioms as this was done, for example, later by Paul Bernays with regard to Hilbert's axiom of completeness ([Ber55]).

Models of axiom systems are conceived as *substrates.* A substrate is "a system of relations between the elements of a domain" (ibid., 2). It is here that Zermelo clearly distances himself from Hilbert's programme of consistency proofs. With regard to the still open question of the consistency of arithmetic it says (ibid., 2):

> A substrate is *presupposed,* as, for example, in the case of the series of real numbers. Something is presupposed which transcends the perceptual realm. The mathematicians must have the courage to do this, as Zermelo states it. In the assumption or postulation of a substrate, the freedom of contradiction of the axioms is presupposed [...]. Zermelo differs from Hilbert on this. It is Hilbert's view that it must be proved.[81]

Substrates are "governed" by logical postulates – "e. g., the law of contradiction, of excluded middle, etc." (ibid., 2). Unlike ordinary axioms, logical postulates (ibid., 3)

---

[80] Hilbert's axiom of completeness states that "the elements of geometry (points, lines, planes) form a system of things which cannot be extended such that all the other axioms remain valid."

[81] The scientific estrangement from Hilbert may have led Zermelo to feel also a personal one. On 31 December 1924 he writes to the Farber brothers (UBA, Box 25, Folder 25) "that for a long time I am no more 'a little darling' *there* [in Göttingen]. So are they all." The "all"-phrase here refers to the "bigwigs," the influential professors in mathematics who do not pay due acknowledgement to Zermelo. Perhaps Zermelo took it from Shakespeare's drama "Julius Caesar." In his funeral speech for Caesar, Antonius uses it, thereby referring to Brutus and the conspirators around him.

cannot be negated and still give 'sense'. Besides, they are presupposed explicitly or implicitly, in all deductive reasoning. [...] To present them completely is obviously difficult.

In particular, the validity of the principle of the excluded middle is ensured when arguing in a substrate (ibid., 7):

[It] holds in every substrate. That lies in the meaning of a substrate. It is really an analytical judgement, to say this.

So there is a decisive difference from intuitionism. In fact, "Brouwer does not see the distinction between an axiom system and a substrate" (ibid., 7). Without explaining in detail, the notes state that "Brouwer is wrong in disputing the law of excluded middle entirely" (ibid., 4), "[his] treatment of the law of excluded middle is not good, but it is to be taken more seriously than Weyl's position" (ibid., 6).

During the winter semester 1926/27 Marvin Farber was staying again in Freiburg. Presumably it was then that Farber and Zermelo agreed on envisioning a monograph on logic. Initial drafts were to be written by Farber and then discussed with Zermelo. The Farber papers house a typescript ([FarZ27]) of 34 pages entitled *The Foundations of Logic: Studies concerning the Structure and Function of Logic,* consisting of three sections entitled "The Nature of Logic and its Problems," "The Nature of Judgement," and "Assertion and Negation." In several letters from this time Farber reports on progress, promising Zermelo to send typewritten drafts. However, Zermelo obviously never received them. In a postcard to Farber of 23 April 1928[82] it says:

Moreover, first of all I would like to have seen a sample of your logic elaboration in order to decide whether a *joint* continuation of the work would be really justified and promising. If you send me something up to August, I could work on it during my holidays, and if you then came here perhaps in the summer of 1929, the base for a fruitful cooperation would be already laid. (OV 4.08)

Nine months later[83] he refers to heavy teaching duties which Farber had reported,[84] and concludes that "your 'Logic' will only make very slow progress," but nevertheless still shows a strong interest in seeing parts of the first version. After that the correspondence ends. The project was not realized. As the three sections which Farber had drafted in 1927 were never seen by Zermelo, and faced with the sceptical remarks in the preceding quotation, it is an open question as to what extent Farber's presentation really meets Zermelo's views. On 10 April 1928 Farber touches on this point when informing Zermelo about an address "Theses Concerning the Foundations of Logic" which he had given at the 1927/28 Chicago meeting of the American Philosophical Association:[85]

[82]UBA, Box 25, Folder 25.
[83]Postcard to Marvin Farber of 15 January 1929; ibid.
[84]Letter to Zermelo of 28 December 1928; UAF, C 129/29.
[85]Letter to Zermelo; UAF, C 129/29.

The paper will be published. In it I acknowledge indebtness to you, which does not commit you to responsibility for the statements therein contained [...]; on the other hand, it will be clear when our work appears that the ideas belong to both of us.

4.3.3 Cantor's Collected Papers and the Relationship to Abraham Fraenkel

Around 1926/27 Zermelo had plans to edit a German translation of the works of the Harvard mathematician and philosopher Alfred North Whitehead (1861–1947), according to Marvin Farber's report an enterprise Whitehead highly appreciated.[86] However, a competing project finally came to realization: "Already before Christmas [1926], at the suggestion of members of the Philosophical Seminar of the University [of Freiburg], in particular of Oskar Becker," Zermelo asked the Springer Verlag whether they would be agreeable to publishing Cantor's collected works.[87] The reasons for the edition are presented in Zermelo's preface to the published collection ([Zer32d], iii):

> In the history of science it certainly rarely happens that a whole scientific discipline of fundamental importance is due to the creative act of one single person. This case is realised in Georg Cantor's creation, set theory, a new mathematical discipline. Developed in its main features over a period of about 25 years in a series of papers by one and the same researcher, it has become a lasting property of science since then. Thus all later research on this field has to be conceived only as complementing expositions to his basic thoughts. Even apart from their historical importance, Cantor's original papers are still of immediate interest for the present-day reader. In their classical simplicity and precision they are appropriate for a first introduction and in this respect are still unsurpassed by any newer text book, and because of their underlaying wealth of ideas they provide pleasantly inspiring reading for the advanced person. The still growing influence set theory has on all branches of modern mathematics and, in particular, its eminent importance for present-day foundational research, has created the desire in mathematicians and philosophers to read and study in their natural context the papers which are otherwise scattered over various periodicals and are partially difficult to access.

When he got a positive answer from the Springer Verlag, he sent a list of those papers of Cantor which he had chosen for publication. Springer responded with the information that Abraham A. Fraenkel had asked them whether they would publish a biography of Cantor which he was preparing for the Deutsche Mathematiker-Vereinigung.[88] In a letter of 25 March 1927 Zermelo offered to Fraenkel a collaboration either by taking part in the editorship or

---

[86] Letter from Marvin Farber to Zermelo of 9 June 1927; UAF, C 129/29.

[87] Letter from Zermelo to Fraenkel of 25 March 1927. I thank Benjamin S. Fraenkel for providing his father's letters and for the permission to quote from them.

[88] It appeared as [Fra30].

by contributing his Cantor biography in a suitable form. Already on 30 March Fraenkel sent an answer,[89] expressing his pleasure and emphasizing that he felt himself honoured by the possibility of a collaboration. He suggested plans on how to organize it, but left the decision to Zermelo. Finally, the second alternative was agreed upon. In October Fraenkel sent a first version of the biography. Zermelo's answer, dated 30 October and signed on 9 November, a typewritten letter of five pages, was to become the origin of a growing estrangement. In a mixture of objective comments, ironical side remarks, harsh criticism, and points which nevertheless show an inherent sensibility, he urges Fraenkel to adopt his view of how to write a biography. We quote some parts:

Unfortunately, the first perusal already led to the conclusion that the text of your presentation still needs considerable changes, especially with respect to its style, before it can fulfil its purpose and be ready for printing [...]. In a biography the name of the hero, the person presented, should be written out in full everywhere if it cannot be replaced by a circumscription; it is respect for the person which requires it [...]. "The nature-lover C." (p. 10) makes an absolutely grotesque impression. As if at a festive party a guest (or even the host himself) were to run around with a cloakroom number [...]. On page 41 there is a "fusion" between mathematics and philosophy. It seems, as if current events in the world of banks played a role here. Surely you meant an "internal relationship." We would rather see science preserved from both "fusion" and "confusion" [...]. Intimate letters of family members which were given to the biographer in confidence may turn out extremely valuable for him because they may provide additional features of the personality in question. However, with respect to publication one cannot be restrained enough; whenever a publication is unnecessary, it also appears tactless [...]. The events following the Heidelberg Congress in 1904 should either be described in their total context or left out [...]. I believe that ideological confessions of the biographer are also out of place here. For example, somebody who is of the opinion that a logical penetration is a bad thing also in science and should be replaced there by a mystic "intuition," has lots of opportunities today to express this fashionable conviction in spoken or written words; but it need not happen just in the biography of a man who as a mathematician struggled for logical clarity throughout his life. For example, we should rather leave the problem whether Dedekind's "logicistic" influence on Cantor has been useful or harmful (p. 10) to the future that will pass over many a question which is noisily discussed these days and proceed to the agenda.

That's all for today, although I still have to say a good many further things of principal importance. But I hope that you may accept my point of view on the whole and that you will finally be convinced of the necessity that your manuscript still needs an essential revision with respect to form and contents before it can be printed. There only remains the question whether you will do the revision yourself or leave it to me. (OV 4.09)

---

[89]UAF, C 129/33.

Obviously Fraenkel complains about the offensive style in a letter now lost. On 27 November Zermelo replies with a mixture of excuse and defence:

> I am glad that inspite of all you are willing to undertake the necessary changes in your manuscript, and so seem to acknowledge the points I have exposed as being justifiable at least partially. And just this was what I wished; otherwise I would not have had to go to the trouble of a detailed explanation. Of course it would have been much easier for me to just accept your paper – and why read it then? – but such a procedure appeared to me incompatible with the duties of an editor. And you should have noticed that I read your paper very carefully and thoroughly and that my criticism is only on an objective basis in the interests of the work. Now, as far as the "tone" of my remarks is concerned which has apparently hurt your feelings, I really do not know what you mean. Maybe the jocular way [. . .] in which I satirized the "fusion"? Isn't it actually suitable and natural to express the infelicity of a term by recalling the associations which come to one's mind? Well, there are people who view *every* joke as an insult because they lack a sense of humour; but nobody would like to be counted as such an unfortunate person [. . .]. But no harm meant! The main thing is that you are willing to undertake the necessary changes. (OV 4.10)

This response was only partially successful. It is obvious that Fraenkel did not accept Zermelo's suggestions in matters of style. On 12 December 1929,[90] more than two years later, he writes in a cool and distant manner:

> With regard to the factual and in many ways stylistic changes you propose, I will not be able to agree to the majority of them, because I do not want to disown either my scientific views or my style. (OV 4.11)

Traces of this conflict can even be found in the published collection: In the preface, despite giving a detailed motivation for the selection of the contents and in contrast to an earlier version, Zermelo does not mention Fraenkel's biographical contribution.[91] On the other hand, Fraenkel does not follow Zermelo's wish with regard to a more elaborate description of the events during the 1904 International Congress of Mathematicians in Heidelberg (2.5.3). When commenting on them ([Can32], 473), he simply writes that "soon Cantor had the satisfaction of seeing the incorrectness of König's conclusion confirmed." A more detailed description could have given information to what extent Zermelo had really been involved, thereby perhaps also providing access to the way he approached the proof of the well-ordering theorem.

---

[90]Letter written from Jerusalem; UAF, C 129/33.

[91]The earlier preface (UAF, C 129/256) ends as follows: "The concluding part consists of a biography of Cantor from the experienced pen of the mathematical author who is well-known to wide circles because of his 'Einleitung in die Mengenlehre'." Zermelo even intended to mention Fraenkel's contribution in the title. A suggestion was "The Collected Works of Georg Cantor with Explanatory Comments and a Biography by A. Fraenkel Edited by E. Zermelo" (ibid.).

The relationship between the two pioneers of set theory was never to recover, even when after the appearance of the Cantor edition the letters lost their frostiness. For Zermelo there was still another sore point: Fraenkel had started to criticize seriously the notion of definiteness as early as 1922, characterizing Zermelo's explanation of 1908 even as leading to concerns "on the same level as Richard's paradox" ([Fra27], 104), just the paradox which Zermelo had intended to avoid. As it seems, Zermelo finally got the impression that Fraenkel was systematically working against him, that he "stabbed him in the back at *every* opportunity" ([Zer31b]).

Instead of Fraenkel, Reinhold Baer and Arnold Scholz became partners in the Cantor enterprise. A large number of letters which Baer wrote to Zermelo around 1931[92] treats editorial questions and gives evidence of Baer's substantial contribution. Questions concerning philosophical aspects were discussed with Oskar Becker.

When Cantor's collected papers appeared in the spring of 1932, Zermelo gave free copies to the mathematics departments in Freiburg and Warsaw, the mathematical institutions he felt obliged to. Among the individuals he took into account for receiving a complimentary copy we find Jean Cavaillès in Paris.[93] Cavaillès had provided transcriptions of a part of the Cantor-Dedekind correspondence and had shown a vivid interest in the foundations of set theory in general and in Zermelo's views in particular which resulted in two major publications ([Cav38a], [Cav38b]). He intended to edit his transcriptions, regarding a publication in the *Jahresbericht der Deutschen Mathematiker-Vereinigung*.[94] Five years later he published them together with Emmy Noether in a separate edition.[95] Emmy Noether had originally proposed incorporating the correspondence into the Cantor collection:[96]

> I would be very pleased if the Cantor-Dedekind correspondence were published in your Cantor edition. [...] The Cantor-Dedekind correspondence ought to appear in a Cantor edition, not in a Dedekind one, because here it is Cantor who makes the new scientific statements.

Zermelo dismissed her proposal,[97] selecting only those parts of the correspondence which formed "an essential and indispensable supplement" ([Can32], 451). The passages chosen concern Cantor's "final conceptions about the system of *all* ordinals and the system of *all* cardinals together with consistent and inconsistent totalities" (ibid., 451), thus throwing light on the history of

---

[92]UAF, C 129/4.

[93]Letter from Zermelo to the Springer Verlag of 17 May 1932; UAF, C 129/136.

[94]Letter to Zermelo of 12 July 1932; UAF, C 129/21.

[95][NoeC37]. For additional letters and drafts of the correspondence and for further information see [Gra74]. Cf. also [Gue02]. An English translation of a large part of the correspondence is in [Ewa96], 843–878 and 923–940.

[96]Letter to Zermelo of 12 May 1930; UAF, C 129/86.

[97]Cf. letter from Zermelo to the Springer Verlag of 17 May 1932; UAF, C 129/136.

the paradoxes.[98] Moreover, they give in print Dedekind's now classical proof of the Schröder-Bernstein equivalence theorem.[99]

During the editorial work Zermelo became involved in a serious foundational controversy. On the one hand, it was rooted in Gödel's and Skolem's finitary approach to mathematics with its inherent limitations, on the other hand, it resulted from the endeavour of intuitionism to propagate a kind of mathematics which had been deprived of actual infinity and classical logic. In some kind of personal uprising which we will discuss later, Zermelo fought for what he called "true mathematics," Cantor's set theory being an essential part of it. So we are not astonished to see him addressing this matter of concern in the Cantor edition, too. Using a remark of Cantor's about a certain Chevalier de Meré, "a man of reputation and intellect," who had tried to reduce probability theory to absurdity, thereby enhancing its development by Pascal and Fermat,[100] he concludes the preface ([Can32], v):

> May the present book [...] promote the awareness and the understanding of Cantor's lifework in accordance with its creator and in the spirit of genuine science, independent of trends of time and fashion, and unperturbed by the attacks of those who in timid weakness would like to force a science which they cannot master any longer to turn back. However, as Cantor says, "it may easily occur that just at the spot where they try to give science its deadly wound, a new branch, more beautiful, if possible, and more promising than all former ones, will burst into bloom before their eyes – like probability theory before the eyes of the Chevalier de Meré."

The work was reviewed in about a dozen journals. Whereas the *Zentralblatt* announces only the title[101] and Ludwig Bieberbach's review in *Jahresbericht der Deutschen Mathematiker-Vereinigung* ([Bie33]) gives only the contents, the other reviews judge the editing from "sober" to "excellent." In *The Bulletin of the American Mathematical Society* Eric Temple Bell ([Bel33]) concludes that "on the whole the editing appears to be as impartial as is fitting in collected works." In *The Mathematical Gazette* Dorothy Maud Wrinch ([Wri32]) congratulates Zermelo "very heartily on his editorship" and offers him "on behalf of the many English admirers of Cantor [...] most cordial thanks." For the *Jahrbuch über die Fortschritte der Mathematik* Erika Pannwitz states that the annotations Zermelo provides realize the intended aims "in an excellent manner" ([Pan32], 44). The edition was also taken note of in philosophical circles. In *The Philosophical Review* Paul Weiss ([Weis34]) writes:

---

[98]The selection comprises parts of the letters which Cantor wrote to Dedekind on 28 July, 3 August, 28 August, 30 August, and 31 August 1899; ibid., 443–450. The second letter is given as a part of the first one.

[99]I.e., the corresponding part of Dedekind's letter to Cantor of 29 August 1899 and Cantor's answer of 30 August 1899; ibid., 449–450.

[100]Cf. [Can73], 36, or [Can32], 359.

[101]*Zentralblatt für Mathematik und ihre Grenzgebiete* **4** (1933), 54.

In connection with both infinitude and continuity, Cantor is dealing with topics which have little to do with those discussed by philosophers, even though they are designated by the same names. Cantor is a great mathematician. I cannot, however, see that he has much to teach philosophers besides some mathematical technicalities, the wisdom being clear and rigorous, and the virtue of persisting in the development of a new concept despite the scorn and abuse received from more respected contemporaries.

Criticism concerning editorial care was put forward by Ivor Grattan-Guinness; in particular, it aims at the edition of the 1899 letters of Cantor's correspondence with Dedekind, pointing to the treatment of Cantor's letter of 3 August 1899 (cf. fn. 98).[102]

The royalties Zermelo got from Springer were highly welcome. As we shall see in a moment, his financial situation had become worse in the meantime. In a letter to Richard Courant of 4 February 1932[103] we read that after the publication of the Cantor edition

I would be ready and willing to take on any literary-scientific work if there were an opportunity to do so. My financial situation makes such a secondary income a necessity. [...] I have no chance of getting a paid lecturership or anything like that in Freiburg or in Karlsruhe. [...] If you hear of any opportunity for translations or similar things, I would appreciate any corresponding information. Of course, in a time of general unemployment one should not have illusions of this kind. But just the same one can't afford to miss any opportunity that is offered. [...] Here I am thinking first of all of Hessenberg's *Grundbegriffe der Mengenlehre*[104] which [...] is out of print now, but would surely merit a new edition. Should I write to the publishers in this respect if they still exist? [...] Well, there is no hurry; however, I would not like somebody to beat me to it. (OV 4.12)

4.3.4 The *Notgemeinschaft* Project

During his years in Freiburg, Zermelo regularly complains about the insufficiency of his pension. The fees students had to pay for his courses did not essentially improve the shortage of income. He therefore decided to apply for a fellowship of the Notgemeinschaft der Deutschen Wissenschaft, the Emergency Community of German Science. The community had been founded in 1920 by the leading German academies and scientific associations in order to prevent a collapse of research in the economically disastrous time after the First World War. In 1929 it was renamed "Deutsche Forschungsgemeinschaft," a name still used today.[105] Support for individual projects was based

---

[102]See, for example, [Gra00a], 548; cf. also [Can1991], 406. For corrections concerning Zermelo's edition, cf. [Gra74], 134–136.

[103]UAF, C 129/22.

[104][Hes06], a book Zermelo also recommended in his courses on set theory.

[105]For measures supporting research in the post-war Germany of the 1920s see, e. g., [Zie68] and [Mars94].

on reports by external experts. Zermelo succeeded in obtaining a three-year fellowship 1929–1931 for his research project "On the Nature and the Foundations of Pure and Applied Mathematics and the Significance of the Infinite in Mathematics."[106] Probably the positive decision was mainly due to the assessment and engagement of Lothar Heffter who had already shown a strong commitment in awarding Zermelo the honorary professorship.[107] The continuation of the fellowship for 1930 and 1931 was granted on the basis of reports which Zermelo delivered in December 1929[108] and in December 1930 ([Zer30d]). They contain valuable information about his foundational work between 1929 and 1931. The letter accompanying the 1930 report describes his financial situation, summing up the details as follows:

> Hence, without further support from the Notgemeinschaft I would be dependent next year on my entirely insufficient pension [...], whereas on the other hand I could not expect an improvement of my economic situation from a state post like younger researchers. (Ov 4.13)

The grant for 1931 was lowered and given without the prospect of a further renewal. Altogether the support equalled about sixty percent of his pension. It improved his financial situation to such an extent that in the summer of 1930 he was able to afford a cruise along the Norwegian coast.[109] However, in the end it left him with the situation described in the letter to Courant quoted above.

Scientifically the years of the *Notgemeinschaft* project form the kernel of his Freiburg period of intensive research activities with the focus on foundational questions. His work on set theory and logic was centered around one aim: to provide a foundation for "true mathematics." According to his firm belief, the emerging formal systems of logic with their inherent weakness in expressibility and range that were to dominate the 1930s, went in the wrong direction. Full of energy and without shying away from serious controversies, he set himself against this development. His commitment let him even postpone work on a book *Set Theory (Mengenlehre)* which was planned to appear with Akademische Verlagsgesellschaft in Leipzig.[110] The next seven sections describe his engaged fight. It starts with a prelude in Warsaw.

---

[106] "Wesen und Grundlagen der reinen und der angewandten Mathematik, die Bedeutung des Unendlichen in der Mathematik." Letters of the *Notgemeinschaft* of 2 February 1929, 20 December 1929, and 3 January 1931; UAF, C 129/140.

[107] Letter from Heffter to Zermelo of 14 March 1929; UAF, C 129/42.

[108] UAF, C 129/140.

[109] Aboard the "Oceana;" cf. letter to "a colleague" of 25 May 1930; UAF, C 129/279.

[110] Correspondence with the publisher (March to April 1928); UAF, C 129/49.

4.4 Warsaw 1929

4.4.1 The Prehistory

On 20 February 1929 the Faculty of Mathematics and Sciences of the University of Warsaw, following an application by Wacław Sierpiński and Stefan Mazurkiewicz, officially invited Zermelo to give a series of talks during the spring term. They were able to provide 220 US-Dollars for the expenses. As Sierpiński wrote the next day,[111] the Faculty had agreed unanimously to this invitation. Bronisław Knaster, a member of the Warsaw logic group, offered Zermelo to stay in his apartment.[112] Zermelo accepted. Plans for the journey had first been discussed at the 1928 International Congress of Mathematicians in Bologna and were already very definite at the beginning of 1929.[113] In April Knaster suggested paying visits also to Cracow and Lvov.[114] The final programme covered the months from May to July; it led to Halle,[115] Breslau, Cracow, and Lvov on the way to Warsaw and to Gdansk, Greifswald, Hamburg, and Göttingen on the way back; the stay in Warsaw was scheduled for about four weeks in May/June.

These plans caused problems in Freiburg: Zermelo had already announced a course on differential geometry for the summer semester to come. When the plans for Warsaw became definite, he had cancelled it with the agreement of Loewy, then Dean of the Faculty. Apparently Loewy had failed to take care for a substitute. Heffter, being very careful with respect to teaching matters, reacted with an excited letter[116] characterized by a condescending and patronizing style together with a consciousness of position:

Via detours [...] the rumour has got through to me that you have been invited to give some talks in Warsaw, that you, therefore, are not willing to give the course on differential geometry which you have announced for the summer semester, that you, moreover, have already informed the Dean about this. May I politely ask you to let me know to what extent this is true and, in case it is, what steps you have taken to prevent the word you have given in our announcements for the summer semester from being broken?

You know, of course, that we always fix the programme for the next semester in a cooperative[117] way. [...] Neither colleague Loewy nor I assert any right of age, except for the courses for beginners which for the time being are shared by us.

---

[111]Letter to Zermelo; UAF, C 129/110.
[112]Letter to Zermelo of 4 March 1929; UAF, C 129/62.
[113]Letter from Hasso Härben to Zermelo of 29 January 1929; UAF, C 129/38.
[114]Letter to Zermelo of 18 April 1929; UAF, C 129/62.
[115]Following an invitation by Helmut Hasse; UAF, C129/40. Hasse had employed Zermelo's elegant proof of the uniqueness of prime number decompositions (cf. 4.5).
[116]Letter to Zermelo of 7 March 1929; UAF, C 129/42.
[117]The word is underlined by Zermelo.

Zermelo and Bronisław Knaster with Lvov mathematicians
Auditorium of Lvov University[118]

Being highly sensitive to criticism intolerable in style and attitude, Zermelo answered in a likewise excited form, thus starting a controversial correspondence from door to door:

> As a lecturer without official lectureship I can by no means be made responsible for the mathematical teaching programme; however, I am ready to speak at any time with any colleague who wishes to discuss questions concerning him, but in the spirit of collegiality and based on equality and mutuality. (OV 4.14)

Without following the escalation, we conclude with an interjection of Heffter[119] which reveals another side of Zermelo: a certain helplessness in everyday life, a feature also confirmed by his later wife Gertrud and heavily contrasting to the self-confidence he usually showed in scientific affairs:

> *You,* who otherwise came to me with quite simple questions for advice and help – I remind you of the intended cut in your Zurich pension which could be prevented by a single word.

---

[118]Kazimierz Kuratowski (1), Bronisław Knaster (2), Stefan Banach (3), Włodzimierz Stożek (4), Eustachy Żylinski (5), Stanisław Ruziewicz (6), Hugo Steinhaus (7), Stefan Mazurkiewicz (8).
[119]Letter of 14 March 1929; UAF, C 129/42.

Zermelo with Warsaw mathematicians[120]

In any case, Zermelo went to Warsaw. The mutual relationship between the combatants soon returned to what it were before. Heffter continued to appreciate Zermelo's scientific quality, excusing his sometimes strange behaviour by a nervous character. Conversely, Zermelo joined the educational programme of the Friends of the University which Heffter had organized.

4.4.2 The Warsaw Programme

Between 27 May and 10 June 1929, Zermelo gave nine one hour talks in the Warsaw Mathematical Institute.[121] The topics he treated range from a justification of classsical logic and the infinite to a critical discussion of intuitionism and Hilbert's proof theory, but he also dealt with specific topics such as a new notion of set or the von Neumann ordinals. In chronological order the talks were entitled as follows:[122]

---

[120]Front row from left to right: Wacław Sierpiński, Zermelo, Samuel Dickstein, Antoni Przeborski; second row from left to right: Jan Łukasiewicz, Stanisław Leśniewski, Bronisław Knaster, Jerzy Spława-Neyman, Franciszek Leja.

[121]On 24 May he gave an additional talk "On the Logical Form of Mathematical Theories" ([Zer30b]) which will be discussed below, and on 7 June an additional talk on "Reflection in analytical curves" (cf. *Comptes Rendus des Séances de la Société Polonaise de Mathématique, Sect. Varsovice* **8** (1929), 323).

[122]UAF, C 129/288. For the titles in German see OV 4.15.

T1 What is Mathematics? Mathematics as the Logic of the Infinite.
T2 Axiom Systems and Logically Complete Systems as a Foundation of General Axiomatics.
T3 On Disjunctive Systems and the Principle of the Excluded Middle.
T4 On Infinite Domains and the Importance of the Infinite for the Whole of Mathematics.
T5 On the Consistency of Arithmetic and the Possibility of a Formal Proof.
T6 On the Axiomatics of Set Theory.
T7 On the Possibility of an Independent Definition of the Notion of Set.
T8 The Theory of "Basic Sequences" as a Substitute for "Ordinal Numbers."
T9 On Some Basic Questions of Mathematics.

This list deviates from Zermelo's original plans. At the end of April he had sent a list of seven topics to Knaster.[123] Another list, probably even older, gives six titles.[124] The *Nachlass* contains notes of six talks ([Zer29b]), again with partly deviating titles:[125]

W1 What is Mathematics?
W2 Disjunctive Systems and the Principle of the Excluded Middle.
W3 Finite and Infinite Domains.[126]
W4 How Can the Assumption of the Infinite be Justified?
W5 Can One "Prove" the Consistency of Arithmetic?
W6 On Sets, Classes, and Domains. An Attempt to Define the Notion of Set.

Talk T9 may have addressed ideas related to the last two entries on the Knaster list, namely

- Logic and intuition in mathematics. Are the mathematical axioms based on "intuition"?[127]
- Pure and applied mathematics. The application of the mathematically infinite to the empirically finite. In general, the relationship between thinking and being and between sentence and fact.[128]

The topic on logic and intuition had been particularly favoured by Sierpiński. Notes of this talk have not been preserved. Hints on the epistemological convictions expressed there can be drawn from the second appendix to Zermelo's 1930 report ([Zer30d]) to the Notgemeinschaft. The report lists five topics of

---

[123]UAF, C 129/62.

[124]UAF, C 129/288. Zermelo may have sent this list to Sierpiński in March; cf. Sierpiński's letter to Zermelo of 25 March 1929 (UAF, C 129/110).

[125]For the titles in German see OV 4.16.

[126]Numbered as the fourth talk in [Moo80], 136.

[127]"Logik und Anschauung in der Mathematik. Sind die mathematischen Axiome auf 'Anschauung' gegründet?"

[128]"Reine und angewandte Mathematik. Die Anwendung des mathematisch Unendlichen auf das empirisch Endliche. Allgemein das Verhältnis von Denken und Sein, von Urteil und Sachverhalt."

intended investigations; the last one, "On the Relationship Between Mathematics and *Intuition*,"[129] may provide basic features to the effect that mathematics starts beyond intuition:

> Mathematical science starts only when treating intuitively given material in a logical-infinitary manner; it, hence, *cannot* itself be based on "intuition." Also in geometry the preference of "Euclidean geometry" does *not* root in its "intuitive actuality," but solely in its logical-mathematical simplicity.

Talk T8 certainly treated Zermelo's version of the von Neumann ordinals under the name of *basic sequences*. On the Knaster list the corresponding talk is entitled "Basic Sequences and Ordinal Numbers: An Attempt to Give a Purely Set-Theoretic Definition of the Transfinite Ordinal Numbers."[130] Zermelo here really seems to have in mind identifying ordinal numbers as types of well-orderings in the sense of Cantor with the von Neumann ordinals. Roughly a year earlier, in his course on set theory in the winter semester 1927/28,[131] he had still been indifferent, using basic sequences *instead of* ordinal numbers and *as* ordinal numbers as well.

The notes W1 to W6 form the main source for Zermelo's foundational views during the late 1920s. They serve as a platform from which he develops new insights when faced with the pioneering results of Skolem and Gödel. It therefore seems worthwhile having a closer look at them.

4.4.3 The Nature of Mathematics

Notes W1 start right away with a clear description of the axiomatic method and its constitutive role for mathematics:

> In order to exhaust its entire extent, mathematics must *not* be characterized by its object (such as space and time, forms of inner intuition, theory of counting and measuring), but only by its characteristic method, the proof. Mathematics means to systematize the provable, and thus is applied logic; its task consists in the systematic development of "logical systems," whereas "pure logic" only investigates the general theory of logical systems. Now, what does "proving" mean? A "proof" is a deduction of a new proposition from other given propositions which follows general logical rules or laws, thereby ensuring the truth of the new proposition by the truth of the given ones. Therefore, the ideal of a mathematical discipline would be given by a system of propositions which already contains all propositions which are purely logically deducible from it, i. e., a "logically complete system."[132] A

---

[129] "Über das Verhältnis der Mathematik zur *Anschauung*."

[130] "Grundfolgen und Ordnungszahlen: Versuch einer rein mengentheoretischen Definition der transfiniten Ordinalzahlen."

[131] Lecture notes "Allgemeine Mengenlehre und Punktmengen;" UAF, C 129/157.

[132] In his talk of 24 May ([Zer30b]), Zermelo characterizes the representation of a mathematical theory as a logically complete system of sentences as "being better

"complete system" is given, for example, by the totality of all logical consequences which are deducible from a given system of basic assumptions or by an "axiom system." However, not all "complete systems" are necessarily specified by a finite number of axioms. One and the same complete system may be given by several, even infinitely many different axiom systems, for instance Euclidean geometry or the arithmetic of real or complex numbers. Thus a complete system is something like an "invariant" of all equivalent axiom systems, and the question about the "independence" of the axioms has nothing to do with it. A complete system is related to any of the axiom systems by which it is determined as a "field" is related to its "basis," and the investigation of such complete systems may promise similar advantages of greater generality and lucidity as the transition from algebraic equations to algebraic fields.

When reading this passage, one might think that Zermelo has in mind some kind of formal rules underlying the deductions which constitute the proofs. A year earlier David Hilbert and Wilhelm Ackermann had presented a formalization of propositional logic in their *Grundzüge der theoretischen Logik* ([HilA28]). Moreover, Gödel had already finished his Ph. D. thesis which contains a proof of the completeness theorem for first-order logic. However, in none of the talks does Zermelo give any indication about a formal system of logic he refers to. Even if he should have had in mind a specific one, he distanced himself from all of them a year later when he learnt about the weakness coming with them such as the existence of countable models of first-order set theory or the consequences of Gödel's first incompleteness theorem.

4.4.4 Intuitionism *versus* Mathematics of the Infinite

Already some years before, in discussions with Marvin Farber, Zermelo's criticism of intuitionism had condensed in the principle of the excluded middle. From the point of view of intuitionism, an (intuitionistic) proof of a disjunction $(\varphi \vee \psi)$ consists in either a proof of φ or a proof of ψ, and a proof of a negation $\neg\varphi$ consists in reducing φ to an absurdity. Hence, a proof of a disjunction $(\varphi \vee \neg\varphi)$ requires either a proof of φ or a reduction of φ to an absurdity. If, for example, φ is the Goldbach conjecture (*Every even natural number* ≥ 4 *is the sum of two prime numbers*), then $(\varphi \vee \neg\varphi)$ is not yet proved, because even in traditional mathematics neither a proof nor a refutation of φ have been found so far. Hence, the principle of the excluded middle cannot be accepted as being universally valid.[133]

---

suited for certain questions than the representation by an axiom system." In a letter to Helmuth Gericke of 2 October 1958 (MLF), Paul Bernays points out that the term "system of sentences" ("Satzsystem") for a mathematical theory can also be found in the papers of Paul Hertz ([Hert22], [Hert23], [Hert29]) which were available "partly earlier, partly at the same time." – For Hertz's work cf. [SchrH02].

[133] Note that the principle of the excluded middle has not been refuted by this argument, because it has not been reduced to an absurdity. – For precise definitions and further details cf., for example, [Hey56], Ch. VII, or [TroD88], 8seq.

Notes W2 argue against intuitionism by taking up the notion of "substrate" and its role as discussed with Farber: All mathematical systems developed so far, and hence, mathematical reasoning in general, refer to an existing or hypothetical domain of objects such as the domain of natural numbers or the domain of real numbers. As the sentences which are true in some model, form a *disjunctive* system, i. e., a logically complete system which for any sentence φ either contains φ or $\neg\varphi$, mathematicians are therefore right if they use the principle of the excluded middle:

> In fact, it is always this kind of application of the general logical principle (limited to hypothetical models) that plays such an essential role in the proofs of classical mathematics of all disciplines, and is indeed, in my opinion, indispensable. Mathematics without the principle of the excluded middle (understood in the right manner) as the "intuitionists" request and believe they are able to offer, would no longer be mathematics. [...] Just "realizability" by models is the basic assumption of all mathematical theories; without it also the question for the "consistency" of an axiom system loses its genuine meaning. For the axioms themselves do not do harm to each other, as long as they are not applied to one and the same (given or hypothetical) model. "Consistency" only makes sense, if the possible inferences are based on a closed circle of logical operations and principles. The principle of the excluded middle is an essential part of them. (OV 4.17)

Notes W3 strengthen the indispensability of models by asserting that true mathematics is necessarily based on the assumption of *infinite* domains. This also holds for theories of finite structures such as the theory of finite groups or the theory of finite fields. As they treat structures of arbitrarily high finite cardinality, they

> are actually developed only in the framework of a comprehensive, infinitary arithmetic. Purely "finitistic" mathematics in which, as a matter of fact, nothing is left to be proved because everything could be verified already by a finite model, would no longer be mathematics in the true sense. Rather, true mathematics is genuinely infinitistic and based on the assumption of infinite domains; it can directly be called the "logic of the infinite."

4.4.5 The Justification Problem

Of course, Zermelo does not deny the necessity of justifying the existence of (infinite) models (W4):

> Couldn't just this apparently so fertile hypothesis of the infinite have brought inconsistencies into mathematics, thereby completely destroying the essence of this discipline which is so proud of its logical correctness? (OV 4.18)

However, "the infinite as such can nowhere be made obvious." So, in general it is impossible to prove consistency, i. e. to exhibit a model. What can be done in this situation? Notes W4 and W5 provide an answer. At a first glance, one might try to give a "logico-mathematical" proof of consistency with the aim of excluding that both a sentence and its negation can be proved, that means, one might try to realize the respective part of Hilbert's programme. For several reasons Zermelo is sceptical about whether such a justification can be given. First, logic as applied in mathematical proofs is so complex that there is no hope of taking into account all modes of inferences (W4):

> Such a proof – if possible – had to be based on a thorough and complete formalization of all of logic relevant for mathematics. Any "incompleteness" of the underlying "proof theory," any forgotten possibility of an inference, for example, would jeopardize the whole "proof." However, as apparently such a "completeness" may never be guaranteed, no possibility of formally proving consistency remains. (OV 4.19)

Second, such a consistency proof would have to deal with an infinite set of sentences and, therefore, would merely result in a reduction of one question concerning an infinite domain to another one, thus leading to a *regressus ad infinitum.* Therefore, so his résumé (W5),

> it will not do to base the formalism on a formalism: at one point something has really to be thought, to be stipulated, to be assumed. And the simplest assumption which can be made and which is sufficient for the foundation of arithmetic (as also of the whole of classical mathematics), is precisely the idea of "infinite domains;" it necessarily suggests itself to logical-mathematical reasoning, and is actually the basis of the whole of our science as it has developed historically. (OV 4.20)

Hilbert's proof theory being doomed to fail, is there left any possibility of justifying the assumption of the infinite? The answer, for exemplification given in the case of arithmetic, is short and definite (W4):

> Such an assumption can only be justified by its success, by the fact that it (and it alone!) has made possible the creation and development of the whole of present arithmetic which is essentially just a science of the infinite. (OV 4.21)

These quotations lead to obvious questions. As mentioned above, Hilbert and Ackermann had already published their *Grundzüge* book containing a complete system of rules for propositional logic, and Gödel had just obtained his first-order completeness result ([Goe29]). Moreover, there might still have been hope that higher order logics also admitted complete calculi – a possibility that was only lost when Gödel found his incompleteness results ([Goe31a]). But Zermelo remained sceptical. In a letter to Marvin Farber of 24 August 1928 he had already formulated his doubts in a plain way:[134]

---

[134]UBA, Box 25, Folder 25.

Hilbert's "Logistic" which has just appeared [i. e., [HilA28]] is more than meagre, and moreover, I no longer expect something overwhelming from his "foundations of arithmetic" which he announced again and gain. (OV 4.22)

Why did Zermelo feel so sure that it was impossible to catch the richness of the rules of logic, why these deep doubts about the possibility of consistency proofs and, hence, in Hilbert's programme?

Concerning the methods of metamathematics, Hilbert had commented on the fundamental character of the combinatorics involved:[135]

> Mathematics as any other science never can be founded by logic alone; rather, as a condition for the use of logical inferences and the performance of logical operations, something must already be given to us in our imagination (Vorstellung), certain extralogical concrete objects that are intuitively present as immediate experience prior to all thought. [. . .] This is the basic position which I regard as requisite for arithmetic and, in general, for all scientific thinking, understanding, and communication. And in mathematics, in particular, what we consider are the concrete signs themselves, whose shape, according to the conception we have adopted, is immediately clear and recognizable. This is the very least that must be presupposed; no scientific thinker can dispense with it, and therefore everyone must maintain it, consciously or not.

Hilbert identifies the combinatorial ability with a third source of cognition besides experience and pure thought, relating it to Kant's *a priori* intuitive mode of thought.[136]

One could argue now that for Zermelo the constructs of logic have the same epistemological status as the other objects of mathematics such as numbers, points, and sets. Therefore it does not make sense "to base a formalism on a formalism." This point of view comes together with his refusal to adopt a strict distinction between formulas and the objects they are about, i. e. between syntax and semantics, an attitude that did not change even when such a distinction became common in the mid-1930s.

As far as his doubts about the adequateness of logical systems developed so far are concerned, one may remember his remark about the insufficiency of any "textbook-like representation" of mathematics:[137]

> Geometry existed before Euclid's "Elements," just as arithmetic and set theory did before Peano's "Formulaire," and both of them will no doubt survive all further attempts to systematize them in such a textbook manner.

The richness of mathematics – as also that of logic – cannot be caught by some kind of finitary systematic approach.

In the autumn of 1930, Gödel proved his incompleteness theorems. They showed that axiomatized mathematics suffered from an essential weakness and

---

[135][Hil28], 65–66; [vanH67], 464-465 (modified).
[136][Hil31], 486; [Hil96d], 1149–1150.
[137][Zer08a], 115; [Zer67b], 189.

Hilbert's proof theory as originally formulated had failed. Sufficiently strong axiomatic theories such as arithmetic or set theory – if they are consistent – are incomplete and do not allow proofs of consistency even if one admits all means they provide. Nevertheless, classical mathematics flourished and intuitionism became the conviction of a minority, tolerated by the majority, but also recognized as an enrichment in mathematical studies. Faced with this situation, Zermelo should have been satisfied: His doubts about the realizability of Hilbert's proof theory and his scepticism of the scope of the axiomatic method had been justified and his concern that intuitionism might harm classical mathematics had been disproved by reality. However, as we have already indicated earlier and will learn in more detail in the next chapters, matters developed in a different way.

And a last point: Given Zermelo's argument for the impossibility of a consistency proof for arithmetic, a consistency proof for set theory should be impossible as well. But set theory had suffered from paradoxes. So there was no justification by success. To overcome this dilemma, Zermelo pleads for grounding set theory on principles that are conceptually more convincing. In notes W6 he starts a search for such principles. Locating the roots of the paradoxes in "Cantor's original definition of a set as a well-defined comprehension of objects where for each object it is determined whether it belongs to the comprehension or not,"[138] he pleads for a new definition of set free from such deficiencies. He finds it in a set theory distinguishing between sets and classes – von Neumann's set theory being the example he may have had in mind – together with a definition of sets which allowed their separation from classes without employing the usual variety of set existence principles (4.6.2).

In his 1930 report to the Notgemeinschaft[139] Zermelo writes that he appreciated his time in Warsaw not only because of the hospitality he had enjoyed, but also because of the intensive discussions with interested colleagues, among them "in particular the gentlemen Knaster, Leśniewski, and Tarski." Part of these discussions concerned his notion of definiteness from 1908. We have already remarked that in the beginning 1920s this concept had become an object of serious criticism. The Warsaw logicians certainly expressed doubts as well. The discussions sharpened Zermelo's picture to such an extent that he decided to lay it down in his definiteness paper ([Zer29a]).

The harmony of Zermelo's stay was especially due to Bronisław Knaster who took care of everything. Knaster's letters from the following years give evidence of a mutual esteem that had grown during the time in Warsaw. They also support the impression shared by those who came into closer contact with Zermelo. In Knaster's letters of 5 August 1929 and 1 October 1932[140] we read:

> I have grown so fond of your scientific points of view and of your further remarks and aphorisms that I miss you. –

---

[138] Referring to [Can82], 114–115; [Can32], 150.
[139] UAF, C 129/140.
[140] UAF, C 129/62.

Zermelo in Warsaw

My *very dear* Professor Zermelo! [...] How would I love to hear again your lectures and your wise and astute remarks about the world and the people, which I still remember.

4.5 Foundational Controversies: An Introduction

The Warsaw talks provide a picture of Zermelo's views on foundational questions as they had formed during the 1920s and had become visible in the discussions with Marvin Farber around 1925. We have noted several characteristic features:

- Mathematics – or, more precisely, logic as the main tool for proofs – is essentially based on the principle of the excluded middle. In other words, mathematics is classical mathematics and intuitionism contradicts its true character.
- Genuine mathematics is concerned with infinite structures.
- As a science of the infinite, mathematics cannot be justified by a proof theory in Hilbert's sense; it can only be justified by its success.
- Having suffered from contradictions, set theory should be based on more convincing principles in order to enhance the conviction that the paradoxes have been overcome.

Apparently, Zermelo still views logic as dealing with customary finitary mathematical propositions, typical examples being his own axioms of set theory. The notion of definiteness in his definiteness paper ([Zer29a]) essentially amounts to the identification of definite properties with second-order definable ones, and is thus also of finitary character. But soon things were to change: All of a sudden some features of the Warsaw notes such as the reservations against intuitionism will experience a much sharper formulation, and some features which were only lightly addressed there such as infinitary languages will play an important role. We shall next sketch this development.

After having written his definiteness paper, Zermelo turned to the last item on the list above, to the search for a more convincing conceptual base of set theory. Early in 1930 a proposal was ready for publication: the so-called *cumulative hierarchy* which he regarded as *the* way to overcome the set-theoretic paradoxes.

During the publication process of the corresponding paper ([Zer30a]) Thoralf Skolem published a critical and somewhat polemical answer ([Sko30a]) to the definiteness paper where he pointed to his own first-order version of definiteness (cf. [Sko23]). Using this version, Zermelo's axiom system from 1908 becomes a countable system of first-order sentences. Presupposing its consistency, it therefore – as Skolem had shown, too – admitted a countable model. Moreover, Skolem stated, any finitary axiomatization of set theory would be subject to the same consequence.

Zermelo now faced a principal shortcoming of finitary axiom systems: Any such system seemed to be unable to describe adequately the Cantorian universe of sets with its endless progression of growing cardinalities. So the decisive role of infinite totalities which he had emphazised so strongly in Warsaw, could not be mirrored adequately in a finitary language.

To overcome this situation, Zermelo focused now on infinitary languages, maintaining that any adequate language for mathematics had to be infinitary. Of course he knew that in order to verify this claim, he first had to develop a convincing theory of infinitary languages together with some kind of infinitary proof theory. Indeed, he started such an enterprise, publishing his results in three papers between 1932 and 1935.

One might wonder why Zermelo did not acknowledge Skolem's approach at least for pragmatic reasons; for Skolem had demonstrated that first-order definiteness was sufficient to carry out all ordinary set-theoretic proofs. Instead, he developed a strong feeling that Skolem's arguments and the epistemological consequences Skolem drew from his results, meant a severe attack against mathematics and that he, Zermelo, was in charge of fighting back in order to preserve mathematical science from damage. His former opponents, the intuitionists, could right away be included among the adversaries.

In 1931 Zermelo got knowledge of Gödel's first incompleteness theorem: Under mild conditions which are satisfied in mathematical practice, any consistent finitary axiom system of sufficient strength is incomplete in the sense that it admits propositions which are neither provable nor refutable in it. So Skolem's results about the weakness of finitary axiom systems had got a companion, and Zermelo a further opponent.

It should have been clear very quickly to any well-informed observer that Zermelo could not win his fight. First, he did not strive for a presentation of his counterarguments with a standard of precision as was exercised by Gödel and Skolem; despite conceptual novelties, his papers on the subject are somewhat vague and, probably for this reason, did not unfold direct influence on further research. Secondly, his epistemological engagement prevented him from considering the results of Gödel and Skolem in an unprejudiced way as mathematically impressive theorems, striving for their technical understanding, and only then pondering their epistemological meaning and utilizing the analyzing power they provide. As it was these results which shaped the discipline of mathematical logic in the 1930s, he placed himself outside the mainstream of mathematical foundations.

If one takes a closer look at the development sketched so far, the lines exhibited dissolve into a finer network. For instance, by propagating infinitary languages and infinitary proofs as the only adequate means of performing mathematics, Zermelo distances himself from Hilbert even more than he had with his doubts in consistency proofs. When presenting this web, we shall concentrate on its main points, describing each point in a somewhat self-contained way, at the same time trying to keep intersections and repetitions to a minimum. The points chosen are:

- Definiteness revisited: How the controversies started.
- The cumulative hierarchy: A conclusive picture of the set-theoretic universe.
- The Skolem controversy: A fight for true set theory.
- Infinitary languages and infinitary logic: Indispensible prerequisites of genuine mathematics.
- The Gödel controversy: A fight against the rise of formal logic.

In contrast to his increasing isolation in foundational matters Zermelo received recognition in the German mathematical community. On 18 December 1931 he was elected a corresponding member of the mathematical-physical class of the Gesellschaft der Wissenschaften zu Göttingen.[141] On 2 November 1934[142] he presented his prime number paper ([Zer34]) to the academy.

The first part of the paper gives a simple argument for the uniqueness of prime number decomposition which does not use the notions of the greatest common divisor or the least common multiple.[143] We give the proof in detail.

For *reduction ad absurdum,* let $m \geq 2$ be the smallest natural number for which the uniqueness of prime number decomposition fails. Let

---

[141]Letter from the academy of 22 December 1931; UAF, C 129/113.

[142]UAF, C 129/24.

[143]In a letter to Zermelo of 22 December 1934 (UAF, C 129/11), Erich Bessel-Hagen points out that a similar but not quite as simple version was given by Gauß in his *Disquisitiones arithmeticae,* art. 13. He finds it "very strange that later authors, in particular Dirichlet who knew Gauss' *Disquisitiones* so well, returned to the much more involved Euclidean algorithm."

$$m = pp'p'' \cdots = pk \text{ and } m = qq'q'' \cdots = ql$$

be two different prime number decompositions of m with p the smallest prime number dividing m and q the smallest prime number among $q, q', q'', \ldots$. Then $p \le q$. As l is uniquely decomposable, p and q are different, i. e., $p < q$. Hence,

(∗)   all prime divisors of l are $> p$.

Let $n := m - pl$. Then $n = p(k - l) = (q - p)l \ne 0$ and $p \mid (q - p)l$. By (∗) and as n is uniquely decomposable, $p \mid (q - p)$. Let $(q - p) = pj$. Then $(j + 1) \mid q$ and $2 \le j + 1 < q$, a contradiction.[144]

On 16 February 1933 Zermelo was also elected an extraordinary member of the class for mathematics and the sciences of the Heidelberger Akademie der Wissenschaften; the academy followed a suggestion of the geometrician Heinrich Liebmann and the analyst Artur Rosenthal, both professors in Heidelberg and members of the academy.[145] Liebmann and Zermelo had got to know each other when Liebmann was an assistant in Göttingen from October 1897 to October 1898. On 2 March 1933[146] Zermelo expressed his thanks for the appointment and promised to take part in the activities of the academy. However, there are no documents giving evidence that he really participated in the academy's life. Obviously this was due to the fact that Liebmann, Rosenthal, and Zermelo himself experienced serious difficulties with the Nazis and were forced out of their professorships, Liebmann in October 1935 ([Dru 86], 164), Rosenthal in December 1935 (ibid., 223–224), and Zermelo in March 1935.

In August 1932, Heinz Hopf invited Zermelo to the International Congress of Mathematicians in Zurich,[147] asking him for two talks, one in the section of philosophy and one in the section of mechanics and mathematical physics. For the first talk he offered double the time scheduled for a contribution. Besides Zermelo only Paul Bernays enjoyed this privilege. For the second talk he suggested speaking about the topic of the "sugar paper" ([Zer33a]). Zermelo did not accept, although Fraenkel's talk ([Fra32a]) on separation and the axiom of choice and Bernays' talk ([Ber32]) on Hilbert's proof theory would have offered him an opportunity to express his opposition.

Zermelo's letter to Courant of 4 February 1932[148] reveals that it was Courant who had recommended his membership in the Göttingen academy. It also shows that despite the signs of acknowledgement he had experienced, Zermelo was concerned that his scientific work had not found the recognition he had hoped for:

---

[144]Already as early as 1912 Zermelo had exhibited his proof to some number theorists, among them Adolf Hurwitz and Edmund Landau. Kurt Hensel had asked for permission to refer to it in his lectures (letter to Zermelo of 1 August 1926; UAF, C 129/44). In fact, it initiated Helmut Hasse's generalization of the uniqueness theorem to division rings (cf. [Has27], footnote 1 on p. 3 and p. 6).

[145]Files of the academy.

[146]Letter to the academy; MLF.

[147]Letter to Zermelo of 12 August 1932; UAF, C 129/54.

[148]UAF, C 129/22.

Indeed, I very much appreciated the Göttingen appointment. I did so even more because, as you know, I did not after all enjoy much recognition. At first I thought that I owed the appointment mainly to Hilbert, my first and sole teacher in science. I therefore wrote to him straightaway at Christmas. But so far I haven't received an answer. [...]

I did not travel to Göttingen for Hilbert's birthday,[149] as the distance is too great. Moreover, I did not know whether I would have been really welcome there. Perhaps, however, I will get another opportunity to see my "home university" again. (OV 4.23)

Zermelo might have had in mind a newspaper article[150] on the occasion of his 60th birthday about half a year before where he was attributed "highest achievements of modern mathematics," but was characterized as a follower of Hilbert's axiomatic method who only tried to go his own way in the foundation of mathematics in recent years. He had underlined the words "only" and "tried," adding an exclamation mark to the latter. And the new own ways? The controversies with Gödel and Skolem were to show that only his friend Arnold Scholz was ready to accompany him. Faced with this isolation, his impression of failing recognition was to increase during the next years and end in a state where he feared falling completely into oblivion.

His impression of being perhaps not really welcome in Göttingen may have its roots in his awareness of an estrangement from Hilbert. He had written in this sense already in 1924 to the Farber brothers. Of course, he may also have thought of Hermann Weyl. Weyl had become Hilbert's successor in 1930. Sympathizing with Brouwer's intuitionism and showing a deep scepticism against the axiom of choice, he was a prominent target of Zermelo's criticism.

4.6 Definiteness Revisited

After his stay in Warsaw Zermelo spent some time at the seaside resort of Sopot near Gdansk where he wrote down his definiteness paper ([Zer29a]), completing it on 11 July 1929. It is the first set-theoretic paper to appear after a break of sixteen years. Planned already before his visit to Warsaw, but finalized during discussions with the Warsaw logicians, it is intended to meet criticism of his 1908 notion of definiteness. Its likewise critical reception had a lasting impact on Zermelo's foundational work. Our discussion starts by recalling some background facts (cf. 2.9.4).

4.6.1 Definiteness Until 1930

An essential feature of Zermelo's axiom system of set theory from 1908 is the restriction of unlimited comprehension to comprehension inside a set,

---

[149]Hilbert's 70th birthday on 23 January 1932.
[150]Vossische Zeitung Nr. 178, 28 July 1931; UAF, C 129/126.

formulated in the axiom of separation and saying that for any set S and (certain) properties $E(x)$ there exists the set $\{x \in S \mid E(x)\}$ of the elements of S having property E. In Zermelo's terms, separation does not deal with properties, but with propositional functions, i. e., descriptions of properties.

The limitation of comprehension to *sets* of discourse is intended to exclude inconsistent formations such as the set of all sets or the set of all ordinals, and, of course, the set $\{x \mid x \notin x\}$ responsible for the Zermelo-Russell paradox. In order to exclude Richard's paradox, Zermelo restricts the propositional functions allowed for separation to those which he calls "definite." He does not provide a precise definition, but merely gives his well-known vague description:[151] A propositional function is definite if *the basic relations of the domain concerned, by means of the axioms and the universally valid laws of logic, determine without any arbitrariness whether it holds or not.*

We detail the criticism which this notion experienced until the late 1920s by considering the objections put forward by Fraenkel, Skolem, and Weyl.[152]

In the first edition of his *Einleitung in die Mengenlehre* Fraenkel in principle still accepts Zermelo's description of definite propositional functions ([Fra19], 8–9, 140); in his words, given a set S and a propositional function $E(x)$, separation is allowed if for each element $x \in S$ it is "internally determined," whether $E(x)$ holds or not. Apparently he follows Cantor who had demanded[153] that for a set S it should be "internally determined" for any x whether $x \in S$ or not. Some years later, however, Fraenkel changed his mind: he calls Zermelo's description ([Fra22b], 231) "the weakest point or rather the only weak point" of his axiomatization and calls it ([Fra23], 2) "an essential deficiency." In the second edition of his *Einleitung in die Mengenlehre* of 1923 (p. 196–197) he speaks of a mathematically not well-defined concept that might become a new source of concern. In his *Zehn Vorlesungen*, in a subsection entitled "The Necessity of Eradicating the Notion of Property," he characterizes Zermelo's explanation as a mere comment that would lead to difficulties "lying on the same level as Richard's paradox" ([Fra27], 104); in order to "eradicate this sore spot" he then gives the version of definiteness which he had already published before[154] and which he also uses in the second edition of his *Mengenlehre*. It is given by an inductive definition which starts with certain basic propositional functions and describes how to gain new propositional functions from propositional functions already obtained.

In his address ([Sko23]) to the 1922 Helsinki Congress of Scandinavian Mathematicians, Thoralf Skolem speaks of Zermelo's notion of definiteness as an unsatisfactory and very imperfect point. Having expressed his astonishment that, according to his knowledge, nobody had solved the "easy" task of giving a precise definition, he proposes identifying definiteness with first-order

---

[151] [Zer08b], 263; [Zer67c], 201.
[152] For related material cf. [Moo82], 260–272, and [Ebb03].
[153] [Can82], 114–115; [Can32], 150.
[154] [Fra22a], 253–254; [Fra25], 254.

definability.[155] He thus gives the Zermelo axioms what was to become their final form.

Some years earlier, in *Das Kontinuum*, Hermann Weyl had characterized Zermelo's definition as an "apparently unsatisfactory explanation" ([Wey18], 36). When trying to clarify this point he had been led to a notion of a definable set of objects of a "basic category" (such as the category "point of a space" or "natural number") which strongly resembles first-order definability. Weyl does not intend to give a precise axiomatization of Cantorian set theory. Sharing the intuitionistic doubts in classical mathematics, he is interested in a safe foundation of analysis. However, if formally transferred to the basic category "set," definiteness in his sense coincides with first-order definability according to Skolem's approach.[156]

In his definiteness paper Zermelo reacts to this criticism by arguing against specific points and providing a more precise version of definiteness as well. Commenting first on the minor role which logic plays in the formulation of his axiom system from 1908, he emphasizes that a generally accepted system of logic had been missing, as it was still missing "at present where each foundational researcher has his own system" – obviously a side remark against the variety of systems such as intuitionism, formalism, and logicism that dominated the foundational discussions in the 1920s. He then criticizes Fraenkel's inductive definition of definite propositional functions as a distortion: it is "constructive" or "genetic," aiming not at the propositional function itself, but rather at its genesis. In particular, the inductive way uses the notion of finite number, namely by presupposing that the inductive steps are performed only finitely many times. However, the notion of number should not serve to establish set theory, it should be an *object* of set-theoretic foundations. He does not mention the discussion in *Das Kontinuum* ([Wey18], 36–37), where Weyl very distinctly describes his former efforts to avoid the use of the notion of finite number as "hunting a scholastic pseudo-problem."[157] This "long lasting and unsuccessful endeavour" had led him to the conviction that "*the imagination of iteration, of the series of natural numbers, should form the ultimate foundation of mathematical thinking*" and should not itself be subject to a set-theoretic foundation ([Wey18], 36seq.), a view that is also shared by Skolem ([Sko23], 230):[158]

> Set-theoreticians are usually of the opinion that the notion of integer should be defined and the principle of mathematical induction should be proved. But it is clear that we cannot define or prove ad infinitum; sooner or later

---

[155]Hence, the propositional functions E allowed in the axiom of separation are those which can be built up from basic expressions $x \in y$, $x = y$ by use of propositional connectives and quantifications of the form $\forall x$, $\exists x$, where x, y range over sets.

[156]See also [vanH67], 285 for this view.

[157]Cf. also [Wey10] and its discussion in [Moo82], 164.

[158]English translation [vanH67], 299.

we come to something that is no further definable or provable. Our only concern, then, should be that the initial foundations are something immediately clear, natural, and not open to question. This condition is satisfied by the notion of integer and by inductive inferences, but it is decidedly not satisfied by set-theoretic axioms of the type of Zermelo's or anything else of that kind; if we were to accept the reduction of the former notions to the latter, the set-theoretic notions would have to be simpler than mathematical induction, and reasoning with them less open to question, but this runs entirely counter to the actual state of affairs.

Zermelo acknowledges the version of definiteness that comes with von Neumann's functional approach ([Neu25]), where the basic objects are functions and sets are defined via functions. Because of the latter feature he criticizes von Neumann's way as being too complicated, offering his own version of definiteness as a simpler one. He emphasizes that it had been in his mind already when he wrote his axiomatization paper ([Zer08b]) and, in fact, had been implicitly present in the proofs of definiteness given there. Its constructive or genetic version "lies at hand" ([Zer29a], 342):

A [propositional function] is "definite" for a given [axiom] system, if it is built up from the basic relations of the system exclusively by use of the logical elementary operations of negation, conjunction, and disjunction as well as quantification, all these operations in arbitrary but finite repetitions and compositions.

Once more criticizing the inherent finiteness condition, he ends up with the intended "axiomatic" version that is given here in an abbreviated form: The totality of definite propositional functions

(1) contains the basic relations and is closed under the operations of negation, conjunction, disjunction, and first-order and second-order quantification and
(2) does not possess a proper part satisfying (1).

To be more exact, the second-order quantifications seem to range only over definite propositional functions, i. e., over the properties to be defined, a problematic point which Skolem will address in his reply.

What is gained against the version of definiteness as given in 1908? Overall, there is a shift from the conceptual side to the syntactical one. By explaining definiteness as the closure of basic relations under propositional connectives and quantifications, he surely gains clarity. Nevertheless, there is no sharp distinction between language and meaning.[159] In fact, Zermelo will never be clear on this point. His negligence will have a major impact in later discussions with Gödel. The separation of syntax and semantics or – more adequately – its methodological control became the watershed that separated the area of

---

[159] For instance, Zermelo does not distinguish between properties and their descriptions ([Zer29a], 342) or between basic relations and atomic formulas.

the "classic" researchers such as Zermelo and Fraenkel from the domain of the "new" foundations as developed by younger researchers, among them Gödel, Skolem, and von Neumann. The acknowledgment of the syntactic structure of language comes together with an unquestioned acceptance of inductive definitions of formulas.

After the definiteness paper had been published, Skolem immediately wrote a reply ([Sko30a]), criticizing Zermelo's notion of second-order quantification, and – more momentous – informing him about his own first-order version of definiteness in his 1922 Helsinki address and, hence, about the existence of countable models of set theory: The first-order version of definiteness leads to a first-order version of Zermelo's axiom system; by the Löwenheim-Skolem theorem (4.8.1) this system – if consistent – has a countable model. It was this consequence that roused Zermelo's resistance. In fact, his further scientific endeavour in foundational matters aimed at ruling out the existence of a countable model of set theory. He probably never checked the correctness of Skolem's arguments or considered the results as an interesting mathematical theorem. He always pointed to the weakness of Skolem's axiomatization, seeing the reason for its deficiency in its finitary approach, henceforth advocating that no finitary axiom system was able to catch the richness of mathematics. As a consequence, any genuine axiom system had to be infinitary.

At a first glance his reaction is astonishing because his own way of having made definiteness precise less than a year earlier in the definiteness paper also leads to a finitary axiom system of set theory. Apparently the striking insight into the weakness of finitary axiomatizations as exemplified by the existence of countable models of first-order set theory, all of a sudden showed the indispensability of infinitary languages and paved the way for his infinitary convictions.

4.6.2 Incorporating Definiteness Into the Notion of Set

Infinitary languages were not the only means Zermelo employed to overcome Skolem's consequences. He also worked increasingly towards a conceptual reformulation of set theory. This enterprise had already started earlier, but became interwoven now with the question of definiteness.

In the *Grenzzahlen* paper ([Zer30a]) Zermelo conceives inexhaustibly long hierarchies of models of Zermelo-Fraenkel set theory. They are versions of the von Neumann hierarchy where the ordinals contain an unbounded sequence of strongly inaccessible cardinals. In the version of the paper which he submitted to *Fundamenta Mathematicae*, the axiom of separation is still restricted to definite propositional functions. Under the impression of Skolem's results and as a first reaction, he removed the definiteness condition, now explicitly emphasizing that separation should be applicable for *arbitrary* propositional functions. In his own words ([Zer30d], 5):

> SKOLEM wants to restrict the formation of subsets to special classes of defining functions whereas I, according to the true spirit of set theory, admit a

free separation and postulate the existence of all subsets in whatever way they are formed.

However, footnote 1 of the published version indicates that this stipulation did not eliminate definiteness forever:

> I reserve for myself a detailed discussion of the "issue of definiteness," following my last note in this journal[160] and the critical "Remarks" of Mr. Th. Skolem.[161]

Hence, the *ad hoc* changes in the *Grenzzahlen* paper may be regarded as some kind of emergency brake. Various pieces from the *Nachlass* give more evidence that he still stuck to limited separation, trying to combine a reasonable limitation with the exclusion of countable models. In particular, he discussed the problem with Reinhold Baer. On 27 May 1930, Baer writes with respect to the *Grenzzahlen* paper:[162]

> Nevertheless, it still may be an interesting question, how many "reasonable" possibilities there are to elucidate definiteness. To be called reasonable, first of all a notion of definiteness should ensure the usual set-theoretic conclusions. Only in the second place would I require cardinalities to be invariant under transitions to larger domains of sets. – Anyway, there is one fact that should be observed: a subset of a set belonging to some domain of sets may exist even if there is no definite function according to the axiom of separation; just that it need not exist in this case.

The way in which Zermelo pursued the notion of definiteness deviates from the way he went in the *Grenzzahlen* paper. Aiming at Skolem and his "set theory of the impoverished,"[163] the concepts of the *Grenzzahlen* paper are built on the ideas of "free separation" and "unbounded progression" ([Zer30a], 47) and lead to a hierarchical universe that may even contain large cardinals. When pursuing definiteness, Zermelo built on the ideas of the Warsaw notes W6 and conceived a universe which is composed solely of definable sets and where any hierarchical aspect seems to have disappeared.[164]

Without providing details, Zermelo proposes taking as sets those classes which allow a categorical definition, i. e., a definition which up to isomorphism has exactly one domain as a model. Examples are the set of natural numbers and the set of real numbers, categorically defined by the Peano axioms and by Hilbert's axioms for completely ordered fields, respectively. In the Warsaw notes W6 he argued in favour of this proposal as follows:

---

[160][Zer29a].

[161][Sko30a].

[162]UAF, C 129/4; reprinted in [Ebb04], 83.

[163]Letter from Baer to Zermelo of 24 May 1931; UAF, C 129/4.

[164]For the apparent incompatibility between the "dynamic" *Grenzzahlen* universe and the seemingly "static" universe of definable sets cf. [Ebb03], 208–211.

In any case, a set in the sense of set theory has to have a "cardinality" which is uniquely determined by its definition, that means, any two "classes" or "domains" which fall under the definition have to be "equivalent" in the sense of Cantor, i.e., it must be possible to map one domain in a one-to-one way onto the other. Certainly, this is the case for domains which are given by a "categorical" system, for instance, for domains which satisfy a categorical axiom system just as the "countable" sets satisfy the Peano postulates. Hence, it suggests itself to generally define "sets" as "domains of categorical systems."

Early in 1931 he started to work out these ideas. In a diligently typewritten note ([Zer31f]) and in a fragment entitled *Mengenlehre* ([Zer32c]), probably part of a version for a textbook on set theory, he gives a more detailed elaboration. In the fragment he defines:

A "set" is a finite or infinite domain of well distinguishable things or objects that is characterizable by a categorical system of postulates.

The formulation takes up Cantor's well-known definition of a set as – so Zermelo at the beginning of the notes W6 – "a well-defined comprehension of objects where for each thing it is decided whether it belongs to it or not." In fact, he is fully convinced that his notion of set is in accordance with that of Cantor ([Zer31f]):

It exactly coincides with what Cantor really means by his well-known definition of a "set" and may everywhere and consistently be treated as a "set" in all purely mathematical considerations and deductions.

The categorical systems defining a set may use additional basic relations such as the successor relation with the Peano axioms. But there is no explanation concerning the nature of these relations and of the isomorphisms involved; they obviously have to be regarded as part of the background universe of classes and functions. Moreover, Zermelo does not comment on the language in which the definitions are to be given. Examples such as the Peano axioms for the natural numbers and Hilbert's axioms for the real numbers suggest that second-order definitions should be allowed.

As remarked by Taylor, the 1908 axiomatization already shows that "an intuitive concept of mathematical definability is at the root of Zermelo's concept of set" ([Tay93]), 549) and that "absolutely all mathematical objects are capable of (predicative) 'determination' and, hence, are more or less definable" (ibid., 551). Faced with the introductory sentence of the *Mengenlehre* fragment ([Zer32c]) that "set theory deals with the mathematically defined infinite totalities or domains which are called 'sets,'"[165] definability has now become *the* dominant property of a set. Remembering Zermelo's credo that no finitary language is able to capture all of mathematics, it would have come

---

[165] "Die Mengenlehre hat es zu tun mit den mathematisch definierten unendlichen Gesamtheiten oder Bereichen, die als 'Mengen' bezeichnet werden."

with a surprise if in this context he had definitely fixed a language such as that of second-order logic. After 1930 any possible candidate had to be an infinitary one.

The definition of the notion of set comes with a shift concerning the role of axioms: Hitherto, Zermelo had emphasized that the universe of sets should be given in the axiomatic way, i. e. by "providing a system of conditions which should be satisfied by the [intended] domain" ([Zer32c], 1), the axioms then forming the basis of mathematical conclusions. Now, he *defines* the universe as the totality of sets obeying the condition of categorical definability. In order to ensure that the usual mathematical considerations may be performed, he therefore has to show that the axioms of set theory are satisfied or, in other words, that the universe of categorically definable sets is a model of the Zermelo-Fraenkel axiom system ZFC; in technical terms, that the universe of definable sets is an *inner model* of ZFC inside the universe of classes.

According to the *Grenzzahlen* paper, Zermelo takes it for granted that the axiom of choice is valid (as a "general logical principle"), and he sketches proofs for the axioms of separation, power set, union, and replacement. Of course, a detailed argumentation would have to provide a precise notion of categorical definability together with an explicit list of the assumptions on classes that are needed. As Zermelo does not do so, his considerations cannot really be followed.[166] Apart from this deficiency, in particular the axiom of separation has become a provable fact. In the lecture notes for the course on set theory which Zermelo gave in the winter semester 1933/34,[167] this 'theorem of separation' has the following form:

Each (categorically defined) part of a set is itself a set.[168]

Later in the notes it is re-formulated in the even simpler form:

Each *part* of a set is itself a set.[169]

This is the last version of separation which can be found in the *Nachlass*. So in the end the notion of definiteness has been absorbed into the notion of set.

4.7 The Cumulative Hierarchy

In 1929 von Neumann gave the first definition of the hierarchy which is now called the *von Neumann hierarchy* ([Neu29]). As his basic objects are functions instead of sets, the definition becomes rather involved. When arguing, say,

---

[166] For details, see [Ebb03].

[167] UAF, C 129/164.

[168] "Jeder (kategorisch bestimmte) Teil einer Menge ist selbst eine Menge."

[169] "Jeder *Teil* einer Menge ist selbst eine Menge."

in the Zermelo axiom system together with the axiom of replacement, the hierarchy can be defined more elegantly in set theoretic terms by iterating the power set operation along the ordinals, starting with the empty set and taking unions over all preceding levels at limit points:

$$V_0 = \emptyset$$

$$V_{\alpha+1} = \text{ power set of } V_\alpha$$

$$V_\beta = \bigcup \{V_\alpha \mid \alpha < \beta\} \text{ for limit ordinals } \beta.$$

The hierarchy is *cumulative* in the sense that for any ordinals α and β with $\alpha \le \beta$, V_α is a subset of V_β. Therefore the von Neumann hierarchy is frequently called the *cumulative hierarchy*. For any set S from the union class V of the V_α there exists the smallest ordinal β such that $S \in V_{\beta+1}$, its rank $\mathsf{rk}(S)$. It is easy to see that for $S \in V$ and $S' \in S$ one has $S' \in V$ and $\mathsf{rk}(S') < \mathsf{rk}(S)$. Hence, as any strictly descending sequence of ordinals is finite, V cannot contain infinite descending $\in$-chains $\ldots \in S_2 \in S_1 \in S_0$. In other words, V satisfies the axiom of foundation that just excludes such sequences. Equivalently,[170] the axiom can be stated more simply as to ensure that any non-empty set S contains an element S' such that $S \cap S'$ is empty. Von Neumann shows that V satisfies not only the axiom of foundation, but all the Zermelo axioms together with the axiom of replacement. In syntactic terms he thus proves that the addition of the axiom of foundation to these axioms preserves consistency.

Both, the axiom of foundation and the cumulative hierarchy, were already conceived more than a decade earlier by Dimitry Mirimanoff ([Mir17]) with the intention of giving a set-theoretic analysis of the Burali-Forti paradox.[171] However, despite its "striking relation to the now standard idea of the von Neumann cumulative hierarchy, one should be wary of seeing too much modernity" ([Hal84], 189–190); the axiom of foundation is introduced as a working hypothesis and the hierarchy does not appear in explicit terms.

4.7.1 Genesis

In his *Grenzzahlen* paper ([Zer30a]) Zermelo developed his own picture of the cumulative hierarchy. Whereas the hierarchy had served von Neumann as a technical means to prove the relative consistency of the axiom of foundation,[172] Zermelo viewed his results as a conceptual answer to "decisive" questions which occurred to his mind when he returned to foundational problems in the mid-1920s:

---

[170]With respect to the other axioms of Zermelo-Fraenkel set theory together with the axiom of choice.

[171]Cf. the detailed discussion of Mirimanoff's work in [Hal84], 185seq. Cf. also [Fel02c], 208seq., and [Ferr99], 370seq.

[172]Von Neumann introduces the axiom of foundation in order to approach a categorical axiom system of set theory ([Neu25], 39; [Neu61], 55). For an earlier discussion of this point cf. [Sko23], §6.

What has a "domain" of "sets" and "urelements" to look like in order to satisfy the "general" axioms of set theory? Is our axiom system "categorical" or is there a multiplicity of essentially different "set-theoretic models"? Is the notion of "set" as opposed to a mere "class" absolute, determinable by logical characteristics, or only a relative one, depending on the set-theoretic model taken as a basis?[173]

Moreover, the "lucid" ([Kre80], 189) presentation in the *Grenzzahlen* paper provides an impressive and independent alternative to Skolem's finitary restrictions. Akihiro Kanamori ([Kan04], 521) puts its character in the following words:

Zermelo in his remarkable *On Boundary Numbers and Set Domains* offered his final axiomatization of set theory as well as a striking, synthetic view of a procession of natural models of set theory that would have a modern resonance. Appearing only six articles after Skolem [1930][174] in *Fundamenta*, [it] is ostensibly a response, more informal and rough around the edges than his writings decades earlier, but its dramatically new picture of set theory reflects gained experience and suggests the germination of ideas over a prolonged period. The article is a *tour de force* which sets out principles that would be adopted in the further development of set theory and focused attention on the cumulative hierarchy picture, dialectically enriched by initial segments serving as natural models.

In March 1930 Zermelo informed Knaster and, later, Sierpiński in Warsaw that the paper was nearly finished, telling them that it "concerned the foundations of set theory and promised a satisfying explanation of the so-called antinomies." He gave a short description of his results and proposed a publication in *Fundamenta Mathematicae* or in the *Annales de la Société Polonaise de Mathématique*.[175] On 27 March Knaster asked him to send the finished manuscript to Sierpiński for publication in *Fundamenta*.[176] The next day he wrote a second message,[177] informing Zermelo that Alfred Tarski, when he had been shown the results, had told him that he had conceived "similar thoughts," but that he, having not published them, made no claims of priority. Knaster urged Zermelo to submit the manuscript "as soon as possible."

The incident shows that Zermelo had probably not discussed his ideas when he was in Warsaw less than a year earlier. Nevertheless, Kanamori may be right in stating that the final paper suggests the germination of ideas over a prolonged period. There is a strong argument for this: On 21 February 1928 Reinhold Baer submitted a paper ([Bae28]) which contains results and remarks taking up basic features of Zermelo's work. For example, he states (ibid., 382) that "a consistent set theory can always be extended in such a

---

[173][Zer30d], 2.
[174]I. e., after Skolem's answer ([Sko30a]) to the definiteness paper.
[175]Letter from Zermelo to Sierpiński of 26 March 1930; copy in UAF, C 129/110.
[176]UAF, C 129/62.
[177]UAF, C 129/62.

way that the totality of its sets is a set in the extended set theory," a property which represents an essential feature of Zermelo's cumulative hierarchy. As Baer was Zermelo's closest scientific partner around that time, the two may have discussed these issues in the late 1920s. When Baer left Freiburg in the autumn of 1928, the discussions could no longer be continued from door to door. According to remarks in letters from Baer written in 1930[178] it was Zermelo who really brought the seeds to fruition.

4.7.2 The Zermelo-Fraenkel Axiom System

Zermelo bases his investigations on an axiom system which he calls "the extended ZF-system"[179] or "ZF'-system" – "ZF" for "Zermelo-Fraenkel" – and which consists of the axioms from 1908 without the axiom of infinity (as it would not belong to "general" set theory), namely the axioms of extensionality, pairing,[180] separation, power set, and union, together with the axiom of replacement and the axiom of foundation; the axiom of choice is not explicitly included, but taken for granted as a "general logical principle." The axiom of foundation is formulated in the two equivalent versions given above.[181] Only here, in the *Grenzzahlen* paper, is it introduced as an essential axiom of set theory, at the same time getting its name ([Ber41], 6).

During the process of publication of the paper Zermelo was confronted with Skolem's first-order approach ([Sko23]) which admitted countable models of set theory, a fact totally alien to his idealistic point of view (4.8.1). As a reaction he cancelled the condition of definiteness that was still present in the axiom of separation as formulated in the submitted version ([Zer30e]), now allowing "*arbitrary* propositional functions." Putting aside his former scepticism (cf. 3.5), he also allowed arbitrary functions in the axiom of replacement (ibid., 30), altogether arriving at an axiom system of second order. Although he was informed then that Skolem had formulated the axiom of replacement nearly as early and for similar reasons as Fraenkel, he stuck to the names ZF and ZF'. He thus followed von Neumann who used the name "Zermelo-Fraenkel set theory" for the first time ([Neu28a], 374). Later, the abbreviation ZF was generally accepted for the first-order version of Zermelo's ZF' including the axiom of infinity, but not the axiom of choice; ZFC now meant ZF together

---

[178]Letters to Zermelo of 27 May 1930 and of 12 July 1930; UAF, C 129/4.

[179]Extended by the axiom of foundation.

[180]It essentially equals the axiom of elementary sets from 1908, postulating for any sets x and y the existence of the set $\{x, y\}$.

[181]Literally: "Every chain of elements (going backwards), in which each term is an element of the preceding term, breaks off with finite index at an urelement. Or equally: Every subdomain T contains at least one element t_0 none of whose elements t is in T." ("Jede rückschreitende Kette von Elementen, in welcher jedes Glied Element des vorangehenden ist, bricht mit endlichem Index ab bei einem Urelement. Oder, was dasselbe ist: Jeder Teilbereich T enthält wenigstens ein Element t_0, das kein Element t in T hat.")

with the axiom of choice. It thus happened that Skolem's initial does not appear in the name of the system which he, Skolem, had formulated for the first time.[182] The system ZFC is *the* purely set-theoretic axiom system modern set theory is based upon. The competing von Neumann-Bernays-Gödel system on classes and sets is of equal strength as regards purely set-theoretic statements.

4.7.3 Cumulative Hierarchies, Large Cardinals, and the Paradoxes

Having presented the axiom system, Zermelo first explicates the fundamental facts about the von Neumann ordinals which he calls "basic sequences" as in his first investigations of this concept around 1915 (3.4.3). He introduces the cumulative hierarchy essentially as described above. More exactly, the hierarchy starts with an arbitrary nonempty domain of urelements (i. e. objects without elements) instead of the empty set.[183] The following discussion nevertheless refers to the usual von Neumann case without urelements, pointing to urelements only in case they are essential.[184]

Zermelo shows that any model of the axiom system ZF′ – in his terminology any *normal domain* – coincides with the cumulative hierarchy defined in it and that a level V_α is a normal domain just in case α is a *boundary number* (*Grenzzahl*), i. e. α equals the first infinite ordinal ω (if the negation of the axiom of infinity is true) or is a *strongly inaccessible* cardinal, that means:

- α is uncountable.
- α is regular, i. e., it does not contain an unbounded subset of order type less than α.
- For any cardinal κ less than α, also 2^κ is less than α.

In order to prove that a normal domain coincides with its cumulative hierarchy, the axiom of foundation is crucial.[185] In this sense the axiom ensures the cumulative-hierarchical structure of a normal domain and of the set-theoretic universe itself, i. e., it ensures that kind of structure which has become a basic constituent of the concept of set.

Zermelo represents the union V of all levels V_β of a normal domain as a level, too, namely as the level V_π belonging to the supremum π of the ordinals in V, by this "inspired move having π outside V, but in set theory" ([Kan04],

---

[182]Some of the editors of [Haus2002] use the term "Zermelo-Skolem-Fraenkel set theory" and the abbreviation ZSF.

[183]It should be emphasized that urelements are essential for Zermelo. On a postcard to Fraenkel of 20 January 1924 (provided by Fraenkel's son Benjamin S. Fraenkel) Zermelo explicates: "It seems questionable to demand that each element of a set should be a set *itself*. Well, formally this works and simplifies the formulation. But how is it then with the *application* of set theory to geometry and physics?" (OV 4.24)

[184]In particular, we write the customary "V_α" instead of Zermelo's "P_α."

[185]Cf. Zermelo's *Hilfssatz* in [Zer30a], 35. – In fact, the coincidence is equivalent to the axiom of foundation.

522). He speaks of π as the *characteristic* of V. He shows that normal domains over equivalent domains of urelements are determined up to isomorphism by their characteristics.[186]

Some of Zermelo's arguments depend on the second-order point of view. We will not discuss this here,[187] but only refer to a special aspect which is frequently employed – for example, in the proof of the isomorphism result just mentioned – and will become important for his later research: the absoluteness of the power set operation. Oskar Becker, in a letter of 31 December 1930,[188] touches on this question, explicating in an informal way the possibility that, for a fixed strongly inaccessible cardinal α, nonisomorphic models V_α that are based on "thin" and "thick" power sets, respectively, might be conceivable. Zermelo will never accept this possibility. For him the concept of a full power set is indispensable ([Zer31f]):

> Any mathematical consideration must be founded on the notion of "allness" or "quantification" as a basic category of logic which cannot be subject to further analysis whatever. If in the special case [the formation of the power set] one would like to restrict allness by special conditions, this would have to be done again by quantifications. We would hence arrive at a *regressus ad infinitum*.

As a consequence, he will refuse to think in models of set theory with restricted power sets; later, this point of view will lead him to try to refute the existence of countable models (4.8.4, 4.12.2).

The concluding section of the *Grenzzahlen* paper comments on the non-categoricity of ZF′ and touches on the problem of consistency, in particular discussing the problem of the existence of strongly inaccessible cardinals. Zermelo observes that ZF′ (extended by the axiom of infinity) is not strong enough to prove the existence of such cardinals. Essentially he argues that otherwise the model V_α with α the smallest strongly inaccessible cardinal satisfies the axioms, but does not obey the consequence of containing a strongly inaccessible cardinal. Following now the lines of the Warsaw notes W6 to the effect that a "categorically defined domain" such as that comprising a normal domain "should also be interpreted as a set in some way," he is led to the conclusion that such a cardinal really should exist.[189] By iterating this argument, he is

---

[186]Namely, given two normal domains $\mathcal{D}_0$ and $\mathcal{D}_1$ over domains U_0 and U_1 of urelements, respectively, both normal domains with the same characteristic π, and given a bijection f from U_0 onto U_1, there is exactly one isomorphism ι from $\mathcal{D}_0$ onto $\mathcal{D}_1$ which extends f. It can be defined from level to level in π steps by transfinite induction, starting with f, in a successor step extending the part of ι defined so far, which maps some level $V_\alpha(U_0)$ of $\mathcal{D}_0$ onto the corresponding level $V_\alpha(U_1)$ of $\mathcal{D}_1$, in a natural way to the respective power sets, and taking unions at limit stages.

[187]For details see, e. g., [Moo82], 269seq., or [Kan04], 519seq.; for a systematic treatment see [She52].

[188]Letter to Zermelo; UAF, C 129/7. It is published in [Pec05c].

[189]If no such cardinal exists, all models of ZF′ together with the axiom of infinity coincide up to isomorphism with the single normal domain beyond V_ω. As a categor-

led to postulating "a new axiom of 'meta set theory'," namely the existence of an unbounded sequence of strongly inaccessible cardinals. When doing so, he continues, the antinomies disappear because any inconsistent set of a normal domain, such as the set of all sets of that domain or the "Russell set" of all sets of that domain not being an element of themselves, are now sets in the next normal domain. By a similar argument the Burali-Forti paradox vanishes as well, as the ordinals of a normal domain V_α just form the first new ordinal, namely α, of the next normal domain. He, therefore, sees the antinomies defeated:[190]

> Scientific reactionaries and anti-mathematicians have so eagerly and lovingly appealed to the "ultra-finite" antinomies in their struggle against set theory. But these are only apparent "contradictions" and depend solely on confusing *set theory itself*, which is not categorically determined by its axioms, with individual *models* representing it. What appears as an "ultrafinite non- or super-set" in one model is, in the succeeding model, a perfectly good, valid set with both a cardinal number and an ordinal type and is itself a foundation stone for the construction of a new domain.

Moreover, in its natural unfolding and by the fruits it provides, the "unbounded series of Cantorian ordinals" along which the cumulative hierarchy develops with its neverending succession of well-behaved set-theoretic universes, mirrors a basic property of the human spirit. This correspondence with human thinking is drawn in a "grand" ([Kan04], 529) and stylistically gripping *finale* ([Zer30a], 47):

> The two diametrically opposite tendencies of the thinking spirit, the idea of creative *advance* and that of collection and *completion*, ideas which also lie behind the Kantian "antinomies,"[191] find their symbolic representation and their symbolic reconciliation in the transfinite number series based on the concept of well-ordering. This series reaches no true completion in its unrestricted advance, but possesses only relative stopping-points, just those "boundary numbers" which separate the higher model types from the lower. Thus the set-theoretic "antinomies," when correctly understood, do not lead to a cramping and mutilation of mathematical science, but rather to an, as yet, unsurveyable unfolding and enriching of that science.

---

ically defined domain, this normal domain is a set and, hence, a level V_α of the von Neumann hierarchy with strongly inaccessible α, a contradiction. See also [Tai98] in this respect.

[190] [Zer30a], 47; English translation from [Ewa96], 238–239.

[191] Also in Cantor's collected papers ([Can32], 377) Zermelo refers to a parallelism between the set-theoretic antinomies and Kant's antinomies of pure reason. Even if rejecting Kant's theory of mathematics as based on "reine Anschauung," he felt himself forced to admit that Kant's theory of antinomies "expresses a deeper insight, an insight into the 'dialectical' nature of human thinking." According to the philosopher Gottfried Martin ([Martg55], 54) Zermelo also used to treat this point in his lectures. Martin attended Zermelo's course on set theory in the winter semester 1933/34. Cf. also [Hal84], 223seq.

Zermelo's chain of thoughts, starting with the principle that categorically defined domains should be sets and ending in a boundless unfolding of ordinals and set-theoretic universes, found its admirer in Oskar Becker. In the letter mentioned above Becker writes:

> The principle to consider any categorical domain also as a set appears to me much more transparent and satisfying than the principle saying that, finally, ordinal numbers with a given property $\mathcal{E}$ exist if this property $\mathcal{E}$ does not contradict the essence of ordinal numbers. In the end it may be a matter of taste or of one's point of view. Your principle has something mystical; it holds the belief that in the sky of the remote part of the ordinal numbers all non-contradictory desiderations of the mathematicians will be finally fulfilled.

Non-categoricity, Zermelo emphasizes, is more than compensated by the richness of models which guarantees "an unlimited applicability of set theory" ([Zer30a], 45). Moreover, as the models of set theory formed an ascending chain of V_α's and as the V_α's were fixed up to isomorphism, some benefits of categoricity were preserved; for example, as the power set of a set S is already fixed in the smallest normal domain V_α containing S, the generalized continuum hypothesis GCH "does not depend on the choice of the model, but is once and for all determined (as true or false) by our axiom system" ([Zer30d], 3).

4.7.4 Continuation

As announced at the end of the *Grenzzahlen* paper and evident from notes in the *Nachlass*, Zermelo continued to work on the subject. In a fragmentary note entitled "Über das mengentheoretische Modell" ("On the Set-Theoretic Model") ([Zer30f]), probably from September 1930, he defines the von Neumann hierarchy not inside a set-theoretic model, but inductively from below, representing the new subsets of a level in the next level not by themselves, but by "arbitrary" elements. In several notes from around 1930 to 1933 he tries to elaborate the considerations about large cardinals in a stricter manner. For instance, a carefully typewritten one page manuscript entitled "Proof of the Existence of Exorbitant Numbers,"[192] probably from late 1930 or early 1931, elaborates the argument for the existence of strongly inaccessible cardinals mentioned above. And a note "The Unbounded Series of Numbers and the Exorbitant Numbers" ([Zer33b]) from November 1933 serves the same purpose, explicitly stating in the form of four axioms what is needed.[193] However, this work never grew into a successor of the *Grenzzahlen* paper. In a letter to the Notgemeinschaft of 3 December 1930[194] Zermelo explains:

> Other ones [of the investigations ensuing the *Grenzzahlen* paper] which I also consider promising, are still causing serious difficulties and will probably

---

[192] "Beweis der Existenz der *exorbitanten* Zahlen," part of [Zer31e].
[193] For details see [Ebb06a].
[194] It accompanies [Zer30d] and is in UAF, C 129/140.

require even longer intensive research until they are ready for publication. (OV 4.25)

We have emphasized elsewhere that the main part of Zermelo's energy during the early thirties was absorbed by the development of infinitary languages. In contrast to the finitary systems treated by Skolem and Gödel, they alone seemed to be able to allow an adequate formulation of "genuine mathematics." However, this endeavour did not restrain him from sticking to the *Grenzzahlen* theme. For him both directions of research – that on the cumulative hierarchy and large cardinals and that on infinitary languages – were closely related to each other by being based on the same essential principle: the axiom of foundation. It lies at the root of the cumulative hierarchy and at the same time allows a corresponding hierarchy of infinitary languages to be defined (4.9.4). In a letter to Baer Zermelo writes ([Zer31b]):

> The question of the antinomy of Richard and of the Skolem doctrine *must* eventually be discussed *seriously*, now, where frivolous dilettantism is again at work to discredit the whole area of research. [...] I believe I have finally found in my *"principle of foundation"* (*"Fundierungsprinzip"*) the right instrument for explaining whatever is in need of elucidation. But nobody understands it.

The parallelism between the hierarchy of sets and the hierarchy of infinitary languages could also have been used to build a bridge between the two competing concepts of the universe: on the one hand the cumulative hierarchy developing in an unbounded, inaccessibly long sequence of set-theoretic models, on the other hand the seemingly static universe of categorically definable sets. However, Zermelo finally failed to accomplish a reconciliation. Moreover, his fight against the "Skolem doctrine" was also lost: Instead of following his idealistic conceptions, logicians joined Skolem and Gödel, soon accepting their finitary points of view. So the *first-order* Zermelo-Fraenkel axiom system ([Sko23]) became *the* axiom system underlying modern set theory. According to Kanamori ([Kan94], XV),

> Skolem's move was in the wake of a mounting initiative, one that was to expand set theory with new viewpoints and techniques as well as to invest it with a larger foundational significance.

And soon the respective results of the emerging giant Gödel left their mark on the future development of mathematical logic and set theory.

4.7.5 The Reception of the *Grenzzahlen* Paper

Zermelo considered his *Grenzzahlen* paper as very important. Firstly, the subtitle "Neue Untersuchungen über die Grundlagen der Mengenlehre" takes up the title "Untersuchungen über die Grundlagen der Mengenlehre I" of his 1908 axiomatization paper which contains his first axiom system and whose second part never appeared. Secondly, he intensively discussed the work with various people, among them, as we know already, Reinhold Baer, but also Arnold Scholz and Oskar Becker. After it had been published, he eagerly waited for reactions, showing discouragement when they did not come. About a year after its appearance he complains ([Zer31b]):

> Nobody understands [my "principle of foundation"], just as nobody has yet reacted to my Fundamenta article – not even my good friends in Warsaw. After all, Hilbert once said that it takes fifteen years before a paper is read. Now Hilbert himself had quick success with his "geometry," but was it really such a great achievement?

Did Zermelo's *Grenzzahlen* paper have an impact on the developments initiated by Skolem and strongly influenced by Gödel?[195] When describing the essence of the natural extension of Russell's simple theory of types into the transfinite which in its cumulative variant amounts to the cumulative hierarchy ([Goe33]), Gödel does not mention it, although according to Solomon Feferman ([Fef95], 37) "he could have referred to it for the first clear statement of the informal interpretation of axiomatic set theory in the cumulative hierarchy." With respect to an early version of Gödel's constructible hierarchy from 1935, Alonzo Church regarded Russell's conceptual influence as being too strong;[196] according to his impression Gödel's procedure "could be considerably simplified if it were made to refer to the Zermelo set theory instead of to the system of Principia Mathematica." Robert Solovay, when commenting[197] on lecture notes by Gödel ([Goe39], [Goe40]) and referring to the cumulative hierarchy, the constructible hierarchy, and the axiom of separation, is "struck by the paucity of references to Zermelo and the subsequent developments leading to modern set theory." This paucity contrasted to Gödel's frequent references to Russell, which "strike him as much too generous," and he supports this impression with a number of examples. "Perhaps," he continues, "Gödel took [Zermelo's] axiomatization for granted as the only sensible way to develop set theory." Dreben and Kanamori conclude ([DreK97], §4) that apparently Gödel may have been more influenced by Russell than by Zermelo. As a matter of fact, Gödel already read the *Grenzzahlen* paper shortly after its appearance and suggested to Zermelo[198] discussing "various concerns" with him. Zermelo, however, caught in his fight against finitism, did not react.

Seen from a later point of view, there is evidence that the *Grenzzahlen* paper had a conceptual influence. By invoking the axiom of foundation, Zermelo succeeded in representing the essence of set theory in the cumulative hierarchy. He thereby "presented a rationale for his choice of axioms that had been lacking in his system of 1908" ([Moo82], 270) and that is now considered the

---

[195] For Gödel's decisive role cf., e. g., [FloK06].

[196] Letter from Church to Paul Bernays; ETH-Bibliothek Zürich, Nachlaß Bernays, Hs. 975.796.

[197] [Goe86], Vol. III, 118–120.

[198] In the letter [Goe31b].

strongest conceptual argument for set theory to have overcome the paradoxes. According to Kanamori ([Kan04], 521–522) it was Zermelo who

> substantially advanced this schematic generative picture with his inclusion of foundation in an axiomatization. [...] Indeed, it can be fairly said that modern set theory is at base the study couched in well-foundedness, the Cantorian well-ordering doctrines adapted to the Zermelian generative conception of sets.

The cumulative hierarchy picture served explicitly to motivate the axiom system in Joseph Shoenfield's handbook contribution [Sho77]. It initiated axiomatizations of set theory such as that of Dana Scott ([Sco74]).[199]

Taking up Zermelo's considerations about large cardinals, Kanamori emphasizes a second point ([Kan96], 28):

> In synthesizing the sense of progression inherent in the new cumulative hierarchy picture and the sense of completion in [...] the [strongly] inaccessible cardinals, he promoted the crucial idea of internal models of set theory.

The conceptual value of large cardinals which Zermelo had emphasized so strongly, was later also acknowledged and shared by Gödel. In his treatise on Cantor's continuum hypothesis he states with explicit reference to Zermelo's "meta-axiom" of the existence of strongly inaccessible cardinals ([Goe47], 520):

> These axioms show clearly, not only that the axiomatic system of set theory as known today is incomplete, but also that it can be supplemented without arbitrariness by new axioms which are only the natural continuation of the series of those set up so far.

Altogether, the *Grenzzahlen* paper deserves to be the last word Zermelo published on set theory *sui generis* and, certainly, it fulfils the expectations he had of it, not with respect to his hope to overcome first-order set theory, but in a sense that could have satisfied him in a deeper way, namely to provide a view of the set-theoretic universe now so familiar and convincing that one can hardly imagine an alternative.

4.8 The Skolem Controversy

In his definiteness paper ([Zer29a]) Zermelo gave a version of definiteness that essentially corresponds to second-order definability. In essence, the axiom of separation gained the following form:

---

[199]The discussion here does not include the so-called reflection principles. For the root of their role in this context see [Levy60], and for a detailed discussion of reflection in connection with the *Grenzzahlen* paper see [Tai98].

Thoralf Skolem at the time of the controversy
Courtesy of Jens Erik Fenstad

For each second-order formula $\varphi(z, x, x_1, \ldots, x_n)$ and any sets $a, a_1, \ldots, a_n$ there exists the set $\{b \in a \mid \varphi(b, a, a_1, \ldots, a_n)\}$.

The paper had probably been written without knowledge of Skolem's 1922 Helsinki address ([Sko23]) which contains a similar definition with first-order definable properties instead of second-order ones. Skolem justifies his choice for two reasons: Firstly, it replaces Zermelo's vague and unsatisfying notion of definiteness from 1908 by a precise version, and secondly, it suffices for all set-theoretic arguments.[200]

Let us recall (cf. 4.5) that with first-order separation Zermelo's axiom system becomes a countable set of first-order sentences and, hence, subject to applications of first-order model theory. Leopold Löwenheim showed for single first-order sentences that satisfiability yields satisfiability over a countable domain ([Loe15]). Skolem proved a generalization to countable sets of first-order sentences, the well-known Löwenheim-Skolem theorem ([Sko20]): Every countable set of first-order sentences which has a model, has a countable one. Hence, assuming that the (first-order) Zermelo axiom system has a model, it has a countable one. It is therefore not strong enough to fix the intended

---

[200]See also [Sko58] for this point.

set-theoretic universe – a consequence that roused Zermelo's sharp opposition
and would strongly influence his further foundational research.

4.8.1 The Löwenheim-Skolem Theorem

The Löwenheim-Skolem theorem suggests itself to those having some famil-
iarity with first-order languages. In order to support this assessment, we shall
give the (simple) proof for a special case that already exhibits the main idea.
Consider countably many first-order sentences

$$\forall x \exists y \varphi_0(x, y), \ \forall x \exists y \varphi_1(x, y), \ldots$$

with quantifier-free $\varphi_i(x, y)$ containing only relational symbols (and no symbols for elements and functions) and being true in the structure $\mathcal{A}$ with domain
A. Then one can obtain a countable model as follows: one starts with a countable non-empty subset A_0 of A and closes it in countably many steps under
the existence of elements as required by the existential quantifiers. To do so,
one successively constructs an ascending chain

$$A_0 \subseteq A_1 \subseteq A_2 \subseteq \ldots$$

of *countable* subsets of A such that for $n = 0, 1, 2, \ldots$ one has:

> For all $a \in A_n$ and all i, if $\exists y \varphi_i(a, y)$ is true in $\mathcal{A}$,
> then there is a $b \in A_{n+1}$ such that $\varphi_i(a, b)$ is true in $\mathcal{A}$.

The union of the substructures of $\mathcal{A}$ with domains $A_0, A_1, \ldots$ is a countable
model of $\forall x \exists y \varphi_0(x, y), \ \forall x \exists y \varphi_1(x, y), \ldots$.

As Skolem shows, the general case can be reduced to a case similar to that
just considered using likewise simple arguments.[201]

4.8.2 Skolem's Paradox

Coming back to set theory and assuming that the first-order Zermelo axiom
system Z has a model, the Löwenheim-Skolem theorem implies that Z has a
countable model. It was a particular consequence of this result that caused
troubles in foundational circles: Let φ_{unc} be a sentence in the first-order language of set theory saying that there exists an uncountable set, and let $\mathcal{M}$
be a countable model of Z. As φ_{unc} is provable from Z and as $\mathcal{M}$ satisfies all
consequences of Z, it satisfies φ_{unc}. We thus have: If there is any model of
Zermelo set theory at all, then there is a *countable* model $\mathcal{M}$ that contains an
uncountable set. The phenomenon became known as "Skolem's paradox."

Skolem treated the paradox in a precise and definitive manner. He clearly
realized that the meaning of a sentence or a property may depend on the

---

[201] For details cf. [Ebb06b].

structure it refers to or, in other words, need not be *absolute* under the change of models. For example, being the largest city is false for New York if we refer to the structure consisting of the cities of the two Americas together with the ordering according to size, but becomes true if we refer to the United States.

In the present case, let a be a set in $\mathcal{M}$ witnessing φ_{unc} in $\mathcal{M}$. To concentrate on the main aspect assume that the element relation in $\mathcal{M}$ is the "real" one and that also the natural numbers in $\mathcal{M}$ are the "real" ones, given as the finite von Neumann ordinals. As $\mathcal{M}$ is countable, so is a, i.e., there is a one-to-one function g which maps a into the set ω of natural numbers. How can this be compatible with the uncountability of a in $\mathcal{M}$? Going back to the definition of uncountability, φ_{unc} may be formulated as a first-order sentence of set theory saying

There is an x such that there is no mapping f which is an injection from x into the set ω of natural numbers.

Hence, the uncountability of a in $\mathcal{M}$ simply means that there is no f in $\mathcal{M}$ which is a one-to-one function in the sense of $\mathcal{M}$[202] mapping a into the set ω of natural numbers. Hence, no function g witnessing the countability of a will be an element of $\mathcal{M}$ witnessing countability of a in $\mathcal{M}$. Thus, the paradox vanishes. In particular, the argument shows that countability (or being of a certain cardinality) in a model of set theory depends on the functions available in it.

Skolem's conclusion was that the notions of set theory are *relative* to the universe of sets under consideration ([Sko23], 224):

The axiomatic founding of set theory leads to a relativity of set theoretic notions, and this relativity is inseparably connected with every systematic axiomatization.[203]

Later Skolem even strengthened his opinion by what can be regarded the essential motto of the so-called "Skolem relativism" ([Sko29], 48):

There is no possibility of introducing something absolutely uncountable, but by a pure dogma.

Because of this relativistic attitude he avoided traditional set theory and became negligent of problems concerned with semantical notions. He thus may have been prevented from recognizing that his paper of 1923 ([Sko23]) contains a nearly full proof of the first-order completeness theorem – six years before Gödel obtained it.

---

[202] That is, an element obeying in $\mathcal{M}$ the (first-order) definition of a one-to-one function.

[203] Strictly speaking, he proved this only for first-order axiomatizations; however, according to [Hen50] the argument for the first-order Löwenheim-Skolem theorem can be generalized to higher-order logics, if one also relativizes the scopes of higher-order quantifiers.

The reaction to Skolem's results was split. For example, Fraenkel ([Fra27], 57) was not sure about the correctness of the proof of the Löwenheim-Skolem theorem, and he seems to have had difficulties in analysing the role of logic with sufficient rigour to understand Skolem's paradox, say, along the clarification sketched above.[204] On the other hand, von Neumann instantly recognized the importance of the results, but reacted with scepticism about the possibility of overcoming the weakness of axiomatizations they reveal ([Neu25], 240):

> At present we can do no more than note that we have one more reason here to entertain reservations about set theory and that for the time being no way of rehabilitating this theory is known.

4.8.3 A "War" Against Skolem

What about Zermelo? When faced with the existence of countable models of first-order set theory, his first reaction was not the natural one, namely to check Skolem's proof and evaluate the result – it was immediate rejection. Apparently, the motivation of ensuring "the valuable parts of set theory" which had led his axiomatizations from 1908 and from the *Grenzzahlen* paper had not only meant allowing the deduction of important set-theoretic facts, but had included the goal of describing adequately the set-theoretic universe with its variety of infinite cardinalities. Now it was clear that Skolem's system, like perhaps his own, failed in this respect. Moreover, Skolem's method together with the epistemological consequences Skolem had drawn from his results, could mean a real danger for mathematics like that caused by the intuitionists: In the Warsaw notes W3 he had clearly stated that "true mathematics is indispensably based on the assumption of infinite domains," among these domains, for instance, the uncountable continuum of the real numbers. Hence, following Skolem, "already the problem of the power of the continuum loses its true meaning" ([Zer30d], 5).

Henceforth Zermelo's foundational work centred around the aim of overcoming Skolem's relativism and providing a framework in which to treat set theory and mathematics adequately. Baer[205] speaks of a real *war* Zermelo had started, wishing him "well-being and victory and fat booty,"[206] at the same time pleading for peace:

> Fraenkel, von Neumann, etc.[207] have an axiom system that, because of their version of "definite," allows for "absolutely" countable models. Actually, this is no disaster; for the usual set-theoretic conclusions are not hurt by this fact.

---

[204] [Fra19], 3rd edition 1928, 333.
[205] In a letter to Zermelo of 27 May 1930; UAF, C 129/4.
[206] "Heil und Sieg und fette Beute," a slogan of the German Youth Movement.
[207] Of course, also Skolem!

However, peace was not to come. In fact, many harsh formulations such as the following one ([Zer32a], 85) give a vivid impression of Zermelo's uncompromising engagement, at the same time also revealing his worries:

> It is well known that inconsistent premises can prove anything one wants; however, even the strangest consequences that Skolem and others have drawn from their basic assumption, for instance the relativity of the notion of subset or equicardinality, still seem to be insufficient to raise doubts about a doctrine that, for various people, already has won the power of a *dogma* that is beyond all criticism.

Details of Zermelo's immediate reaction to Skolem become apparent from letters which Reinhold Baer wrote to Zermelo in 1930/1931.[208] They reveal that in the beginning Zermelo had doubts about the correctness of Skolem's proof. However, apparently he did not check its technical details. Maybe the insufficiency of first-order logic which Skolem had discovered, prevented him from a thorough examination. Instead, he tried right away to disprove the existence of countable models. Already on 4 December 1921,[209] in connection with Fraenkel's way of making definiteness precise, he had asked questions that anticipate the pattern of his method:

> Isn't it possible that your restriction [of definiteness] is too *narrow*? For in each case it would yield only *countably* many subsets, wouldn't it? Wouldn't it, thus, lead to the (countable!) Weyl-Brouwer pseudo-continuum? This question seems worth a more detailed discussion. But maybe the uncountability can be obtained from a combination with *other* axioms. (OV 4.26)

Starting with a countable model $\mathcal{M}$, he invoked various methods to obtain a contradiction by constructing sequences of subsets of ω in $\mathcal{M}$ of length uncountable in $\mathcal{M}$. As Baer immediately recognized, he had to fail because of the non-absoluteness of uncountability.[210] In a letter from 2 June 1930 Baer may have convinced him at least in principle that Skolem's results are technically correct, because his later letters do not come back to this point. Baer writes:

> I [do not] believe that your objection to Skolem's proof can be elaborated in such a way that one might be successful in finding a gap; for as far as I know the proofs for the existence of absolutely countable models of set theory, difficulties always stem from the axiom of separation, whereas the axiom of choice and the axiom of replacement are harmless as for any given set they only claim the existence of a single set with a certain property; hence, in general they do not concern the cardinality of the model.

---

[208] UAF, C 129/4. Zermelo's part of the correspondence is lost up to [Zer31b]. The letters concern mainly the edition of Cantor's collected papers, but also address Skolem's work. The parts quoted here are published in [Ebb04], 83–85.

[209] Postcard to Fraenkel, provided by Benjamin S. Fraenkel.

[210] Cf. [Ebb04] for details. – Zermelo had to fail also because the set-theoretic principles which he used can be formulated in the first-order language of set theory.

> By the way, despite all my endeavour, I have not been able to find a sound
> objection to the applicability of the surely true theorem of Löwenheim to
> the usual axioms of set theory.

Besides his endeavour to refute the existence of countable models of set theory by providing a concrete inconsistency, Zermelo proceeded more systematically. The tone of the resulting papers – both the published and the unpublished ones – is harsh. Perhaps they were influenced by a certain sharpness
in Skolem's criticism ([Sko30a]). However, such an explanation may fall too
short: Zermelo regarded Skolem's position as a real danger for mathematics
and, therefore, saw "a particular duty" ([Zer31c]) to fight against it; hence,
from his point of view, the "war" was by no means a war of aggression, but
was a defensive one.

4.8.4 Modes of Defence

There are at least two strategies of defence: a systematic introduction of infinitary languages and a conceptual reformulation of set theory.

Infinitary languages. As mentioned already, Zermelo may have been sure that
the weakness of finitary languages which Skolem had exhibited also applied
to his axiom system of 1929 with second-order separation and to the axiom
system of the *Grenzzahlen* paper. So he had to take into account that these
systems did not adequately describe the set-theoretic universe as well and that
he was in need of stronger means of formulating set-theoretic principles. His
remedy consisted of infinitary languages. When faced with Skolem's results,
their indispensability all of a sudden became obvious to him. Skolem had
considered such a possibility, too, but had discarded it because of a vicious
circle ([Sko23], 224):

> In order to get something absolutely uncountable either the axioms them-
> selves would have to be present in an absolutely uncountably infinite number
> or one would have to have an axiom which could provide an absolutely un-
> countable set of first-order sentences. However, in all cases this leads to a
> circular introduction of higher infinities, that means, *on an axiomatic basis
> higher infinities exist only in a relative sense.*

Similar arguments were put forward by Zermelo himself on various occasions.
For instance, in the Warsaw notes W4 he had ruled out the possibility of
a consistency proof for arithmetic, as such a proof would merely reduce a
statement about an infinite domain – that of numbers – to a statement about
another infinite domain – that of propositions. Nevertheless, he started the
enterprise to develop infinitary languages and an infinitary logic (cf. 4.9).
The addressees *expressis verbis* will always be Skolem and, having published
his incompleteness theorems, Gödel as well because of his likewise finitary
approach.

A conceptual reformulation of set theory. When Zermelo heard about Skolem's results, his *Grenzzahlen* paper was in the process of publication. In the expanded version of the cumulative hierarchy developed therein, the Cantorian universe of sets is deprived of its distinguished role; it is substituted by a boundless sequence of *normal domains*, set-theoretic universes ordered by inclusion and coming to existence as those levels of the cumulative hierarchy which belong to strongly inaccessible cardinals. Having been confronted with Skolem's results, Zermelo instantly tried to use these ideas as a tool against Skolem's first-order approach which "relativized the notions of 'subset' and 'cardinality', thereby contradicting "the true spirit of set theory" ([Zer30d], 4–5). In a letter from 25 May 1930 to "a colleague" in Hamburg[211] he shows confidence in a conceptual success:

> By "relativizing" the notion of set in this way [namely by allowing non-sets of one normal domain to be true sets in the next one], I feel able to refute Skolem's "relativism" that would like to represent the whole of set theory in a *countable* model. It is simply *impossible* to give *all* sets in a constructive way [...] and any theory, founded on this assumption, would by no means be a theory of sets. (OV 4.27)

Zermelo kept working in this direction. Finally, his belief in the necessity of full power sets which had accompanied his conception of the cumulative hierarchy, led him to a (wrong) refutation of the existence of countable models of set theory. The *Nachlass* contains a note from as late as 4 October 1937 entitled "Relativism in Set Theory and the So-Called Theorem of Skolem."[212] Neglecting the non-absoluteness of the power set operation, Zermelo shows therein that in a countable model of set theory the power of the continuum has to be finite or absolutely uncountable. The latter alternative being impossible, it has to be finite.

In order to illustrate how far Zermelo is away from treating models of set theory adequately, we describe his argument in detail. Let $\mathcal{M}$ be a countable model of set theory.[213] Furthermore, let N be the set of natural numbers (in $\mathcal{M}$ and in the standard model), $\mathcal{C}^{\mathcal{M}}$ its power set in $\mathcal{M}$ (Zermelo's "lean continuum") and $\mathcal{C}$ its power set in the standard model (Zermelo's "fatter continuum"). Zermelo argues in the standard model. He assumes implicitly that $\mathcal{C}^{\mathcal{M}}$ is closed under set-theoretic differences and explicitly that it is closed under *arbitrary* intersections and unions, i.e.,

(∗) for any *standard* set $S \subseteq \mathcal{C}^{\mathcal{M}}$, $\bigcap S \in \mathcal{C}^{\mathcal{M}}$ and $\bigcup S \in \mathcal{C}^{\mathcal{M}}$.

It is here that he neglects relativization to $\mathcal{M}$. He defines

$$N_a := \bigcap \left\{ s \in \mathcal{C}^{\mathcal{M}} \mid a \in s \right\} \text{ for } a \in N$$

[211]UAF, C 129/279. Probably the addressee is Emil Artin.
[212]UAF, C 129/217; published in [vanDE00], 155–156.
[213]Zermelo treats $\mathcal{M}$ as a transitive submodel of the standard model.

and states that the set $\mathcal{T} := \{N_a \mid a \in N\}$ is a subset of $\mathcal{C}^{\mathcal{M}}$ and a partition of N. With $\sim$ meaning equivalence and $P(\mathcal{T})$ the power set of $\mathcal{T}$, he then obtains

$$P(\mathcal{T}) \sim \mathcal{C}^{\mathcal{M}},$$

where the bijection is defined as to map any $S \subseteq \mathcal{T}$ to $\bigcup S$.[214] If $\mathcal{T}$ is finite, $\mathcal{C}^{\mathcal{M}}$ is finite. Otherwise it is routine to show that $P(\mathcal{T}) \sim \mathcal{C}$ and, hence, $\mathcal{C}^{\mathcal{M}} \sim \mathcal{C}$.[215] So, altogether, the lean continuum is finite or its cardinality equals that of the fatter continuum $\mathcal{C}$. It, hence, is finite.[216]

Having obtained the apparent contradiction, Zermelo sees Skolem defeated, and his feelings do not spare intuitionism, formalism, and Hilbert's proof theory either:

> "Skolem's theorem" thus leads to the interesting consequence that *infinite* sets can be realised in finite models – a consequence that would not be more paradoxical than several other consequences already obtained from this nice theorem. In this way also the ideal aim of intuitionism, the abolition of the infinite from mathematics, would be brought nearer to realization together with the goal of "formalism" which, as we know, strives for the proof of consistency. For, as we know, one can prove *anything* from absurd premises. Hence also the consistency of an arbitrary system of sentences.

With his refusal to think in terms of models of set theory and to acquire the technical ability to treat questions of absoluteness, Zermelo was not alone ([vanDE00], 157):

> It took some time before the insight that set theory allowed a rich variety of models became commonplace. The next step, how to exploit this insight in practice, took even longer. It was only a small group that appreciated the possibilities of the model-theoretic aspects of set theory; Skolem, von Neumann, and Gödel led the way in this respect.

Finally the first-order Zermelo-Fraenkel axiom system became the dominating axiom system underlying modern set theory and the handling of models of set theory one of the major driving forces. It is a tragic coincidence that Zermelo, although having contributed crucial model-theoretic features in the *Grenzzahlen* paper, completely missed this development.

4.9 Infinitary Languages and Infinitary Logic

4.9.1 The Infinity Theses

Zermelo's *Nachlass* contains a one-page typewritten manuscript ([Zer21]), dated "17 July 1921" and entitled "Theses About the Infinite in Mathematics"

---

[214]Note that, by (∗), $\bigcup S \in \mathcal{C}^{\mathcal{M}}$.

[215]Strangely enough, Zermelo does not use that $\{a\} \in \mathcal{C}^{\mathcal{M}}$ and, hence, $N_a = \{a\}$ for all $a \in N$. Under this additional assumption, (∗) yields immediately that $\mathcal{C}^{\mathcal{M}} = \mathcal{C}$.

[216]See [vanDE00] for further details.

which can be considered the essence of and a clue to his infinitary conceptions as they developed in the early thirties. The theses are:[217]

(1) Each genuine mathematical proposition has an "infinitary" character, i.e. it refers to an *infinite* domain and has to be regarded as a comprehension of infinitely many "elementary propositions."

(2) The infinite is neither physically nor psychologically given to us in reality, it has to be comprehended and "posited" as an idea in the Platonic sense.

(3) Since infinitary propositions can never be derived from finitary ones, the "axioms" of all mathematical theories have to be infinitary as well, and the "consistency" of such a theory can only be proved by exhibiting a corresponding consistent system of infinitely many elementary propositions.

(4) By its nature, traditional "Aristotelian" logic is finitary and, hence, not suited for a foundation of the mathematical sciences. Therefore, there is a necessity for an extended "infinitary" or "Platonic" logic which rests on some kind of infinitary "intuition" – as, e.g., with the question of the axiom of choice –, but which, paradoxically, is rejected by the "intuitionists" out of habit.

(5) Every mathematical proposition has to be conceived as a combination of (infinitely many) elementary propositions, the "basic relations," via conjunction, disjunction, and negation, and each deduction of a proposition from other propositions, in particular each "proof," is nothing but a "regrouping" of the underlying elementary propositions.

Of course, taken literally, the pithy language leaves several open questions. Heinrich Scholz sent Zermelo a list of such questions in 1942;[218] they concern, for example, the nature of consistency proofs for infinitary axiom systems and the meaning of proofs as regroupings of the underlying elementary propositions. The context shows that Zermelo had sent the theses to Scholz sometime between March and August 1942.

It is not clear when the theses were conceived.[219] If they were really written in 1921, they would anticipate in an astonishingly far reaching way the infinitary programme that Zermelo started around 1930, "a theory of axiomatic systems [...] according to which truly adequate formulations of mathematical theories are infinitary in almost every respect" ([Tay02], 479). If they were written considerably later, they could be considered Zermelo's infinitary

---

[217]Original version under OV 4.28.

[218]Letter of 3 October 1942; UAF, C 129/106.

[219]Above the year 1921 given on the manuscript, there is the figure "42" written in pencil. Furthermore, the *Nachlass* contains a handwritten version of the theses among notes on loose sheets of a writing pad (UAF, C 129/238), two notes being dated 13 June 1926 and 3 February 1937, respectively. The version is less detailed, and thesis (5) is part of thesis (3). Thesis (1) appears in an almost identical formulation in a letter to an unknown addressee written in the summer of 1932 (UAF, C 129/268).

legacy. For his work on infinitary languages and infinitary logics as it is documented in his last publication, the *Satzsysteme* paper ([Zer35]), had not been completed there; in the last paragraph it says (ibid., 146):

> The preceding representation is only the *beginning* of a still unfinished investigation which aims at establishing an "infinitistic" truly mathematical syllogistic and proof theory.

Faced with this uncertainty, we will view the theses as a leitmotiv and a description of the final goal Zermelo intended to realize.

4.9.2 Infinitary Languages

Indications of Zermelo's infinitary convictions can be found in the Warsaw notes ([Zer29b]). They aim at both, the significance of infinite structures for mathematics and the infinitistic character of mathematical propositions. The first point is discussed in detail. In the second point Zermelo does not go as far as the infinity theses suggest. In particular, he does not touch the necessity of an infinitary logic. There is only a single remark touching infinitary languages, a remark saying that mathematical disciplines deal with propositions "which comprehend infinite multiplicities of simple propositions," recalling thesis (5). It was only after his encounter with Skolem that he became convinced that infinitary logical constructs offered a promising way to overcome the deficiencies of finitary formalisms.[220]

In 1931 Zermelo started to develop a corresponding theory. His infinitary languages are bound to a domain D together with relations and functions over D, i. e., to a mathematical structure with domain D. Their propositions are built up from so-called "basic relations" ("Grundrelationen").[221] In the case of the real numbers with the usual ordering $<$ and usual addition $+$, these basic relations are of the form $s < r$ or $r + s = t$ with real numbers r, s, t. Zermelo allows different truth values to be attributed to basic relations, speaking of "distributions of truth."[222] Hence, basic relations are treated as propositional variables or constants.[223]

Now, given a mathematical structure, the propositions of the language coming with it are the basic relations together with those propositions that can be obtained by forming the negation $\neg\varphi$ of a proposition φ or the conjunction $\bigwedge \Phi$ or the disjunction $\bigvee \Phi$ of a *possibly infinite* set Φ of propositions.[224] The

---

[220]Cf. also the discussion in [Tay02], 504seq.

[221]In the infinity theses Zermelo speaks of elementary propositions.

[222]"Wahrheitsverteilungen;" cf. [Zer32a], 86, and [Zer35], 142.

[223]In another context (cf. [Zer35], §5) they are considered as first-order atomic propositions where the elements $d \in D$ serve as first-order variables or constants. The assignments admitted are just the permutations of D. (We shall speak of this restriction as *domain relatedness*.) Obviously, there is no clear distinction between syntax and semantics.

[224]Zermelo writes $\overline{\varphi}$, $\mathfrak{K}(\Phi)$, and $\mathfrak{D}(\Phi)$ for $\neg\varphi$, $\bigwedge \Phi$, and $\bigvee \Phi$, respectively.

propositional combinations can be repeated along the ordinal numbers. For a distribution of truth, the meaning of a proposition is fixed as follows: basic relations are either true or false as determined by the distribution, and the propositional connectives have their usual meaning.

Zermelo does not admit quantifiers. However, he speaks of the formation of conjunctions and disjunctions as "quantifications" ([Zer32a], 86). This may indicate that quantifications are tacitly present via infinitary descriptions.[225] For example, in the case of the real numbers the proposition $\exists x \, (x < s)$ can be represented as $\bigvee \{ t < s \mid t$ a real number$\}$, which means, the disjunction over all propositions of the form $t < s$ for t a real number; analogously, the proposition

$$\bigwedge \{ \bigvee \{ t < s \mid s \text{ a real number} \} \mid t \text{ a real number} \}$$

means that for any real number, say t, there is a bigger one, say s.

Of course, this interpretation of quantifiers is bound to the underlying structure. Without a given structure even a simple first-order proposition such as $\exists x \exists y \, (x < y)$ has no counterpart among Zermelo's infinitary propositions.

The definition of quantifiers as (possibly infinitely long) conjunctions or disjunctions is not new. It can be found already with Charles Sanders Peirce ([Pei85]) and Ernst Schröder ([Schr95a]) and was taken up by Hilbert as early as 1904 ([Hil05a]).[226] Later, in connection with his proof theory, Hilbert argued that it is just this aspect of the quantifiers by which the transfinite enters into logic, bringing with it the danger of antinomies:[227]

If we were always and without concern to apply to infinite sets the processes admissible for finite sets, then we would open the gates to error. This is the same source of error which we know well in analysis. There the theorems valid for finite sums and products can be transformed to the case of infinite sums and products only if the inference is secured by a special examination of convergence. Analogously, we must not treat the infinite logical sums and products, $A(1) \wedge A(2) \wedge A(3) \wedge \ldots$ and $A(1) \vee A(2) \vee A(3) \vee \ldots$, in the same way as finite ones.

Hilbert went on to replace quantifiers with their inherent use of infinitely long formulas by a special construct, the so-called "ϵ-operator" which allows a quantifier-free formulation of mathematics.[228] He thus provided a basic tool for a finitary treatment of mathematics which became the heart of his *Beweistheorie* as it is systematically represented in the *Grundlagen der Mathematik* ([HilB34]).

Even if he had had in mind the infinitary descriptions of quantifiers given above, Zermelo was no longer in Hilbert's tradition: Whereas Hilbert tried to

---

[225]These descriptions are at hand via domain relatedness; cf. fn. 223.

[226]See [Moo97] for details.

[227][Hil23], 155; [Hil96c], 1140. See also [Hil26], 173, English translation [vanH67], 378.

[228]This reformulation makes use of the axiom of choice.

eliminate the infinitary character of quantifiers by special finitary constructs in order to facilitate consistency proofs, Zermelo emphasized the necessity of infinitary languages as the only means of adequately treating the infinite in mathematics. Besides the deviation with respect to the possibility of consistency proofs which had become apparent in his Warsaw talks and is visible also in the third thesis, there now is an additional difference to Hilbert, the deviating view about the role of infinitary languages.

4.9.3 Infinitary Logic

According to the fourth infinity thesis, Zermelo also started to develop infinitary logic, i. e., a logic dealing with infinitary propositions and infinitely long proofs. To provide a rough description[229] we shall first consider a special case. Given an infinitary language and a distribution of truth, one can easily show that any proposition gets a definite truth value. For a proposition φ the inductive argument runs over the subproposition relation inside the set $\mathsf{Sp}(\varphi)$ of subpropositions of φ. In Zermelo's terminology this argument "proves" φ relative to the given distribution of truth, and $\mathsf{Sp}(\varphi)$ is considered the respective proof.

In the general case assume that the proposition φ follows from the proposition ψ in the usual sense, i. e., for any distribution of truth which renders ψ true, φ is also true. Then the set $S = \mathsf{Sp}(\varphi) \cup \mathsf{Sp}(\psi)$ contains φ and ψ and has the property that for any distribution ι all propositions from S get a definite truth value, and there is no such ι for which ψ becomes true and φ becomes false. Similarly as above, any set S of propositions with this property is a *proof* showing that φ follows from ψ:

> For a mathematical *"proof"* is nothing other than a *system of propositions which is well-founded by quantification and which cannot be divided into two classes, the "true" ones and the "false" ones, without violating the syllogistic rules and in such a way that the assumptions belong to the first class and the proposition to be proved to the second one.*[230]

As Kanamori formulates ([Kan04], 530), Zermelo proposed

> proofs not as formal deductions from axioms, but as semantic determinations of truth or falsity of a proposition through transfinite induction based on its well-founded construction from [basic relations].

Of course, Zermelo is aware of the fact that in contrast to his "absolutely infinitary" proofs ([Zer35], 144) proofs in everyday mathematics are finite. They are "*defective* means of our finite intellect to at least gradually approach and master the infinite that we cannot directly and 'intuitively' survey or perceive"

---

[229] Following [Zer32a], 86seq., and [Zer35] 143seq., but neglecting minor differences between these papers.
[230] [Zer32a], 87.

([Zer32a], 85) or, according to Taylor ([Tay02], 479), they are "finitary approximations that serve as convenient vehicles of mathematical practice." There nevertheless remained the question to what extent infinitary proofs could be made intelligible to the "finite human intellect." Having exemplified the power of human understanding by the "quite infinitary" proofs via complete induction in the natural numbers, Zermelo expresses hope that "apparently there are no fixed limits of comprehensibility" ([Zer35], 144–145).

4.9.4 Infinitary Languages and the Cumulative Hierarchy

When submitting the title for a talk on his infinitary theory to the programme committee of the 1931 meeting of the Deutsche Mathematikervereinigung at the small Saxon health resort of Bad Elster,[231] Zermelo speaks of "an inner relationship between mathematical logic and set theory." He thereby refers to a certain parallelism between his infinitary languages and the cumulative hierarchy which we have addressed briefly already in 4.7.4.[232]

For exemplification, let D be the domain of a mathematical structure and E the corresponding set of basic relations. Then there are two hierarchies: There is the cumulative hierarchy starting with D, where for any ordinal number α the level $V_{\alpha+1}$ is the power set of V_α, and for a limit number β, V_β is the union of the preceding levels. Similarly, the infinitary propositions over E can be arranged in a cumulative hierarchy. It starts with $P_0 := E$; for any ordinal number α, the level $P_{\alpha+1}$ results from P_α by adding all negations, conjunctions, and disjunctions of elements of P_α, i.e.,

$$P_{\alpha+1} = P_\alpha \cup \{\neg\varphi \mid \varphi \in P_\alpha\} \cup \{\bigwedge \Phi \mid \Phi \subseteq P_\alpha\} \cup \{\bigvee \Phi \mid \Phi \subseteq P_\alpha\};$$

finally, for a limit number β, P_β is the union of the preceding levels.

Zermelo is convinced that the hierarchical representation of his infinitary languages provides a general framework to treat a variety of mathematical theories ([Zer35], 142):

> In this general form one can represent any mathematical discipline which is based on a certain domain of urelements together with certain basic relations, such as the arithmetic of rational or algebraic numbers and functions, the analysis of real or complex numbers and functions, or likewise the geometry of any space of a given dimension, and also the set theory of "a given normal domain."[233]

The parallelism between the cumulative hierarchy rising above the domain of objects of a specific mathematical theory, and the hierarchy of infinitary propositions rising above the related domain of basic relations, may have

---

[231]Letter to Ludwig Bieberbach of 13 July 1931; UAF, C 129/124.

[232]The parallelism is discussed in more detail in [Tay04].

[233]*Expressis verbis*, but without further explanation, Zermelo excludes theories dealing with several domains such as the general theory of fields.

strengthened his belief in the adequacy of the infinitary languages he had conceived.

The hierarchical structure of both, sets and propositions, is connected with well-foundedness. In fact, a large preparatory part of the *Satzsysteme* paper is dedicated to the study of well-founded relations and provides the methodological framework for inductive definitions and proofs. As Zermelo had already pointed out in his *Grenzzahlen* paper, the corresponding role of well-foundedness is intimately linked to the axiom of foundation. Hence, the similarity between the hierarchy of sets and the hierarchy of propositions is more than a similarity of their definitions – both hierarchies are related by the essential role of well-foundedness as guaranteed by the axiom of foundation. In his own words,[234] the axiom of foundation fulfils the double role

> of fully characterizing the domains of sets that are really useful in mathematics [and] of forming the most natural and safest foundation for a general "infinitistic" and, hence, truly mathematical syllogistics for the systems of propositions. (OV 4.29)

4.9.5 The Failure

Gödel had also announced a talk for the Bad Elster meeting, namely on his new incompleteness results. Zermelo planned to use this coincidence as a forum for his criticism of both, intuitionism and formalism, in general and the finitary point of view as shared by Skolem and Gödel in particular. As he was sure that his infinitary conceptions would cause scepticism and resistance, he tried to get help from people who were already acquainted with his thoughts. In a letter obviously written some weeks before the Bad Elster meeting[235] we read:

> Will you really stay in your summer resort during the entire vacation or would you not prefer to come to Bad Elster for the conference? It would be very important to me to have one or the other in the audience who has read (and understood!) my Fundamenta paper.[236] For I am sure that with my "infinitary logic" I will meet dissent from all sides: neither the "intuitionists" (of course, they are enemies of logic anyway) nor the formalists nor the Russellians will accept it. I only count on the young generation; their attitude towards such things is less prejudiced. After all also the opponents of the "principle of choice" and of the well-ordering have not been proven wrong; but both principles have gradually gained acceptance with the new generation by their natural powers. (OV 4.30)

---

[234]From a one-page manuscript "Fundierende Relationen und wohlgeschichtete Bereiche" ("Well-founding Relations and Well-Layered Domains"), without date, presumably around 1932; UAF, C 129/217. For further details see also [Tay02], 493seq.

[235]Letter without first page to an unknown recipient, perhaps to Reinhold Baer; UAF, C 129/268.

[236]The *Grenzzahlen* paper [Zer30a].

The plans for an extended discussion failed (4.10.1). Even more: Zermelo's hope that in the long run his infinitary logic would gain acceptance, failed as well. Things went a different way. It was Gödel's finitary approach to mathematical theories together with Skolem's first-order approach to set theory which shaped the disciplin of mathematical logic in the 1930s. And it was the young generation as represented by the 25-year old Gödel and the 28-year old von Neumann that promoted this development.

A systematic investigation of infinitary languages started only in the 1950s, finding its first comprehensive presentation by Carol Karp ([Kar64]).[237] The new theory enriched several branches of mathematical logic, some in a way that would have been interesting for Zermelo. From the model-theoretic point of view the restriction to finitary first-order languages means "a severe limitation" ([Kei77], 95) for those parts of mathematics which are outside the scope of first-order axiom systems. As it turned out, infinitary languages provide a promising methodological enrichment in this respect, a fact that suits well with Zermelo's mathematically oriented view. Infinitary *logic* was promoted particularly by the work of Jon Barwise ([Bar67]). It opened new insights into finitary first-order logic, but also interesting connections to set theory ([Bar75]), connections which Zermelo might also have appreciated.

Zermelo did not pave the way for this development. He did not succeed in providing convincing applications of his infinitary languages, for example, by characterizing important mathematical structures or by using his own parallelism between logic and set theory. There are several reasons. On the one hand, his concepts were still far away from mathematical practice; on the other hand and measured against the degree of precision which had been reached meanwhile, they were too vague. So apart from promoting the role of well-foundedness, the publications based on the Bad Elster talk[238] "vanished with hardly a trace" ([Moo97], 117).

Sometimes[239] Zermelo's infinitary languages are identified with the propositional part of the infinitary languages $L_{\infty\omega}$.[240] The latter admit an analogue of the Löwenheim-Skolem theorem.[241] In particular, they cannot fix the relation of equicardinality, one of the key examples in Zermelo's fight against Skolem.[242] Furthermore, according to Edgar Lopez-Escobar ([Lop66]), they are unable to characterize the class of well-orderings. Hence, with this inter-

---

[237]Cf. [Moo97] for the history and, for example, [Dic85] and [Nad85] for the theory.

[238]I. e., [Zer32a], [Zer32b], and [Zer35].

[239]For instance, by Moore (ibid.).

[240]These languages are built up from atomic propositions over a given set of relation symbols, function symbols, and constants like the corresponding first-order languages, but allowing conjunctions and disjunctions of arbitrary length.

[241]Cf. [Dic85], 338seq., or [Ebb06b], 463.

[242]See, e. g., [Zer32a], 85.

pretation Zermelo's main intention is not realized for central notions of set theory.[243]

4.10 The Gödel Controversy

Gödel's celebrated first incompleteness theorem of 1930 ([Goe31a]) marks a turning point in the axiomatic method as it had been paradigmatically shaped by Hilbert's *Grundlagen der Geometrie* ([Hil1899]): A consistent axiom system that is strong enough to capture a certain amount of the arithmetic of natural numbers is incomplete in the sense that there is a proposition in the system which can neither be proved nor refuted by the given axioms. The only further assumption is that the axiom system be given in an explicit form, in modern terms, that it be enumerable, i. e. can be listed by an appropriately programmed computer. All axiom systems that occur in mathematics are of this form, and nearly all of them such as the axiom systems for number theory, analysis, or set theory possess the strength needed in Gödel's theorem and thus turn out to be incomplete. In other words: The established method for recognizing mathematical truth is subject to essential limitations.

When Zermelo heard of this result, his fight against finitary mathematics which up to then had been mainly aimed at Skolem, now also included Gödel. It finally led to a controversy which may be considered a dispute about the future of mathematical logic. For Zermelo it was the third substantial debate. The first one against Ludwig Boltzmann in the late 1890s had seen him stimulating a reconsideration of the foundations of thermodynamics. The second one about the axiom of choice roughly ten years later ended with a full acknowledgement of his point of view and of his achievement in having shaped an essential feature of post-Cantorian set theory. Quite differently, the Gödel controversy was to mark a change in mathematical logic which saw him as a man of the *before*. Kanamori speaks ([Kan04], 531) of a generational transition. As Moore remarks ([Moo02b], 55–56), the situation was mirrored in the differences in age and mode of approaching:

> The old and experienced Zermelo, whose views of logic were mainly formed decades ago, was confronted with the brilliant young Gödel. (OV 4.31)

4.10.1 The Bad Elster Conference

Gödel first announced his incompleteness result during a roundtable discussion which concluded the meeting of the Gesellschaft für empirische Philosophie at Königsberg from 5 to 7 September 1930.[244] The meeting was attended by

---

[243]The identification does not take into account a possible domain relatedness of Zermelo's languages which results in a semantics different from that of $L_{\infty\omega}$. For further reservedness cf. [Tay02], fn. 16.

[244][Daw97], 68seq.; [Wan81b], 654seq.

Kurt Gödel as a young scholar
Kurt Gödel Papers, Princeton University
Courtesy of the Institute for Advanced Study

several logicians, among them Rudolf Carnap, Arend Heyting, and also John von Neumann who immediately and enthusiastically recognized the importance of what Gödel had proved. Zermelo became aware of the result only in the spring of 1931. In a letter of 13 May 1931[245] Reinhold Baer wrote to him:

> The main result of Gödel's paper,[246] about which I informed you in my last letter,[247] can be formulated in a pointed and, hence, not fully correct way as follows: If A is a countable, consistent logical system, for example arithmetic, then there is a proposition in A which is not decidable in A (but may be decidable in extended systems). Hurrah, logicians have also discovered diagonalization! – By the way, you may have a lot of benefit from this "new gentleman;" he is strongly interested in philosophy and the foundations of mathematics, moreover in functional analysis (let him give a talk about this, it is very amusing), summation procedures, etc.

As mentioned before, it was the time when Zermelo had started to work out his theory of infinitary languages in order to overcome the weakness of finitary systems which Skolem had revealed, planning to give a talk about his results

---

[245]UAF, C 129/4, as also the following letters of Baer we quote from. The parts quoted are published in [Ebb04], 85.

[246]Very probably [Goe31a], which had appeared in late March (cf. [Daw97], 74).

[247]The letter is not contained in the Zermelo *Nachlass*.

at the meeting of the Deutsche Mathematikervereinigung at Bad Elster in September. We can imagine that in this situation he regarded Gödel's incompleteness result as further evidence of the inadequacy of finitary approaches to mathematics. Baer's letters show that indeed he doubted its relevance, whereas Baer tried to explain its methodological importance. On 24 May 1931 Baer writes:

> I believe that we have quite the same opinion about Gödel. But actually I think that his work is rather rewarding; for after all it proves that logics of the strength of, say, Principia Mathematica are insufficient as a base for mathematics to the extent we would like to have. In this situation two possible consequences can be drawn:
>
> (1) one surrenders to "classical" logic (as Skolem does) and amputates mathematics;
>
> (2) one declares, as you do, that "classical" logic is insufficient. –
>
> Standpoint (1) is more comfortable; for it gets by on "God's gift" of the natural numbers, whereas standpoint (2) has to swallow the total series of transfinite ordinal numbers, as they are necessary to build up set theory to the extent you have performed[248] and as without this greater "gift of God" one cannot get beyond what Skolem etc. reach – but it is strange that up to now, apart from the continuum problem, the "poorhouse mathematics" of Skolem etc. suffices to represent all that happens in the "mathematics of the rich." In any case, Gödel can be credited for having made clear how far one can get with standpoint (1); it is a matter of faith, whether one now decides in favour of (1) or (2). I suspect that logicians will decide in favour of (1), while mathematicians will rather decide in favour of (2).

Zermelo strongly decided in favour of (2). When he left for Bad Elster, he had made up his mind to fight for his convictions and had taken strategic precautions. After the meeting he reports to Baer ([Zer31b]):

> At Bad Elster I avoided any direct polemic against Gödel both in the lecture itself and afterwards: one should not frighten off enterprizing beginners. I deliberately had Gödel's lecture put on *before* mine and asked that they be discussed *together*. But the sole consequence of my loyalty was that the whole discussion was put *further* back at the unjustified suggestion of Fraenkel (who stabs me in the back at every opportunity!) and then came to naught.

According to Olga Taussky-Todd ([Tau87], 38) the "very irascible" Zermelo and the young Gödel nevertheless had a "peaceful" meeting with intensive discussions during a lunch break – at least after some difficulties, for Zermelo first

> had no wish to meet Gödel. A small group suggested lunch at the top of a nearby hill, which involved a mild climb, with the idea that Zermelo should talk to Gödel. Zermelo did not want to do so and made excuses.

---

[248]In the *Grenzzahlen* paper [Zer30a].

[...] However, when Gödel came to join, the two immediately discussed logic and Zermelo never noticed he had made the climb.

Because of Zermelo's firm intention to fight for his infinitary convictions and against Baer's possible consequence (1), even the most peaceful atmosphere could not lead to a compromise. In fact, immediately after the meeting Zermelo took measures to compensate for the failed public discussion with Gödel ([Zer31b]; cf. 4.10.2):

> Now I am going to do it differently; the gentlemen will *have* to declare their hand finally when I publicly assert that Gödel's much admired "proof" is nonsense.

Besides a plain presentation of his results ([Zer32b]), he prepared an elaboration for the proceedings, the Bad Elster paper ([Zer32a]), where he added a programmatic introduction which describes the motives of his fight (ibid., 85):

> When starting from the assumption that all mathematical notions and the-orems should be representable by a *fixed finite system of signs*, already the arithmetical continuum inevitably leads to the well-known *"paradox of Richard;"* seemingly settled and buried, this paradox has now found a happy resurrection as *Skolemism*, the doctrine according to which *every* mathematical theory, including set theory, can be realized by a *countable* model. [...] However, a sound "metamathematics," a true "logic of the infinite," will only be possible by thoroughly *turning away* from the assumption that I have described above and which I would name the *"finitary prejudice."* Generally speaking, it is *not* the "formations of signs" that – according to the opinion of various people[249] – form the true subject of mathematics, but the *conceptual and ideal relations* between the elements of *infinite varieties* that are set in a conceptual way, and our notational systems are just *defective means* of our *finite* intelligence, varying from case to case, to at least gradually approach and master the infinite that we cannot *directly* and *intuitively* survey or perceive.

He continues that he will present a kind of mathematical logic that, *free* of the finitistic prejudice, will allow the whole of mathematics to be built up without arbitrary prohibitions and limitations.

What does this logic look like? According to the description in the preceding section, it is related to a mathematical structure, say, the domain of the real numbers together with the ordering relation and addition. The propositions arise from the basic relations – such as $2 < 3$ or $2 + 3 = 7$ – by iterating the formation of negations and possibly infinitely long conjunctions and disjunctions. If the basic relations are attributed truth values (say, the natural ones), then every proposition φ becomes either true or false. For a given proposition φ, this can be proved by induction over the subproposition relation. If φ becomes true, the set $\mathsf{Sp}(\varphi)$ of subpropositions of φ, which forms

---

[249]Among them, for sure, also Hilbert.

the frame for the inductive argument, is a proof of φ (relative to the given distribution of truth). In this sense, every proposition φ is decided: As φ becomes either true or false, either $\mathsf{Sp}(\varphi)$ is a proof of φ or $\mathsf{Sp}(\neg\varphi)$ is a proof of $\neg\varphi$. So Zermelo can say ([Zer32a], 87):

> Hence from our point of view each "true" proposition is at the same time also *"provable"* and each proposition also *"decidable."* There are no (objectively) *"undecidable"* propositions.[250]

In essence, a proposition or a proof of it is not a syntactical string of signs, but an ideal (infinitary) object. This is in total contrast to the concepts underlying Gödel's result. With Gödel the finitary representability of the system in question is indispensible. For only then can one code axioms, propositions, and proofs in an effective way by natural numbers, and the presupposed arithmetical power of the system then allows provability to be treated in the system itself. By imitating the paradox of the liar on the syntactical level, thereby replacing truth by provability, one can finally infer the existence of undecidable propositions.

For Zermelo this procedure renders incompleteness a trivial fact: As the set of provable propositions in Gödel's sense is countable and as there are uncountably many true infinitary propositions, there is of course a true proposition which is not provable in Gödel's sense. Hinting also at the fact that undecidable propositions in Gödel's sense may become decidable in an extended system, he concludes ([Zer32a], 87) that Gödel's argumentation can only serve to

> corroborate the deficiency of *any* "finitary" proof theory, yet without providing us with a means to remedy these ills. The real question whether there are absolutely undecidable propositions or absolutely unsolvable problems in mathematics is by no means touched by such relativistic considerations.

4.10.2 The Gödel Correspondence

When conceiving his infinitary languages, Zermelo did not clearly distinguish between a proposition and its meaning, between syntax and semantics. This attitude was widespread among logicians of that time. Grattan-Guinness ([Gra79], 296) reports that according to Barkley Rosser "it was only with Gödel's theorem that logicians realised how careful they needed to be with this matter."

Mixing up a proposition and its denotation, Zermelo believed he had discovered a flaw in Gödel's argument. Immediately after the Bad Elster meeting he informed Gödel ([Zer31c]) about this "essential *gap*," locating the alleged mistake in the "finitary prejudice," namely the "erroneous assumption that

---

[250]If contrasted with the remark from 1908 ([Zer08a], 112; [Zer67b], 187) that "in mathematics [...] not everything can be proved, but every proof in turn presupposes unproved principles," Zermelo's changed views become particularly conspicuous.

every mathematically definable notion be expressible by a 'finite' combination of signs." With the intention of winning Gödel as a comrade, he continues:[251]

> In reality, the situation is quite different, and only after this prejudice has been overcome (a task I have made my particular duty), will a reasonable "metamathematics" be possible. Correctly interpreted, precisely your line of proof would contribute a great deal to this and could thereby render a substantial service to the cause of truth.

Gödel's answer ([Goe31b]) consists of a ten-page letter, giving an extensive description of his method and explaining in detail where Zermelo went wrong with the discovery of the alleged mistake. Moore characterizes this explanation as "patient, how one maybe teaches a schoolboy" ([Moo02b], 60).

In his answer of late October ([Zer31d]), without any reaction to Gödel's explanations, Zermelo mainly explicates the preferences of his infinitary view as compared to what he assumes as Gödel's opinion:

> All that you prove [...] comes down to what also *I* always emphasize, namely that a "finitarily restricted" proof system does *not* suffice to "decide" the propositions of an uncountable mathematical system. Or can you "prove" for instance, that "your" scheme is the "only possible" one? Surely, this can't be done; for what a "proof" really is cannot be "proved" again, but has to be *assumed, presupposed* in some way. And just that is the question: what does one understand by a proof? In general, a proof is understood as a system of propositions that, when accepting the premises, yields the validity of the assertion as being *reasonable*. And there remains only the question of what may be "reasonable." In any case – as you are showing yourself – *not only* the propositions of some finitary scheme that, also in your case, may always be *extended*. So in this respect we are of the same opinion; however, I *a priori* accept a *more general* scheme that *need* not be extended. And in *this* system, really *all* propositions are decidable.

Taking up a remark of Gödel saying that finitary systems do not provide definitions for all, i. e. for *uncountably many*, classes of natural numbers,[252] Zermelo was led to believe that in the end Gödel was also in favour of infinitary systems. The corresponding part of his letter is of a misleading vagueness. Dawson may reflect the interpretation it offers when writing ([Daw97], 77) that Zermelo "continued to be confused about Gödel's results" and "mistakenly thought" that the set of propositions of *Principia Mathematica* "would comprise an uncountable totality." Gödel's impression of technical weakness which had already emerged from Zermelo's first letter and might have led to the scholarly style of his answer, became stronger and, together with Zermelo's sharp tone and uncompromising attitude, added to his decision to end the correspondence.

---

[251]Translation by John W. Dawson ([Daw85], 69).
[252]Ibid., p. 2, fn. 1.

4.10.3 After the Controversy

Despite Zermelo's hurtful style a dispassionate analysis may lead to the conclusion that there was no irreconcilable gulf between him and Gödel. By ways which could have found mutual acknowledgement, both had come to similar results: Gödel had proved in a strict mathematical sense that nontrivial finitary systems are inadequate to catch their richness, and he was open to conceive stronger extensions. Zermelo had gained the insight into the incompleteness of finitary systems on grounds of his infinitary convictions and had conceived an infinitary system that according to his opinion was complete and adequate for mathematics. Kanamori ([Kan04], 535) formulates concisely as follows:

> For Gödel the incompleteness of formal systems is a crucial *mathematical*
> phenomenon to be reckoned with but also to transcend, whereas Zermelo,
> not having appreciated that logic has been submerged into mathematics,
> insisted on an infinitary logic that directly reflected transfinite reasoning.

According to Georg Kreisel ([Kre80], 210) "a man of thought would have used Gödel's theorem to strengthen the case against formal systems." As it is apparent from his first letter, Zermelo in fact tried to win Gödel as a supporter. However, this offer had to fail because of his unconditional insistence as it culminated in the "outburst" ([Kre80], 210) of the Bad Elster paper, and because of the incompetence or superficiality he showed with regard to Gödel's technical procedure. Both reasons may belong together, as the latter one may be rooted in his reluctance to really work through the technical details of a proof that aimed at something sharply contradicting his views. The same pattern can be found in his reaction against Skolem. According to Kreisel ([Kre80], 210) he "simply refused to look at the tainted subjects."

Altogether, "the peaceful meeting between Zermelo and Gödel in Bad Elster was not the start of a scientific friendship between two logicians" ([Tau87], 38). In fact, the events and what followed left both feeling hurt.

Ten years later, in a letter to Paul Bernays ([Zer41]), Zermelo remembered with bitterness what he thought had been a plot between Gödel and Gödel's thesis supervisor Hans Hahn, with whom he had published the encyclopedic contribution on the calculus of variations ([HahZ04]):

> Already at the conference of the Mathematical Union at Bad Elster my talk
> about systems of propositions was excluded from discussion by an intrigue
> of the Vienna School represented by Hahn and Gödel. Since then I have lost
> all interest in giving an address about foundations. Apparently this is the
> fate of everybody who is not backed by a "school" or a clique. (OV 4.53)

Obviously he had also discussed this point with Baer shortly after the meeting. For in a letter from 3 November 1931[253] Baer gave encouragement:

---

[253] UAF, C 129/4.

In his new *History of Logic*[254] H. Scholz very nicely writes that the logicians in Vienna are very pleasant and competent, but that we do not want to obey their dictatorship just as little as any other one.

It may be mentioned here that Rudolf Carnap, one of the leading members of the Vienna Circle, also proposed infinitary logic to overcome Gödel's incompleteness results ([Carn36]).

The young Gödel who unlike Zermelo was not a man of controversies, felt completely misunderstood and attacked by scientifically unjustified arguments. He apparently showed Zermelo's letters to others ([Daw97], 77) and complained about Zermelo's attacks in the Bad Elster paper. On 11 September 1932 he asked Carnap[255] whether he had already read Zermelo's "absurd criticism" in that paper. On 25 September[256] Carnap answered:

> Yes, with astonishment I have read Zermelo's remarks about you in the *Jahresbericht*; despite his astuteness he is sometimes astonishingly wide of the mark.

Coming back to Zermelo, one might assume that his state of health may also have contributed to the style of his reactions. Olga Taussky-Todd reports ([Tau87], 38) that he had suffered a nervous breakdown before the Bad Elster meeting, but had actually recovered in the meantime. In fact, Zermelo's nervous condition seems to have been unstable about that time. Four years later, under personally deplorable circumstances, he would again show signs of nervous tensions.

Eventually, the 1931 debate gave way to calmer reconsiderations. In his elaboration of the Bad Elster paper ([Zer35]) Zermelo avoids any epistemological discussion. The new attitude may have been due to a growing appreciation of Gödel's work. Shen Yuting, a student attending Zermelo's course on set theory in the winter semester 1933/34, recalls ([Wan87], 91):

> When Zermelo spoke of mathematical logic in 1933 with an air of contempt, he added, "Among these people [those who pursue the subject] Gödel is the most advanced."

The course was accompanied by a seminar exclusively devoted to Gödel's incompleteness proof ([Goe31a]), "apparently also for sake of [Zermelo's] own understanding" (ibid.).

4.11 The Loss of the Honorary Professorship

In his letters to Zermelo, Carathéodory usually addressed his friend as "Dear Zermelo." In the letters which he wrote early in 1933, he addressed him as "Dear Zerline." Maria Georgiadou comments on this as follows ([Geo04], 301):

---

[254] [Scholh31], 64seq.
[255] [Goe86], Vol. IV, 346–349.
[256] [Goe86], Vol. IV, 348–353.

"Zerline" [is] the name of the innocent peasant girl in Mozart's *Don Gio-vanni* [...]. The Don, the incarnation of a force of nature without any sense of morality and responsibility, applies his charm to Zerline, who almost yields to his seduction. The implication is clear.

Zermelo will not submit to the Don, a resistance that will lead to the loss of his honorary professorship two years later. According to Sanford Segal he thus "provides [a] case of what in Nazi times was reckoned a bold act with damaging results" ([Seg03], 467).

4.11.1 The University of Freiburg and Its Mathematical Institute After the Seizure of Power

On 30 January 1933 the Nazis seized power in Germany. The universities soon felt the change. Already on 6 April the local government of the State of Baden which was responsible for the University of Freiburg, ordered Jewish professors to be relieved of duty and removed from academia. *Rector designatus* and professor of medicine Wilhelm von Möllendorf, an upright democrat, whose office was to start on 15 April, did not succeed in organizing a supporting movement for the outlawed Jewish colleagues. Even more, during his first week as Rector of the University he already became the target of severe attacks by the Nazi press, apparently arranged together with a group of nationalistic professors who planned to make the philosopher Martin Heidegger the new rector because he seemed to be more open to the national-socialist aims.[257] In order to discuss this situation, von Möllendorf initiated an extraordinary meeting of the Senate on 20 April. At this meeting he resigned and the Senate dissolved itself, ordering a general assembly of professors already on the next day for the purpose of a new election of Rector and Senate. This assembly then elected Heidegger as Rector. Thirteen Jewish members out of 93 professors were excluded from the meeting.[258]

On 23 April 1934 Heidegger resigned after a year in which he had tried to bring together Nazi ideology and his own ideas of a new university, failing for reasons that are still subject to debates.[259] One of Heidegger's decrees – a decree that was to become essential for Zermelo – ordered that lectures be opened with the Hitler salute,[260] i. e., by lifting the outstretched right arm above the horizontal, sometimes accompanying this posture by saying "Heil (hail) Hitler."[261]

---

[257]The description follows [Martb89], 22–24, and [Ott88], 141-145.

[258]From the 56 votes casted, 52 were *pro* Heidegger, 1 *contra* Heidegger, and 3 ones abstentions. Most of those voting for Heidegger were not aware of his party political antanglements, but regarded him as a guarantor for keeping away radical changes from the University.

[259]For a detailed discussion see, for instance, [Ott88] and [Martb89].

[260]Between 1933 and 1945 officially named "German salute."

[261]Heidegger's decree was published in Freiburger Studentenzeitung of 3 November 1933, p. 5. Following a decree of the responsible local government of 19 July

In 1933 the Mathematical Institute had two full professorships, one held by the applied mathematician Gustav Doetsch, who had followed Lothar Heffter, the other one by Alfred Loewy. Each professorship was supported by an assistant position held by Eugen Schlotter with Doetsch and the algebraic number theorist Arnold Scholz with Loewy. Scholz had followed Reinhold Baer who had gone to the University of Halle in 1928, the university of Georg Cantor and Leopold Kronecker.[262] Like Baer, Scholz was interested in foundational questions and developed an increasing friendship with Zermelo of both a personal and a scientific character.[263]

As a Jew, Loewy was suspended by the decree of 6 April 1933 with effect from 12 April. He finally lost his position at the end of the summer semester and died two years later.[264] Doetsch, now the only full professor, organized the successorship for Loewy according to his own wishes, i. e. against algebra. The report of the Faculty,[265] following the recommendations of the search committee, argues along these lines: Among the areas complementary to those represented by Doetsch, algebra and number theory have become totally abstract, as happens with areas under a strong Jewish influence.[266] Being too difficult for average students, these disciplines cultivate a mental arrogance which, as a rule, hides the inability to comprehend concrete and appliable subjects necessary for other disciplines such as physics. As set theory, the theory of real functions, and the calculus of variations are not well represented among younger German mathematicians, the only important area left is geometry, the seminal seed of the whole of mathematics.

At the beginning of the winter semester 1934/35 the geometrician Wilhelm Süss, the later founder of the Mathematical Research Institute Oberwolfach, followed Loewy. Scholz had to leave in order to complete the extinction of algebra.[267] He changed to the University of Kiel in Northern Germany, supported by a fellowship of the Deutsche Forschungsgemeinschaft. When he died in 1942, he had not succeeded in obtaining tenure. His academic career was partly blocked by negative reports from Doetsch and Schlotter where he was

---

1933 concerning the introduction of the Hitler salute at schools and universities (cf. [Schn62], 90–91), Heidegger gave the following order (ibid., 136–137): "*At the beginning* of classes students no longer greet by stamping their feet, but by standing up and lifting their right arm. The lecturer re-greets from the lectern in the same way. The end of classes remains as hitherto." (OV 4.32) Heidegger himself never obeyed this decree (cf. [Wap05], 148–149).

[262] Being of Jewish descent, Baer left Germany early in 1933 because of the anti-Jewish university laws.

[263] For further details concerning Baer, Loewy, and Scholz, cf. 4.1.

[264] For details concerning mathematics in Freiburg during the Nazi period, see [Rem95], [Rem99a], [Rem99b], [Rem99c], [Rem01], [Rem02], and [Seg03].

[265] Cf. [Rem01], 382.

[266] This was a standard position of Nazi ideology in mathematics; cf., for example, [Seg03], Ch. 7.

[267] [Rem99a], 61; [Rem01], 383.

blamed for being personally disorganized and not supporting the Nazi regime ([Rem99a], 61–62). Doetsch also worked behind the scene. When he learnt that Scholz, the "petty bourgeois,"[268] was under consideration for a lecturership at the University of Kiel, he turned to Erhard Tornier[269] in order to foil the appointment:[270]

> I am requiring of nobody that he be a National Socialist; I myself am not a member of the Party. However, in accord with the intentions of the government I would support somebody only if he has at least positive views of the present state. There is nothing of such an attitude with Sch[olz]. Quite the reverse, together with the sole friend he has here, namely Zermelo, he is well-known as the typical belly-acher who would like nothing more than the National Socialist government to disappear as soon as possible which, in contrast to former ones, threatens to pursue such characters as him. – It is, however, out of the question that such a softy like Scholz could have any political influence. [. . .] But if people like Scholz see that they are doing very well even today, despite all announcements of the government,* one need not be astonished that they go on to be up to mischief calmly and even more cheekily. [. . .]
>
> Maybe you understand better now that I was really astonished that this man was to be generously awarded a lecturership at a foreign university just at a time where an out and out war has been declared on such drone-like characters.
>
> * Half a year ago he got a fellowship of the [Deutsche Forschungsgemeinschaft] for himself; the affair was initiated behind my back. (OV 4.34)

When Scholz moved to Kiel, Zermelo lost the colleague who was closest to him. Remarks by Olga Taussky-Todd ([Tau87], 38) and, in particular, the letters Scholz wrote from Kiel, give evidence of a deep friendship that on Scholz's side was accompanied by both interest in Zermelo's foundational work and care for the older friend.

---

[268]So Doetsch in a letter to Süss of 1 August 1934; UAF, C 89/5.

[269]Tornier (1894–1982) was a *Privatdozent* in Kiel, working on the foundations of probability theory, and an eager Nazi. He had been appointed to a professorship in Göttingen and was thinking of Scholz as his successor in Kiel.

[270]Letter to Tornier of 31 July 1934; typewritten copy by Wilhelm Magnus in UAF, C 89/5. Magnus sent this copy to Süss in a letter of 24 June 1946 (UAF, C 89/331), commenting on it with the following words: "I add a letter, more exactly, the copy of a letter, which Doetsch wrote to Tornier in 1934. [. . .] I didn't need to be so familiar with Arnold Scholz to have the urgent wish that the writers of such letters should have more difficulties in the future to libel others in such a way. Scholz has always assured me that he would not have been able to stand life in Freiburg any longer; I now see that he was right to an extent which I could not imagine at that time." (OV 4.33) – Süss' assessment for Kiel (UAF, C 89/075) acknowledges Scholz's achievements, emphasizing that the degree of abstractness of his work did not justify the conclusion that he had worked under Jewish influence. Concerning his distance from the ideal of an academic teacher in the Third Reich, he should be given the opportunity to prove that he satisfied the corresponding requirements.

Arnold Scholz in the 1930s
Courtesy of Nikola Korrodi

In 1933 Zermelo still observed the political events with a certain coolness. In a letter from 7 September 1933 to Heinrich Liebmann in Heidelberg[271] he writes:

> I hope that the University of Heidelberg has weathered the storms of the new Reich without a lot of damage. By the way, Loewy finally had to retire here – though I do not think that this is a misfortune; he would never have retired voluntarily. But unfortunately we have to keep Heffter's successor [i. e. Doetsch]; there is basically only one opinion about this colleague. Only a miracle can help here. Let us hope for the best! (OV 4.35)

Two years later, in the spring of 1935, and at the instigation of Doetsch, Zermelo lost his position as well. The main reason for his dismissal put forward by the Vice Chancellor of the University was that he did not return the Hitler salute when greeted by it, thus causing offence among lecturers and students. Apart from the corresponding papers in the University Archives that witness Zermelo's antifascist attitude, neither the *Nachlass* nor other documents in the University Archives give information on political activities or convictions.

---

[271]UAF, C 129/70.

A possible exception might be seen in his signature below a general decree[272] from October 1933 where former members of the social democratic party and of the communist party had to declare that they no longer upheld relations to these "treasonous" parties. In a letter to Marvin Farber of 24 August 1928[273] Zermelo calls the events around the 8th International Congress of Mathematicians in Bologna[274] a "nationalistic incitement" and those engaged "chauvinists" and "grousers" ("Stänker"), thereby explicitly mentioning Bieberbach and Picard. Bernays reports[275] that in 1936 Zermelo "was on very bad terms with the Nazi regime."

4.11.2 The Zermelo Case

The affair started in early 1935. After the First World War Germany had lost the *Saarland*, a highly industrialized area at the French border. Having been a bone of contention between France and Germany for centuries and also under French government for some time in the nineteenth century, the treaty of Versailles had put it under control of the League of Nations. Economically it had been put under French control. After fifteen years the people were to be allowed to poll about their future status. The plebiscite took place on 13 January 1935, resulting in a majority for a return to Germany. During the following days the Nazis organised meetings to celebrate the result. Such a "Saar celebration" took place on 15 January at the University of Freiburg. At the end participants sang the Horst Wessel song, *the* song of the fascist movement, thereby offering the Hitler salute. Among those attending the rally we find Zermelo, but also Doetsch' assistant Schlotter, an active and eager supporter of the Nazi movement and intermediary Nazi agent at the Mathematical Institute. Three days later, on 18 January, Schlotter reported about what he saw to the teaching staff of the University, at the same time initiating reports from the students' union. To exemplify the infamous style of these denounce-

---

[272]UAF, B 24/4259.

[273]UBA, Box 25, Folder 25.

[274]After the First World War mathematicians from Germany and its former allies were excluded from the 6th International Congress of Mathematicians in Strasbourg (1920) and from the 7th Congress in Toronto (1924). The exclusion had been initiated by the International Mathematical Union, Émile Picard having been one of its strongest advocates. It was abolished for the Bologna Congress. As this congress was to be held under the auspices of the Union, some German mathematicians, among them, in particular, Bieberbach, but also Erhard Schmidt, pleaded for a boycott by Germans. They were supported by Brouwer, but strongly opposed, for example, by Hilbert. In the end, perhaps enhanced by a decrease in the Union's responsibility for the congress, German participants formed the largest group after the Italians. Also Zermelo attended the congress. - For details cf. [Seg03], 349–355, or [vanD05], 587–599.

[275]Letter to Fraenkel of 27 June 1951, Bernays *Nachlass*, ETH-Bibliothek, Hs. 975.1464.

Zermelo in August 1934

ments, we give Schlotter's letter almost completely and the students' letter in parts.[276] Schlotter writes:

> I feel myself obliged to report an incident which happened during the Saar rally on 15 January 1935, and which outraged everyone around me. Professor Zermelo attended this rally in a way that undoubtedly damaged the reputation of the teaching staff of the University. I have to begin with some remarks which may characterize the habits of this gentleman.
>
> A generation ago, by giving a foundation of set theory, Professor Zermelo made himself an immortal name in science. However, since then he never again achieved anything scientific. He rested on his laurels and became a drunkard. Because of his famous name, a benevolent patron provided him with a professorship at the University of Zurich. When he was sure of this position, Zermelo started travelling without bothering about his duties at the university. He lost his professorship and lived on the generous pension granted to him. He repaid his benefactors' hospitality with unsurpassable malice. Once he registered in a Swiss hotel with the following words: "E. Zermelo, Swiss professor, but *not* Helvetian!" This resulted in a considerable reduction of his pension.– He moved to Freiburg, and because of his name he got an honorary professorship. After the national uprising he attracted attention several times by seriously insulting the Führer and the institutions of the Third Reich. Up to now he has been given jester's licence by students and lecturers. Some time ago I called upon him to join the Freiburg corporation of young teachers and pay the enrolment fee corresponding to

---

[276] If not explicitly marked, the documents quoted in this section are from Zermelo's personal files in UAF, B 24/4259.

his salary. He told me that he did not see why he should do so. He received his money from Switzerland. Anyway, he did not intend to be in Freiburg during the next two semesters. Moreover, he would wait until he was called upon to join by someone more authoritative than I appeared to be to him. [...]

During the Saar rally on 15 January 1935, Professor Zermelo thought he had to present himself right in the forefront of the teaching staff. Probably he didn't do so with the intention of showing his nationalist convictions, but in order to have the best view. When the people present started to sing the Deutschland anthem, he noticed that they raised their hands. Timidly he looked around. When he saw no other way out, he held his little hand waist-high and moved his lips as if he were singing the Deutschland anthem as well. However, I am convinced that he does not even know the text of the Deutschland anthem, for the movements of his lips got more and more rare and did not even match the text. This observation incensed me and all the people around me.

When, after this, the Horst Wessel song was sung, he felt himself visibly uncomfortable. First, the little hand did not want to rise at all. He looked around even more timidly until during the third verse he saw no other possibility but to hold his little hand waist-high again with visible reluctance. He gave the impression he was swallowing poison, desperate movements of his lips could be observed more and more rarely. He seemed to be relieved when the ceremony had ended. The appearance of such a wretched figure ruined all our pleasure. After all, any student observing this surely gained the worst possible impression of a teaching staff which still tolerates such elements in it. If somebody attends such a rally not from nationalist conviction but because he is too curious to stay away, then he should at least stay in the background where his hostile behaviour does not cause offence.

I request the Freiburg teaching staff to investigate whether Prof. Zermelo is (still) worthy of being an honorary professor of the University of Freiburg and whether he should be allowed to damage the reputation of the German teaching staff somewhere else during the next term or the next but one.

Heil Hitler!

Schlotter. (OV 4.36)

On 12 February the student union seconds by a letter "concerning damage caused to the reputation of the teaching staff and the students of Freiburg by Prof. hon. Dr. E. Zermelo:"[277]

Since the National Socialist uprising Mr. Zermelo has several times attracted attention by seriously insulting the Führer and by disregarding the institutions of the Third Reich. For example, he refuses to greet with the German salute. He is still able to exist at Freiburg University only because he has been tolerated up to now as being harmless and is taken seriously only in scientific affairs.

However, this tolerance must not be extended so far as to let him cause damage to the reputation of the teaching staff and the students of Freiburg

---

[277] One of the two students who signed the letter, was a Ph. D. student of Doetsch. He completed his thesis in 1939.

with impunity. An event giving rise to general indignation was his behaviour during the Saar rally of the University. [...]

In order to prevent Prof. Zermelo from further damaging the interests and the reputation of the teaching staff and the students in public, we request the Ministry to investigate this state of affairs and to see to it that Prof. hon. Dr. E. Zermelo is removed from the teaching staff. (OV 4.37)

The union sent this letter together with Schlotter's report to both, the responsible Minister and the Rector.

Instantly, on 14 February, the Rector informed the Minister that he intended to take steps against Zermelo with the aim of retracting the *venia legendi*, i. e. the permit to hold lecture courses. Between 20 February and 28 February the Vice Chancellor, the professor of philosophy and educational science Georg Stieler, heard various witnesses, among them Schlotter, some students, participants of the Saar rally, and professors of the Mathematical Institute. The participants of the Saar rally who were summoned did not know Zermelo in person and seem to have been brought in by Schlotter; one of them reported:

> During the Saar rally SS-Oberscharführer Schlotter drew my attention to [...] Professor Zermelo [...]. I could not see Zermelo myself because my sight was blocked.

The witnesses essentially confirmed the accusations that were made in the letters. Schlotter added that Zermelo did not react to a Hitler salute, whereas he answered a usual greeting in a friendly manner. Concerning "most serious insults of the Führer," one of the students – in a letter to the Rector – stated that he had been informed about them by Doetsch:

> The assertion that Prof. Zermelo had several times made insulting comments about the Führer is based on statements by Prof. Doetsch. Prof. Doetsch told me about several cases where Mr. Zermelo attracted attention by insults to the Führer or other criticism. Partly he himself was a witness. Because of noisy abuse the landlord of a Freiburg inn had told Mr. Zermelo not to visit the inn again. So I have to ask you to consult Prof. Doetsch for details concerning these cases. [...]
>
> I would like to add that Mr. Zermelo, when beginning his lectures, only indicates the prescribed German salute by holding his right hand beside his hip and bending his head to the right, smiling in embarrassment. I have observed that he did this. Further witnesses: the participants of his course.

According to the official lists of participants this student had never attended one of Zermelo's courses after the Nazis had seized power.

On 25 February Doetsch had the following statement put on record:

> Zermelo refuses the Hitler salute to me and others in the Institute. He does not do this by accident, but deliberately. For this did not happen now and then and by chance, but *regularly*. Even a loud "Heil Hitler, Mr. Zermelo" directed solely at him when he had ignored the usual salute again, was

answered by him merely with a squashed *Heil*. But also on the following days he again regularly shirked the response to the Hitler salute as if he had not become aware of my entry and my salute at all. This happened around the beginning of February this year. My assistant Schlotter and my colleague Süss have also confirmed the same to me. Although Zermelo is not taken seriously by mature students, I think after all that such political behaviour is dangerous for the younger ones. (OV 4.38)

Wilhelm Süss who had followed Loewy at the beginning of the winter semester gave only a short statement:

> I have been in Freiburg since 10 October 1934 and have a correct relationship with Zermelo. Since then I noticed that Z[ermelo] does not reply to the Hitler salute. Further negative remarks against Führer and movement have not been made in my presence. (OV 4.39)

Lothar Heffter, the *professor emeritus* and predecessor of Doetsch, who had had a serious debate with Zermelo because of Zermelo's leave for Warsaw in 1929, approved the following protocol:

> I myself have no experience with regard to his refusal of the Hitler salute because I meet him only rarely; I also do not know about any remarks of Zermelo against the Führer and the new Reich. I consider Zermelo an extraordinarily ingenious mathematician where the borderline between genius and abnormality may be somewhat vague. (OV 4.40)

Finally, in the morning of 1 March, Zermelo himself was interviewed. The protocol is rather short:

> I did not intend to violate a legal requirement when refusing the Hitler salute, and at the beginning of my lectures I also made the prescribed movement of the hand; when not always responding to the salute in the prescribed form, I only wanted to defend myself against exaggeration.
>
> When reproached: It may be that I have occasionally been careless with critical remarks against today's politics in inns; however, I only made them in private circles, at any rate not in the room for the teaching staff. (OV 4.41)

Already on the next day he wrote the following letter to the Dean:

> Very esteemed Herr Dean!
> I herewith inform you that after a meeting with the Vice Rector I have decided to renounce any further teaching activity at this university, and I ask you to give my best recommendations to my former colleagues.
> With all due respect
> Dr. E. Zermelo
> hitherto full honorary professor at the University.[278] (OV 4.42)

---

[278] When Abraham A. Fraenkel left Germany in the autumn of 1933 to settle in Jerusalem, he informed his colleagues by a card (UAF, C 129/33) which was signed in a similar way, namely by "Dr. A. Fraenkel, hitherto full professor at the University of Kiel" ("Dr. A. Fraenkel, bisher o. Professor an der Universität Kiel").

Freiburg, 2. März 1935.

Sehr geehrter Herr Dekan!

Hiermit teile ich Ihnen mit, daß ich nach einer
Besprechung mit dem stellvertretenden Rektor
mich entschlossen habe, auf eine weitere
Lehrtätigkeit an der hiesigen Universität
zu verzichten, und bitte Sie, mich meinen
bisherigen Herren Kollegen bestens zu
empfehlen.

In vorzüglichster Hochachtung

Dr. E. Zermelo
bisher ord. Honorarprofessor a.d. Universität.

Zermelo's letter of renunciation
Universitätsarchiv Freiburg, Zermelo files, B 24/4259

Gertrud Zermelo – she married Zermelo in 1944 – remembered that after
the interrogation her later husband had been very dejected because he had
been confronted with the alternative of either performing the Hitler salute
or facing dismissal from his professorship, in the latter case also suffering a
ban on publishing.[279] In the letter to his sisters from the middle of April
reprinted below, Zermelo does not mention this alternative, but states that
during the interrogation he had already rejected continuing his lectureship in

---

[279]In the late 1950s, in connection with negotiations of redress (cf. p. 253), Mrs.
Zermelo argued that the dismissal of her husband was accompanied by a ban on
publishing and, hence, by a possible loss of royalties. The case was not cleared,
but was dismissed for formal reasons (letter from the State Office of Redress of 30
September 1959; MLF). According to the files a ban seems to be improbable.

the situation he was facing. In any case, he was anticipatory in making his decision. For on 4 March the Rector, without having received Zermelo's letter to the Dean, informed the Minister about the investigation and his intended measures, giving force to the claim that the real alternative facing Zermelo was either to be asked to give up his position or to be dismissed:

> On the basis of the reports against Professor Zermelo of 18 January and 12 February I initiated an investigation against the accused on 14 February.
>
> The questioning of the witnesses, as far as it has been possible up to now, proves beyond doubt that Z[ermelo] does not respond to the German salute and thus causes offence among Faculty and students. It also seems to be certain that he does so deliberately and on principle.
>
> So far Schlotter's assertion that Z[ermelo] has several times attracted attention by seriously insulting the Führer and the institutions of the Third Reich could not be clearly established; according to vague statements by Zermelo they may have consisted in disparaging criticism made in inns.
>
> In order to judge this case, one has to take into account Zermelo's personality. Privy Councillor Heffter characterizes him as an ingenious mathematician with whom the borderline between genius and abnormality is vague. As a researcher he is world-famous, as a teacher he has apparently deteriorated strongly. Others call him an awkward customer who does not oppose maliciously but on principle, and who gets angry with everything. With regard to his outward appearance considerable signs of neglect can be observed, he seems to have given himself up to the consumption of alcohol. Hence, there may be no desire to have him remain in the Faculty.
>
> With regard to the proven charges preferred against him I am requesting you to ask him to give up further teaching activity. Otherwise legal procedures would have to be initiated against him to withdraw the *venia legendi*. (OV 4.43)

On 13 March, then still without an answer to his letter of renunciation, Zermelo asked the Rector for confirmation:

> I submissively inform Venerable Magnificence that on 2 March I officially announced my renunciation to the Dean's Office of my Faculty without having received up to now any form of confirmation or acknowledgement of receipt. As I am still receiving official notifications concerning the University, I have to conclude that my renunciation from the Faculty staff has not yet been noticed officially. Maybe it has been overlooked because of a temporary absence of the Dean. However, I attach importance to establishing that immediately after the meeting with the Vice Rector I took the necessary steps to clarify my relationship to the University.
>
> With proper and deep respect
>
> Dr. E. Zermelo. (OV 4.44)

Now things went quickly. The Rector received Zermelo's letter of renunciation on 14 March. On the same day he informed the Minister and asked him to order Zermelo's exclusion from the Faculty. On 26 March this was carried out.

4.11.3 After the Renunciation

Altogether, the affair came to an end that was both deplorable and yet without serious consequences beyond his dismissal. On the one hand, Zermelo lost his position under shameful circumstances aggravated by the obstinate and inhuman behaviour of his colleague Doetsch. On the other hand, the Nazi system seemed to have been satisfied with his dismissal and did not pursue the disparagement of its politics. Maybe it was also Heffter's (well thought out) characterization of Zermelo's personality which worked in this direction.[280]

Zermelo closes his letter of renunciation with "best recommendations to [his] former colleagues." The letter to his sisters reprinted below shows that these recommendations were made with a taste of bitterness. In the beginning his relationship to Doetsch, the main addressee, seems to have been correct, but cool and without any liking. However, in early 1935 he felt a "growing dissatisfaction with the situation in Freiburg" as Doetsch "was tyrannizing science."[281]

Zermelo's relationship to Süss was better. On 18 November 1934 Scholz[282] refers to a letter of 4 November where Zermelo had most probably reported on a meeting with the Süss family and had described the get-together:

> It is really fine that right away you have got together to some extent with the Süss family. Surely this acquaintance will last longer than the "sizing each other up" with Doetsch. Well, Süss himself is a nice person. Please give him my regards!

On 15 March 1935 Zermelo had informed Scholz about the procedures against him and his decision to resign. Scholz's answer of 29 March[283] shows that Zermelo was not fully aware of the way his case was handled and still uncertain whether his decision had been the right one. Having consoled his elder friend by pointing to his decreased pleasure in lecturing, Scholz continues:

> I think that for the present your step is the right one and effective. [...] However, *if*[284] the *Faculty invites* you to stay, at the same time *assuring* you of the necessary *protection*, wouldn't you do it then? [...] After all, the Faculty may nevertheless attach importance to including a minimum of distinguished scientists among their members. Of course, they would have to change their behaviour towards you.

As the documents prove, such hopes were doomed to fail.

---

[280]The Rector's report of 4 March which takes up Heffter's characterization, was formulated by Vice Chancellor Stieler. An eager Nazi still in 1933 (cf. [Ott88], 148), Stieler later became an opponent of the Nazi regime, his change of attitude beginning around 1935.

[281]Letter from Arnold Scholz to Zermelo of 9 February 1935; UAF, C 129/104.

[282]UAF, C 129/104.

[283]UAF, C 129/104.

[284]Emphases represent underlinings by Zermelo's hand.

On 19 April 1935 Zermelo wrote a long letter to his sisters (MLF) which impressively documents to what extent the affair had hurt him. Any doubts about the adequacy of his renunciation had disappeared, and there was no room left for any possibility of a resumption of his lecturership. Taking into consideration the sharp condemnation of Doetsch found there, he had certainly become aware in the meantime of Doetsch's decisive role.

Before quoting the letter, some introductory explanations may be useful: In 1935 only Zermelo's sisters Elisabeth and Margarete were still alive. His eldest sister Anna had died in 1917,[285] his youngest sisters Lena and Marie already in 1906 and 1908, respectively. Because of their fragile health, Elisabeth and Margarete had moved to the mild alpine climate of South Tyrol in Northern Italy. The hint about Gessler's hat Zermelo reports upon was politically very daring and risky: In his drama "Wilhelm Tell" about the Swiss hero of liberty, Friedrich Schiller describes how people in one of the later cantons have to pay respect to a hat of the Austrian governor Hermann Gessler. Tell refuses to do so, thereby starting a movement that in the end leads to liberty and independence from Austria.

Now the letter in full.

My dear sisters,
many thanks for your kind card of 28 March giving me news of you again after a long pause. I am so sorry that your health leaves so much to be desired again and that you had to suffer so much. I have finally learnt your new address, but I would like to hear more about how you are now living. My health at least is quite satisfactory, but I am pretty lonely here and the new year began under gloomy forebodings which were to prove me right. I had not announced courses for the summer semester. And now, under the pressure of circumstances, with intriguing colleagues and denouncing students I have definitely given up my lecturership here for which I didn't get a salary, but which used to give me some pleasure. On 1 March, when the semester had just ended and everywhere people were celebrating the return of the Saarland, I was summoned to the Vice Rector, who disclosed to me the following: there was an investigation underway because I had "refused" (i. e. incorrectly returned) the Hitler salute. My answer was that both, within and out of my teaching activities I had tried to follow painstakingly all the legal requirements of the new state; for among other things I share Socrates' opinion that a citizen has to obey his government even if he doesn't approve of its regulations. On the other hand, not being a civil servant, and not receiving a salary from the state, I had by no means felt obliged to continuously express convictions of the National Socialist party which I did not share. The politics of Gessler's hat didn't seem to have paid off as can be read in Schiller. And so I had answered the Hitler salutes in the same way I usually answered greetings. As far as I knew, I had made critical remarks about governmental measures, if any, only in private and never in the rooms of the University. As far as the matter was concerned, I therefore felt entitled to hold my point of view. However, if the University wanted

---

[285]Letter from Zermelo to Elisabeth and Margarete of 8 November 1917 (MLF).

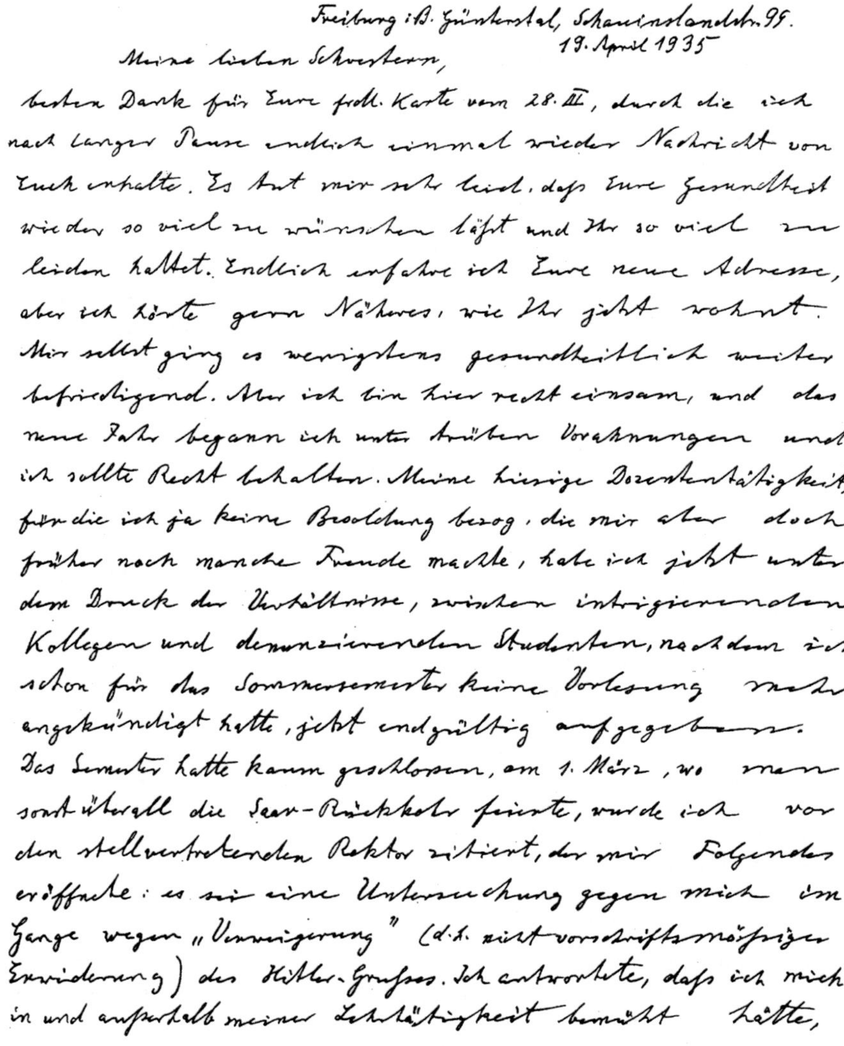

[Handwritten letter in German]

Freiburg i.B. Günterstal, Schauinslandstr. 95.
                                19. April 1935

Meine lieben Schwestern,

besten Dank für Eure fröhl. Karte vom 28. III., durch die ich nach langer Pause endlich einmal wieder Nachricht von Euch erhalte. Es tut mir sehr leid, daß Eure Gesundheit wieder so viel zu wünschen läßt und Ihr so viel zu leiden habt. Endlich erfahre ich Eure neue Adresse, aber ich hörte gern Näheres, wie Ihr jetzt wohnt. Mir selbst ging es wenigstens gesundheitlich weiter befriedigend. Aber ich bin hier recht einsam, und das neue Jahr begann ich unter trüben Vorahnungen und ich sollte Recht behalten. Meine hiesige Dozententätigkeit, für die ich ja keine Besoldung bezog, die mir aber doch früher noch manche Freude machte, habe ich jetzt unter dem Druck der Verhältnisse, zwischen intrigierenden Kollegen und denunzierenden Studenten, nachdem ich schon für das Sommersemester keine Vorlesung mehr angekündigt hatte, jetzt endgültig aufgegeben. Das Semester hatte kaum geschlossen, am 1. März, wo man sonst überall die Saar-Rückkehr feierte, wurde ich vor den stellvertretenden Rektor zitiert, der mir Folgendes eröffnete: es sei eine Untersuchung gegen mich im Gange wegen „Verweigerung" (d.h. nicht vorschriftsmäßiger Erwiderung) des Hitler-Grußes. Ich antwortete, daß ich mich in und außerhalb meiner Lehrtätigkeit bemüht hätte,

First page of Zermelo's letter to his sisters Elisabeth and Margarete

to make me understand that it no longer appreciated my (voluntary and unpaid) lectureship, I was willing to voluntarily give up an activity which couldn't give me any pleasure under these circumstances. The Vice Rector seemed to be satisfied with this and referred me to the Dean to whom I was to address my renunciation; he didn't even try to settle the matter in an amicable way. I *never* received an answer or an acknowledgement of the receipt of my letter addressed to the Dean and ending with "regards to my former colleagues," so that I had to inquire at the Rector's Office two weeks later whether my renunciation had been officially noted. After two weeks I received an official notification from the Ministry for Cultural Affairs in Karlsruhe stating that I had been retired from the Faculty. All this was without any kind of greeting or the slightest phrase of courtesy after I had worked at the University for seven years without any salary. The interests and the reputation of the University seem to have been less important for its official representatives than their wish to recommend themselves to their superiors. Of my colleagues in mathematics and beyond, whom I would count among my friends, not even one has so far tried to talk to me about the matter; I am considered to be suspect, and it's better to stay away from me. But I know which way the wind is blowing and who got me into this pickle: a resentful colleague who has known how to compensate a lack of scientific talent with particularly unpleasant traits of character, and so creeping into a position to which he had no right. But what did he get out of this? Will I now evaluate him more favourably if I tell other colleagues about my experience? And what does the Party get out of it? Did anyone really believe that I could give lectures on number theory or set theory which might endanger the security of the state? It is just regrettable that in these days where the autonomy of the University has been abolished the way is free for all intriguers if they are clever enough to trim their sails to the wind. I don't need the lectureship myself as I can continue my research without it. But whether I will stay in Freiburg remains to be seen. Perhaps I will move to Darmstadt which combines a central location with a mild climate and libraries offering me good possibilities for work; as well as that I am friends with a family living in Jugenheim. I will make up my mind about that in the course of the summer. As far as countries abroad are concerned, apart from Switzerland, which I don't like, I only think of Austria, as Italy fancies itself making noisy threats of war.

Now that's enough from me! Perhaps you can see from this still quite humorous case what is happening here.

With best regards

Your brother Ernst.

Happy Easter! (OV 4.45)

Zermelo's intention to move to Darmstadt became so serious that on 29 June he gave in notice to his landlord that he intended to leave his Freiburg apartment at the end of September. However, he finally decided to stay, and on 6 August the lease was renewed (MLF).

With regard to the behaviour of Doetsch who was obviously the driving force behind the affair, possible explanations have been discussed of the kind that he himself was in danger of being persecuted by the regime because of antinationalist activities in the twenties and was thus demonstrating strict obedience.[286] However, possible excuses are even more out of place when seeing him write the following letter to the Rector on 1 June 1936:

---

[286]Cf. [Seg03], 98seq. – Doetsch did not admit that nationalist views impeded mathematics. In late 1935 he did not prolong the employment of his assistant Schlot-

More than a year ago the former honorary professor at our University, Dr.
E. Zermelo, was induced to lay down his honorary professorship; otherwise
a disciplinary procedure would have been initiated against him because of
his opposition to the Third Reich. Nevertheless, in the membership list of
the Deutsche Mathematiker-Vereinigung Zermelo still has himself listed as
an honorary professor at the University of Freiburg. As this list is generally
and rightly considered to be an authentic list of German mathematical university lecturers, Zermelo is still regarded a member of our University. [...]
Therefore it seems appropriate to me that Zermelo be officially reminded
by the Rector's Office that it had come to their ears that he is still listed as
an honorary professor at the University of Freiburg in the membership list
of the Deutsche Mathematiker-Vereinigung and that this is inadmissible.

I think it very likely that Zermelo still wants to be considered a member
of our University, for even in these days he pushes his way into the meetings
of the Freiburg Mathematical Society – which take place in the University
– without having been invited to them. (OV 4.46)

Two days later, in a letter without any kind of salutation, Zermelo was urged
by the Rector to ensure an immediate correction.

In contrast to Doetsch, Süss tried to maintain a cooperative relationship
with Zermelo. Apparently he continued to invite him to the meetings of the
Freiburg Mathematical Society as is shown by a letter of 7 November 1936:[287]

On next Tuesday a lecture will take place in [our] mathematical society.
Since Herr Doetsch is giving a talk, I have not had an invitation sent to
you. On the one hand, I did not believe you would gladly participate. On
the other, I wish at present to avoid everything that could cause a quarrel
within the society and have as a consequence its going asunder. And so I
hope to have met your consent if I therefore ask you this time to stay away
from the presentation.

Altogether, the formal and practically all personal connections between Zermelo and the University had been cancelled. In private conversations, but
also in sarcastic notes, he worked off his aversion to Nazi politics. On a piece
of paper, probably from around 1936 (MLF), we find several goat rhymes
denouncing the Nazi politics of rebuilding the universities. A sample:

Der Bekehrte

Meinst du, daß Mathematikstunde,[288]
Daß Physik heut und Statik munde?
Auf daß ich meine Kassen runde,
Da treib ich lieber Rassenkunde.

---

ter, arguing that an exaggerated eagerness for the Nazis kept him back from mathematics ([Rem99a], 64).

[287]Letter to Zermelo (UAF, C 129/115); the translation follows [Seg03], 468–469.

[288]Underlinings are given here in order to exhibit the structure of the rhymes which
consists in a chiasmus of consonants.

The Convert

Do you believe that these days one would savour a lesson in mathematics, physics, or statistics? In order to fill my purse, I prefer the study of races.

The rhymes are followed by well-known quotations from Shakespeare's "Hamlet" and Goethe's "Götz von Berlichingen," modified into a composition against Aryan politics: Zermelo has "Hamlet Swastica" ask the question "Aryan or Not-Aryan?" and indicates his answer by "Götz von Berlichingen."[289]

His reformulation of the Ten Commandments (MLF) gives a barely disguised indictment of the Hitler regime:

New Cuneiform Scripts Found in Ninive

The Ten Commands of Baal on Two Clay Tablets

1) I am Baal, Your God. You shall not have any other god.
2) You shall make images of Me as many as you can, and no place shall be without a Baal's image, neither on earth nor under the earth.
3) You shall name Your God at any time and everywhere, for Baal will not hold guiltless those who hide His name.
4) You shall keep the holiday holy and celebrate feasts day for day. Let others have the labour.
5) You shall unconditionally obey the priests of Baal, even against Your father and Your mother.
6) You shall kill all those standing in the way of Baal, men, women, and children, so that the infidels shall be wiped out from earth.
7) You shall not remain unmarried, but bring descendants into the world like sand at the sea.
8) To honour Baal You shall steal as much as You can from those who do not worship Him.
9) You shall bear false witness against the infidels and lie and slander Your head off: Always there is something which gets stuck.
10) You shall covet posts and ranks, those that exist and those that do not, and let no infidel keep his ones. (OV 4.47)

Mathematical logic was to return to Freiburg only thirty years later when in 1966 Hans Hermes became a full professor of mathematical logic and the foundations of mathematics and the first head of a new department for mathematical logic within the Mathematical Institute.

4.12 Retreat

The loss of the honorary professorship and the disappointment over the behaviour of some of his former colleagues with whom he had worked together

---

[289]In a hopeless siege Götz answers the demand to surrender by a "fuck off," thus also showing his contempt.

Zermelo around 1935

for years left Zermelo bitter. Arnold Scholz, whose interest in set theory and the foundations and whose friendship he had enjoyed since the early 1930s, had moved to Kiel. There is evidence that the correspondence with Scholz formed the main body of his scientific contacts in the later 1930s. Maybe even more important, the circumstances of the dismissal procedure had attacked his fragile mental constitution. His later wife Gertrud remembers that the events caused nervous tensions and led to a deterioration of his scientific initiatives.

4.12.1 Günterstal

In a certain sense, Zermelo's growing scientific isolation was mirrored by the circumstances of his life. In March 1934 he had left his apartment in downtown Freiburg and had moved to Günterstal, a nice suburban village-like residential area in the foothills of the Black Forest, where he had already spent the first year after his move to Freiburg. It is separated from the city by wide meadows

Freiburg-Günterstal in the early 20th century, view from Lorettoberg
Stadtarchiv Freiburg, signature M 737/15.09

which extend through a narrowing valley to a tower gate marking the entrance to the monastic area with cloister and churchyard which formed the nucleus of the settlement. A lovely hillside on the right, Loretto Mountain, had seen Ferdinand Lindemann strolling for a walk to Günterstal on the afternoon of 12 April 1882, his thirtieth birthday, finding the conclusive idea for his proof of the transcendence of π and thus refuting the possibility of squaring the circle merely by compasses and ruler ([Lin71], 84).

At the end of the nineteenth century the Berns, an elderly well-to-do Dutch couple, had built an ample country house at the remote end of the settlement on the road leading to the higher mountains and had laid out a large park around it, planted with exotic trees. Feeling disturbed by noisy students who used to visit a nearby inn, and perhaps also missing an open view, they finally left Günterstal, selling "Bernshof" ("Berns Farm") to the city of Freiburg. The city rented it to a couple that in turn rented parts of it to several parties, among them Zermelo since 1934. Until his death in 1953 he lived on the second floor that could be reached through a wide panelled stairwell. The living room, his "knight's hall," where he had his private library, measured more than thirty feet in length, allowing him to walk around while working, a cigarillo in his right hand and a small copper ash tray in his left one.

During the first months after his dismissal in March 1935, scientific activities seemed to continue as usual. Being free of teaching tasks, he intensified earlier plans for a "summer tour" to various colleagues, among them Emil Artin in Hamburg and, of course, Arnold Scholz in Kiel. Scholz instantly tried to arrange an invitation for a talk. The final programme consisted of talks in Bonn on his *Grenzzahlen* paper[290] and in Hamburg on well-founded relations and infinitary languages,[291] and a stay in Kiel. Some days before the talk scheduled there, Scholz had to withdraw his invitation,[292] complaining of the bad treatment Zermelo would experience. Perhaps nationalist colleagues were responsible for the cancellation.

In the spring of 1936 Heinrich Scholz asked him for advice concerning the edition of Frege's collected papers, giving as a reason Zermelo's "exemplary" edition of Cantor's collected works.[293] Scholz also informed him about his plans to edit Frege's *Nachlass.* Zermelo answered immediately,[294] giving pieces of advice, but also offering his assistance:

> By the way, I myself would be ready for any kind of cooperation concerning the intended Frege edition you might wish. For I believe that I have already gained a certain practice in such editorial activities. I cannot but very much welcome this logician of mathematics who has remained unrecognized so long, belatedly coming into his own with the new edition.[295] (OV 4.48)

Bernshof in the early 20th century;
the central windows above the ground floor belong to Zermelo's "Knights' Hall"

---

[290] Correspondence with Hans Beck; UAF, C 129/6.

[291] Copy of announcement in UAF, C 129/129.

[292] Letter to Zermelo of 16 May 1935; UAF, C 129/104.

[293] Letter to Zermelo of 5 April 1936; UAF, C 129/106.

[294] Letter of 10 April 1936, Scholz's *Nachlass,* Institut für Mathematische Logik, Universität Münster.

[295] Scholz's project was delayed by the war, in particular by the destruction of most of the material by an air attack in March 1945, and was completed only after his death; for details cf. [HerKK83].

Zermelo in the garden of Bernshof

Despite these signs of activity, the year 1935 marks the beginning of a decline of Zermelo's mental energy. It was a development in reverse of that in 1926 when he was given the honorary professorship. The second active period of his scientific life around 1930 closely coincides with this professorship, and the second half of the thirties comes with a complete withdrawal from the scene of mathematical foundations. Besides letters from Arnold and Heinrich Scholz there is only one letter in the *Nachlass* written by a well-known researcher: In July 1937 Haskell B. Curry asks for a preprint of the *Satzsysteme* paper ([Zer35]). Similar to the *Grenzzahlen* paper ([Zer30a]) five years before, this paper also promises a further development of its topic. However, there is an essential difference that may indicate the change in Zermelo's scientific activities. Whereas the *Nachlass* contains several notes that might be considered part of the sequel planned for the *Grenzzahlen* paper, there are no such notes concerning the *Satzsysteme* paper. Moreover, apart from the notes on a flawed refutation of the existence of countable models of set theory ([Zer37]),[296] there are no notes in the *Nachlass* which stem from the second half of the 1930s and concern foundational questions or nonelementary questions of other areas of mathematics.

Complementing this development, his daily life concentrated on quiet Günterstal. He started to comment on everyday events with little poems, all written with a sense of humour.[297] But he had not lost his sharp irony. In the pamphlet "Cheeky Little Devil. The Golden Bigwig-ABC or the Art

---

[296] Cf. 4.8.4 or 4.12.2.

[297] An example, the poem "The Easter Egg," written on Easter of 1937 (MLF), is given under OV 4.49.

Winter scenery near Günterstal
Photo by Zermelo

of Becoming a Perfect Bigwig"[298] he lists 23 properties which make a perfect bigwig. For instance, a perfect bigwig should be presumptuous like Toeplitz, a charlatan like Weyl, despotic like Klein, obstinate like Hilbert, business-minded like Courant, mad like Brouwer, and careful about his appearance like Doetsch.

In a letter from Arnold Scholz[299] we read that his Swiss pension had been cut back once again in January 1935. Thus his dream of owning a four-wheeled car could no longer be fulfilled, and the driver's licence he had obtained some weeks earlier remained unused; even worse, his three-wheeled car, which had inspired his development of an electrodynamic clutch (cf. 4.1), had given up the ghost thanks to his ambitious driving style. The letter reveals both Scholz's care about his friend and Zermelo's lack of experience in everyday life:

As your situation will worsen after January 1, I would do two different things: (1) Apply to the tax office. I once heard that somebody who obtains his income from abroad gets a reduction because otherwise he might move his domicile abroad, and then no taxes at all will come in. By pointing this out, you may succeed in being charged only the rate for married people without children. (Analogously this should be the case for all who no longer want to marry.) – (2) Have a friendly talk with [your landlord]. You no longer need a garage. And you cannot drive his car as a substitute as he

---

[298] "Frech wie Oskar. Das goldene Bonzen-ABC oder die Kunst, ein vollkommener Bonze zu werden;" MLF.
[299] Of 11 December 1934; UAF, C 129/104.

Zermelo during a rest

promised, because he apparently now only owns the Maybach.[300] (Certainly such a precious object should be treated with so much care that it is neither fun to drive it nor can you expect to be allowed to drive it.) Moreover, rents have continued to go down, and [your landlord] did not make a difference in the price for a couple and a bachelor. [...] Present him the following offer: Rent starting January 1: 72–75 Marks and 12 Marks for [the cleaning lady]. This is still very generous; for if you gave him notice to quit on 1 March, he would get less.

Each day at about noon he walked for nearly a mile to the restaurant "Kühler Krug" ("Cool Flagon") to have lunch, and in the evening he sometimes walked there again for a glass of red wine. As a rule he also went for longer walks in the afternoon, accompanied by his dog. They often ended up in a little inn higher in the mountains where he ordered his beloved cheese cake together with a glass of beer. As he believed, it was mainly these regular hikes in the forest that stabilized his fragile health.

At some time or other his evening habits changed. In Bernshof there also lived a younger lady, Gertrud Seekamp. After spending her youth in Silesia together with eight brothers and sisters, she had gone to Freiburg via Berlin in the 1920s and in 1929 had settled in Bernshof, becoming the "good spirit" of the landlord's family. More and more frequently it happened that she spent the evening in the knight's hall together with Zermelo and a glass of liqueur. Often these evenings ended with Zermelo reading aloud from classical literature. One sunny summer afternoon in 1944 Gertrud was sitting outside the house in the shade when Zermelo turned up. He very politely asked her to allow him a

---

[300] An expensive luxury car of a high technical standard.

Zermelo with Gertrud Seekamp, his later wife,
in the garden of Bernshof, August 1935

question which then turned out to be whether she was ready to become his wife. Totally surprised she did not find an immediate answer. As a result, however, the wedding took place on 14 October. At this time walks to the "Kühler Krug" had already become more irregular; when the landlady – so Mrs. Zermelo with a smile – had heard of Zermelo's wedding plans, she had stopped all extra attentivenesses. On occasion, Zermelo still addressed his wife by the formal "Sie" instead of the familiar "Du." He then used to excuse himself by pointing out that she and his sisters were the only women in his life he had ever addressed by "Du." He also explained to her that it was his state of health which had put off his decision for a marriage; he simply would not have felt happy with a wife doomed to be married to a sick husband. Interpreting this the other way round, one may conclude that, thanks to his way of life, he felt considerably better in his seventies.

4.12.2 Last Scientific Activities

Zermelo's retreat from the foundational scene did not mean a retreat from mathematics as a whole or from the diversity of his former activities. The

Nachlass contains studies on music theory and papermade slide rules of different kinds to calculate keys and tone intervals, stemming from 1933 and later,[301] and fragments of manuscripts or sketches dealing with various mathematical topics such as puzzle games,[302] permutations of finite and infinite sets,[303] polynomials in division rings,[304] modules and fields,[305] bases of modules,[306] systems of linear equations with infinitely many unknowns,[307] and – probably initiated by the first air attacks of the Second World War – considerations about families of parabolas under the title "Are there places on earth which are protected against bombs?" ("Gibt es Orte der Erdoberfläche, die gegen Fliegerbomben geschützt sind?").[308] However, none of these elaborations can be considered part of an extended project; they may have been something like exercises done without contact to the actual scene of research.[309]

Besides this diversity of topics there are some more coherent themes.

The Ongoing Debate with Skolem

There are two points which we already addressed earlier in full detail (cf. 4.8.4 and 4.9, respectively). They show that Zermelo's "war" against Skolem was not yet over.

On 4 October 1937 he wrote down a flawed proof for the impossibility of the existence of countable models of set theory, finishing it with statements pronouncing the defeat of "Skolemism," intuitionism, and formalism. A letter from Arnold Scholz[310] shows that Zermelo had discussed the argument with him. Scholz can imagine that a countable continuum may be adequate for mathematics because the real numbers explicitly appearing in analysis, those

---

[301] UAF, C 129/218–219 and C 129/285–286.

[302] April 1935; UAF, C 129/251.

[303] April 1935; UAF, C 129/217.

[304] UAF, C 129/296.

[305] November 1937; UAF, C 129/218.

[306] UAF, C 129/218.

[307] January 1940; UAF, C 129/212.

[308] About 1940; UAF, C 129/238.

[309] There is an earlier manuscript of 1933 (UAF, C 129/207), "Über die ähnlichen Permutationen der rationalen Zahlen und das Linearkontinuum und die Grundlagen der Arithmetik" ("On Similarity Permutations of the Rational Numbers and the Linear Continuum and the Foundations of Arithmetic") which obviously forms the beginning of a longer treatise dedicated to the programme of deducing the arithmetical properties of the rationals and the reals solely from properties of the respective groups generated by translations and inversions together with properties of the respective orderings, "in this way gaining a new and natural foundation of the arithmetic of the real numbers which is free of arbitrary conventions as far as possible" (ibid., 4). Already on 31 May 1924 Zermelo had written to Sidney Farber (UBA, Box 25, Folder 25) that he was "eagerly working on a theory of 'groups of transformations of the continuum' as a foundation of arithmetic *and* geometry."

[310] Letter to Zermelo of 3 November 1937; UAF, C 129/104.

numbers "upon which some mathematician has come directly or indirectly," formed a countable set. With respect to Zermelo's point of view he simply states "that we are on different banks of Skolem's paradox."

Besides his endeavour to refute the adequacy of Skolem's first-order approach to set theory, Zermelo had also aimed at an infinitary foundation as being the only way of adequately dealing with the richness of mathematics. There is evidence that he still stuck to this programme – not technically by continuing the investigations of his *Satzsysteme* paper ([Zer35]), but in principle. For as late as 1942 he sent a copy of the infinity theses ([Zer21]) to Heinrich Scholz.[311] Scholz answered[312] by pointing out vague phrases and raising numerous questions. Apparently, however, Zermelo did not react.

A Book on Set Theory

In April 1928[313] Zermelo agreed with the Akademische Verlagsgesellschaft Leipzig to write a book on set theory.[314] Around 1932, perhaps after the edition of Cantor's collected papers had been completed, he worked on the project. The *Nachlass* contains three fragments:[315] (1) an undated fragment of eight pages consisting of one page of the introduction and seven pages of the main text; (2) an undated presentation of the number systems including a proof of the uncountability of the set of real numbers and comprising fourteen pages, each page headed "Mengenlehre;" (3) a fragment of five pages entitled "Mengenlehre 1932" which right away starts with sets as categorically defined domains (cf. 4.6.2).

The introduction of the first manuscript begins with a clear statement: "Set theory deals with the structure of the mathematical sciences: of the mathematical domains and their rules of inference.[316] So set theory should also treat logic. But logic meant infinitary logic where the infinitary propositions are built up parallel to the cumulative hierarchy (4.9.4). Taking into consideration also the categoricity concept of the third manuscript, it becomes obvious that the intended textbook was to reflect important features of Zermelo's set-theoretic views at this time.

Apparently, work on the book was far from being intensive. In a letter of 11 December 1934[317] Arnold Scholz suggested that Zermelo try to renew

---

[311]For Heinrich Scholz cf. 2.4, fn. 60.

[312]UAF, C 129/106.

[313]Correspondence with Emil Hilb, 14 March to 3 April 1928; UAF, C 129/49.

[314]There is evidence that he had an already earlier agreement with the Verlagsgesellschaft: On 10 January 1928 Wolfgang Krull writes (letter to Zermelo; UAF, C 129/66) that he had been asked by them to "explore cautiously and diplomatically when the book would be ready."

[315]UAF, C 129/290.

[316]"Gegenstand der Mengenlehre ist die Struktur der mathematischen Wissenschaften: der mathematischen Bereiche und ihrer Schlußweisen."

[317]Letter to Zermelo; UAF, C 129/104.

the 1928 agreement with the publisher and to resume writing, "now, that you have blown your car and get round less to this sport." However, there still seems to have been no further progress. Even more: In 1941 Zermelo finally gave up, pursuing plans now for a book "Mathematische Miniaturen." When Scholz heard about this, he tried to change Zermelo's mind:[318]

> It does not please me at all that you want to leave set theory now where you have a reasonable agreement [with a publisher]. You can collect [the miniatures] on the side in hours of leisure.

Sharing Zermelo's set-theoretic point of view to a great extent – although more conciliatory – he offered to write the book himself in Zermelo's spirit. On 7 August 1941 he sent detailed proposals for the first two chapters.[319] The cooperation encouraged Zermelo to continue work on the text even after Scholz's sudden death in February 1942. In late 1942, when answering a letter which Zermelo had sent him perhaps in the early summer, Heinrich Scholz wrote:[320]

> I need not tell you that I have partaken with the greatest interest in all you have bestowed on me. For the first time I learn from you that you are working on a book on set theory. It will be highly interesting for us, and I would very much appreciate it if you could tell me without obligation when you expect it to be completed. The same applies to your paper on finite sets. I assume that it will appear in a mathematical journal, and I would like to ask you now to send me at least two copies of it for our seminar. (OV 4.50)

In June 1953 the philosopher Gottfried Martin who had attended Zermelo's course on set theory in the winter semester 1933/34 and since then had stayed in contact with him, wrote to Süss[321] that around 1943 Zermelo had informed him about his search for a publisher of a completed manuscript on set theory. Martin comes back to this in connection with plans for an edition of Zermelo's collected works.[322] As the *Nachlass* does not contain a longer text on set theory and faced with the unsuccessful history of the project, it is doubtful whether Zermelo ever completed the book.

Mathematical Miniatures

In October 1940 Zermelo wrote down a list of sixteen themes which he planned to present in his *Mathematical Miniatures. A Collection of Entertaining Exercises for Mathematicians and Friends of Mathematics*,[323] among them sev-

---

[318]Letter to Zermelo of 10 March 1941; UAF, C 129/104.

[319]Letter to Zermelo; UAF, C 129/104.

[320]Letter to Zermelo of 3 October 1942; UAF, C 129/106.

[321]Letter of 30 June 1953; UAF, C 89/331.

[322]Letter to Helmuth Gericke of 4 July 1960; MLF.

[323] *"Mathematische Miniaturen. Eine Sammlung unterhaltender Aufgaben für Mathematiker und Freunde der Mathematik;"* UAF, C 129/278.

eral topics which he had treated in papers of his own such as a proof of Poincaré's recurrence theorem ([Zer96a]) ("The Eternal Recurrence According to Nietzsche and Poincaré: A Theorem of Dynamics"), the evaluation of a chess tournament ([Zer28]) ("Chess Tournament: Evaluation of its Participants: A Maximality Problem of the Theory of Probability"), or the solution of the navigation problem for airships ([Zer30c]) ("Navigation in the Air: Making Use of Winds in Aeronautics"). Besides one or two topics on properties of infinite sets ("Dancing Class and Hotel Operation Here and in Infinity-Land: One-to-One Mappings of Finite and Infinite Totalities" and maybe also "Rules of Life and Constitution in Infinity-Land") there are no other topics related to foundational questions.[324] After first elaborations of some parts, the book never came to fruition.

4.12.3 The Time of Resignation

When Arnold Scholz died on 1 February 1942, Zermelo not only lost a colleague with whom he had intensively discussed mathematics and whose foundational views were close to his own – he also lost his best friend. Each of Scholz's letters documents care for "Dear Zero." Scholz gave advice in everyday problems and was active in the background to help in scientific matters.[325] The letter of Heinrich Scholz mentioned above mirrors the great loss Zermelo had suffered:

> It moves me in an unusual sense that you miss Arnold Scholz so much. I saw him only once here in Münster; however, it was worth the effort, and his number-theoretic booklet which he brought out shortly after in the Göschen series,[326] shows quite an unusual carefulness and accuracy also in the foundations. Up to now I believed that we foundational researchers in the narrow sense are the only ones who mourn for him. I feel comforted when hearing from you, that you are of the same opinion. I did not understand why he has stayed so much hidden away. (OV 4.52)

Still in the spring of 1941 Arnold Scholz had helped to organize a colloquium in Göttingen on the occasion of Zermelo's 70th birthday. It took place on

---

[324] For the full list (in German) see OV 4.51.

[325] Olga Taussky-Todd reports ([Tau87], 38) that in Bad Elster in 1931 he had been anxious to bring Gödel and Zermelo together.

[326] [Schol39]. Zermelo's *Nachlass* contains two larger fragments of a typewritten manuscript on elementary number theory (UAF, C 129/297 and C 129/268). Obviously, they are an early version of parts of this book, written perhaps around 1936/37 (cf. letter from Arnold Scholz to Zermelo of 14 March 1937; UAF, C 129/104). Entries in Zermelo's hand give evidence that Arnold Scholz had asked him for comments. Years after Scholz's death Bruno Schoeneberg prepared a revised edition ([Schol55]) where the proofs in the introductory chapter which concern metamathematical aspects were eliminated. – Moreover, the *Nachlass* contains a typewritten note about the divisibility of integers of the form $a + bx^2$ (UAF, C 129/226) which probably stems from Arnold Scholz, too.

Zermelo, probably around 1940
Mathematisches Institut der Universität Erlangen-Nürnberg
Sammlung Jacobs, Nr. 165/32

19 July. Among the speakers were Konrad Knopp and Bartel Leendert van der Waerden. Zermelo gave three talks:[327] "Rubber Ball and Lampshade" on the folding and bending of flexible surfaces,[328] "Building Roads in the Mountains" on shortest lines of limited steepness,[329] and "How Does One Break a Piece of Sugar?" on splitting lines of a rectangle.[330] Later he wrote to Süss[331] that "such opportunities have become so rare that they are seized with great pleasure." The birthday party was held in a private circle with a friendly family.

The time that had passed since the loss of his professorship had seen work on various enterprises; however, Zermelo had been lacking the ability to systematically pursue them. This development may have been natural, given his

---

[327] According to an announcement of the Göttingen Mathematical Society; UAF, C 129/129.
[328] "Gummiball und Lampenschirm" (cf. miniature (9)); UAF, C 129/289.
[329] "Straßenbau im Gebirge" (cf. miniature (4)), following [Zer02c].
[330] "Wie zerbricht man ein Stück Zucker?" (cf. miniature (8)), following [Zer33a].
[331] Letter of 18 August 1941; UAF, C89/086.

increasing age and nervous destabilization following the Nazi affair. However, there might still have been another point. As discussed earlier, he was facing difficulties in bringing together diverging concepts of sets and his infinitary languages. So the main foundational task was still unfinished and the "particular duty" to fight against finitism still unsatisfied. Concerning his book on set theory, it may be more than just a hypothesis that it also was the insight into the seemingly unsurmountable difficulties still ahead which caused him to give up. If one remembers the formerly engaged formulations and clear statements about his aims and his convictions, one cannot imagine that he would have been satisfied with writing a book, lacking a conclusive base. In some way, this failure mirrors his overall situation. Facing an increasing weakness and feeling exposed to scientific isolation, at the same time being confronted with the success of a direction of mathematical logic which he was not willing to accept, he finally resigned with a taste of disappointment. In the letter which he sent to Paul Bernays on 1 October 1941 as a response to Bernays' birthday congratulations he writes ([Zer41]):

> I am very pleased that some of my former colleagues and co-workers still remember me, while I have already lost many a friend of mine through death. Well, one always gets more lonely, but therefore is more grateful for each friendly remembrance. [...] With respect to the impact of my intrinsic life's work, as far as it concerns the "foundations" and set theory, I certainly no longer have any illusions. If my name is quoted, it always occurs *only* with respect to the "principle of choice" for which I have never claimed priority.[332] This also happened at the foundational congress in Zurich.[333] I was not invited, but you were so friendly as to send me the report. As far as I could convince myself till now, *no* talk and *no* discussion quoted or took into consideration any of my papers which appeared after 1904 (in particular the two Annalen notes from 1908[334] and the two notes which in the last decade appeared in the "Fundamenta."[335] On the other hand, the questionable merits of a Skolem or a Gödel were rolled out at large. [...] But there may come a time where my papers will also be rediscovered and read. (OV 4.53)

4.13 Twilight of Life

No document in the *Nachlass* shows that Zermelo, now in his seventies, worked on scientific projects after Scholz's death. Everyday problems increased during

---

[332]With this attitude Zermelo follows his conviction "that not mathematical principles, which are common property, but only the proofs based upon them can be property of an individual mathematician" ([Zer08a], 118, first footnote; [Zer67b], 191, footnote 8).

[333]Possibly the Congress *Les fondements et la méthode des sciences mathématiques* in 1938; see [Gon41].

[334][Zer08a], [Zer08b].

[335][Zer30a], [Zer35].

the last years of the war and the time following. It was mainly Gertrud who tried to maintain a modest way of life. Zermelo admired her ability "to make a good meal from nothing." Unfortunately, his vision problems – rooted in a glaucoma – got more serious. In late 1944 plans for an operation were cancelled because it did not promise success. Apparently Zermelo had been too negligent and thus missed having a treatment begun in time. During the next years his powers of vision deteriorated, and in 1951 he was blind.

In May 1945 the Second World War ended together with the Nazi dictatorship. Southwestern Germany came under the control of a French military government which tried to reinstall the necessary structures, at the same time taking measures to ward off the influence of former supporters of the fascist regime. Those professors of the University who were suspected of having been involved in Nazi activities were suspended from duties and could return only after having passed an investigation of their behaviour during the Third Reich. Central Freiburg, where the University was located, had been almost completely destroyed in an air attack by the Allied Forces on 27 November 1944, so regular work in the University could start only step by step.

As far as the Mathematical Institute was concerned, Süss as an office bearer of the "Third Reich" and a former member of the NSDAP, the National Socialist Party, was suspended in the summer of 1945. However, two months later he was reinstalled. During the preceding years he had used his influence to help individuals, and by a skilful cooperation with the Nazis, partly as Rector of the University, he had succeeded in obtaining benefits for university institutions in general and for mathematics in particular. We have already mentioned the establishment of the now world wide renowned Mathematical Research Institute Oberwolfach in 1944. He therefore enjoyed a good reputation not only among German mathematicians and his Freiburg colleagues, but also with the French occupying power.[336]

In contrast to Süss, Doetsch had not been a member of the NSDAP. Therefore he was considered formally innocent. Nevertheless the so-called purification committee accused him of "character defects." He was suspended in winter 1945/46. The purification procedure lasted six years and ended with his re-installment in 1951. As the University refused to cooperate with him, he spent the time until his retirement in 1961 in isolation and with a *de facto* ban from entering the Mathematical Institute.[337]

On 23 January 1946, Zermelo wrote to the Rector's office:[338]

> As a former university professor of mathematics at the University of Zurich,
> by your leave, I would like to inform Your Magnificence of the following.

---

[336]The presentation here and with Doetsch below follows [Rem99a], 81seq. Remmert has a critical look there at Süss' role during the Nazi time.

[337]Cf. [Rem01], 391, or [Seg03], 104. – Doetsch's courses of lectures also took place outside the Institute. They were appreciated among students because of their quality (oral communication from several former listeners).

[338]UAF, B 24/4259.

In 1916 I was forced by illness to temporarily give up my teaching activity and was permanently retired. As a citizen of the German Reich, I soon afterwards took permanent domicile in this university town. As an *honorary professor*[339] I gave regular lectures in pure and applied mathematics for a number of years until I was forced under the Hitler government by political intrigues to give up this activity. Circumstances having changed now, I would readily take it up again to a modest extent as far as my condition allows. I therefore request the University in agreement with the Faculty of Science to look favourably on my reappointment as an honorary professor.

In proper and deep esteem

Dr. E. Zermelo

Retired University Professor. (OV 4.54)

The Faculty supported Zermelo's application as an act of compensation, recommending that he be informed that because of his age the University did not expect him to really teach again. The Senate gave its approval on 13 March. After the necessary formal procedures required by the French military government, on 23 July, some days before his 75th birthday and a little bit more than ten years after his dismissal, the responsible minister of the subordinate German administration appointed him an honorary professor at the University of Freiburg.

It seems that the reinstatement once more motivated him to get back to mathematics. The *Nachlass* contains a list of possible talks from 17 October 1946 with the following items:[340]

(1) On the Significance of the General Well-Ordering Theorem for the Theory of Finite and Denumerably Infinite Sets.
(2) On the Logical Form of Mathematical Propositions and Proofs.
(3) On the Bending of Surfaces and the Emergence of Craters on the Moon.
(4) On the Second Law of Thermodynamics and the Kinetic Theory of Gas.
(5) On a Mechanical Model to Represent Keys and Harmonies.

However, as it seems, his failing physical condition did not admit consistent work. Moreover, his financial circumstances became a real burden. Since the end of the war the Canton Zurich that up to then had paid his pension, stopped the transfer to post-war Germany and left the Zermelo couple without any income. After one year their savings had been spent. Zermelo therefore tried to get permission to settle in Switzerland, hoping to obtain the barred pension there and thus escape the financially depressing situation in Freiburg. A letter which he wrote on 18 January 1947, about half a year after he had applied for the necessary documents, shows the state of resignation caused by his then still unsuccessful efforts and also by his fear of having lost his pension for ever:[341]

---

[339]Emphasis by Zermelo in red letters.
[340]UAF, C 129/278; cf. OV 4.55 for the German version.
[341]Letter to an unknown addressee; MLF.

Best thanks for your friendly postcard from [gap]. Unfortunately, your endeavour was in vain again. After six months of waiting, the local Swiss consulate did not answer my application for admission by a direct decision, but by a notification which might render any further efforts hopeless. I enclose a clipping of it. As I have no relatives in Switzerland and also no invitation from a university or an authority and let alone the willingness of such an authority to pay or to guarantee the costs of my stay, I may consider my application as being definitely rejected. I may also write off my "pension" which had been explicitly approved as "lifelong" when I retired. For all my inquiries concerning it have not been answered. This is what happens if one accepts a however tempting appointment in a however neutral foreign country. Well, in such a case one is entirely without rights, and nothing helps. (OV 4.56)

Fortunately, the local authorities could be brought to give an advance on the money that was accumulating on his Swiss account. Finally, in 1947 and in 1948 he was allowed to spend a couple of weeks in Switzerland, living from the money deposited there. During his first stay he could not be accompanied by his wife because her travelling documents were insufficient. He met Paul Bernays several times and enjoyed the contrast between Zurich and the war-torn Freiburg. On a postcard to his wife of 25 October 1947 (MLF) he writes:

Everywhere one finds good food and good beverages, and trams are going in all directions. It is a pity that I cannot stay here forever. (OV 4.57)

As planned with his wife, he bought urgently needed clothes. In order to avoid customs inspection when crossing the German border, he wore the pieces he had acquired for his wife under his own clothes. He had informed her about his plans days before in a simple postcard. When he left the train on his arrival in Freiburg, he was sweating because of several pullovers under his jacket and hobbling because of the pain caused by the ladies shoes he was wearing. According to his wife, this mixture of helpfulness and clumsiness in things of daily life was a dominant feature of their living together. During the second stay the couple spent six weeks in Switzerland, partly at the lake of Lugano and free of the permanent worries in Freiburg. Mrs. Zermelo enjoyed the hospitality and friendliness of the people, whereas he grumbled about them in a way that may evoke the hotel story discussed earlier (3.1.1, 3.1.5).

In 1948 a monetary reform took place. The Reichsmark, having become almost worthless, was replaced by the Deutsche Mark. Zermelo's advance was re-fixed, but soon turned out to be insufficient again. Mrs. Zermelo therefore applied for an increase. Even though the Rector's Office supported her endeavour in the "warmest" way and the local government benevolently contacted the Swiss authorities responsible, the Swiss clearing office announced reservations because the Swiss pension would not allow an increase. But it offered another stay in Switzerland and was ready to spend 1,000 Swiss francs on the costs. Because of the relief that would result from this amount, the German authorities refused to raise the monthly rate of advance, thus leaving the couple in their insufficient financial situation. When regular payments

The Zermelo couple at Lugano in 1948

from Switzerland were allowed in 1950, the shortage was not remedied, because now the advance had to be paid back. Pieces of paper in the *Nachlass* filled by Zermelo with costs of household goods give evidence of the deficiency. Apparently, Zermelo once more made plans to settle in Switzerland. On 15 September 1950, the Swiss Consulate in Baden-Baden asks for references from persons "also residing in the Canton of Waadt;" they doubt that the Zurich pension was sufficient for the Zermelo couple to live there.

In September 1949 Mrs. Zermelo had already applied for compensation. She had argued that the dismissal of her husband ended their income from students' fees. The legal procedures to clear this question lasted almost eight years. It wasn't until five years after Zermelo's death, that the government of the Federal State of Baden-Württemberg which was now responsible, paid a compensation of about 1,200 German Marks.[342]

The deplorable situation caused by financial need, increasing blindness, and separation from scientific activity Zermelo had to suffer may have intensi-

---

[342]Letter from the State Office of Redress of 21 March 1957; MLF.

fied his feeling of falling into oblivion. On 3 March 1949 he sent his publication list to the Springer Verlag, suggesting an edition of his collected works under the title *Mathematische und physikalische Abhandlungen 1894–1936.*[343] He also may have had in mind the possibility of improving his financial situation by the royalties he could have expected. Springer answered that the momentary circumstances did not allow such an edition.[344] Apparently Zermelo made a second unsuccessful attempt with Keiper, the publisher of Planck's "Erinnerungen," and, on 20 January 1950, a third one[345] with the publisher Johann Ambrosius Barth, where his translation of Gibbs' *Elementary Principles in Statistical Mechanics* had appeared in 1905:

> As I have not found an editor up to now for my own works of mathematical content, the publishing house Keiper has recommended yours to me [. . .] and I therefore ask you for the kind information whether you would perhaps be able to undertake this edition. For reasons of health I am now no longer able to publish new scientific works. (OV 4.58)

Barth answered negatively as well,[346] arguing that the situation had totally changed since 1905. Meanwhile their programme consisted of text books in applied sciences, and under the prevailing circumstances the paper needed for the collected works could not be allocated.

Zermelo's 80th birthday on 27 July 1951 brought congratulations from good friends, among them Heinz Hopf and Reinhold Baer.[347] Baer remembered the pleasant days when he had worked with Zermelo and they had spent hours in discussion and "had made fun of their fellow men." Moreover:

> I have dedicated a little paper on theorems of finiteness[348] to the magician who has put the magic wand of well-ordering at our disposal for travelling in the infinite – this may sound a little bit paradoxical, but is meant not less gratefully and cordially.[349]

Herbert Bilharz dedicated his paper on the Gaussian method for the approximate calculation of definite integrals ([Bil51]) to him. Heinrich Behnke, then editor of the *Mathematische Annalen,* was also among the congratulators. He regretted that lack of time had not permitted the preparation of a special issue, although numerous colleagues would have liked to contribute a paper.[350]

---

[343]UAF; C 129/271.

[344]Letter of 14 March 1949; UAF, C 129/136.

[345]Copy of the letter in UAF, C 129/127.

[346]Letter of 26 January 1950; UAF, C 129/127.

[347]If not stated otherwise, the documents concerning Zermelo's 80th birthday quoted here are in MLF.

[348][Bae52].

[349]On the occasion of Zermelo's 60th birthday Baer and Friedrich Levi had dedicated their paper [BaeL31] to him; it makes essential use of the well-ordering theorem.

[350]Erich Kamke, then president of the Deutsche Mathematiker-Vereinigung, had refused to initiate a special issue of *Mathematische Zeitschrift,* after having discussed

Zermelo in March 1953, some months before his death

The Faculty apologized for sending only a letter, because Zermelo's state of health did not allow a personal congratulation. In his answer[351] Zermelo regrets not having been able to welcome the colleagues in his apartment in order to thank them personally for the gifts. The congratulatory letter of Wilhelm Süss touches upon the unfortunate events of the past:

> Your name has been well-known to mathematicians all over the world for a long time through your fundamental papers on set theory. Fate and human malice later cast many a shadow over the glorious fame of your youth. Whoever had the good fortune we had, to get to know you personally, also knows how your generous heart was able to bear all disagreements.
>
> For many years our Institute was allowed to enjoy your cooperation, until a denunciation forced you to leave the University for years. You did not re-

---

this question with Hellmuth Kneser and Konrad Knopp. He argued that "for years Zermelo has not distinguished himself mathematically." Instead he vaguely brought up an honorary doctorate in Göttingen or in Berlin, but apparently did not pursue this idea (letter from Kamke to Hermann Ludwig Schmid of 24 January 1950; copy in UAF, C 129/58).

[351]UAF, B 15/736.

pay evil with hate here, but were later satisfied with a formal rehabilitation, a further sign of your generous character. (OV 4.59)

Less than two years later and two months before his 82nd birthday, in the morning of 21 May 1953, now totally blind, Zermelo died without suffering from a specific illness. On 23 May he was buried in the churchyard of the monastery in the heart of Günterstal. During the funeral the university flag was hoisted to halfmast. Because of her shortage of money, his widow could only afford a public grave. When it was to expire twenty years later, an initiative supported by the University together with the Faculty of Mathematics took care that he was reburied. He has now found a final resting place in the same churchyard, not far away from the grave of Edmund Husserl. A simple grey stone plate gives his name.

Among the expressions of sympathy addressed to Mrs. Zermelo we find a letter from Bernays of 17 July 1953 (MLF) which in a few, but warm words describes essential features of Zermelo's personality and fate:

> During my first stay in Zurich – at that time [Zermelo] held the Zurich professorship of mathematics and I was *Privatdozent* there – we often sat together exchanging ideas. He had great wit and *esprit.* So I got much intellectual stimulation from him – beyond what I learnt scientifically from him, too. Moreover, I always enjoyed his personal goodwill.
>
> His scientific achievements, in particular in the field of set theory, carry a lot of weight in today's foundational research; I am sure, they will always retain their significance and will keep his name in science alive.
>
> As for you, dear Mrs. Zermelo, you can have the gratifying awareness that you made your husband's life easier in the later years of his life and kept him from desertedness and sad loneliness. All those who with appreciation remember the striking and original researcher personality represented by your husband, will be grateful to you. (OV 4.60)

Zermelo's death left his widow without income. As she had married her husband only after his retirement in Zurich, she did not receive a widow's pension. Being unable to work because of her impaired health, she had to rely on public support. In view of the still unsettled social problems caused by the consequences of the war, it took a great effort from the University to secure a very modest livelihood, which started in August 1954. Because of the shortage of flats in post-war Germany and because of the rent she had to leave the flat which she had shared with Zermelo and, hence, also the knight's hall. She was lucky not to be forced to leave Bernshof, but could move into a smaller apartment at the rear of the building. There she could still enjoy the views of the park and the surrounding mountains. Helmuth Gericke, then professor for the history of mathematics at the University of Munich, who had become acquainted with her husband in the 1930s when he was an assistant to Süss, helped her to sift through Zermelo's *Nachlass* and have it acquired by the University Archives. Together with the philosopher Gottfried Martin from the University of Bonn he started preparations to edit Zermelo's collected papers. First plans for such an edition were already conceived by Martin shortly

Memorial slab on the grave of the Zermelo couple
Photo by Susanne Ebbinghaus

after Zermelo's death.[352] They foresaw two volumes, the first one containing the collected papers, the second one comprising pieces from the *Nachlass* and a biography.[353] Both Gericke and Martin agreed to invite Bernays to participate. On 20 Februar 1956 Martin turned to him (MLF):

> I owe very much to Zermelo in both scientific and personal respect and would like to pay off a little bit of these debts by trying to get his collected works printed. [...] It would be highly desirable for these collected works of Zermelo to have a presentation put at the beginning which sketches the historical situation in which they originate, that furthermore says which significance they had and which developments they initiated. I would highly appreciate it if you, very esteemed Mr. Bernays, as the most competent expert on these connections, would be willing to give such a presentation.

Bernays agreed to write the introduction and to serve as an official editor.[354] Furthermore, the Kant Society promised financial support stemming from a private donation.[355] Work really started, but came to a standstill in 1962 because Gericke who was in charge of the major part, was absorbed by other projects.

---

[352]Letter from Gericke to Bilharz of 19 December 1953; MLF.
[353]Letter from Martin to Gericke of 21 August 1962; MLF.
[354]Letter to Martin of 6 July 1956; MLF.
[355]Letters from Martin to Gericke of 16 November 1956 and 16 April 1958; MLF.

Gertrud Zermelo in August 1999
Photo by the author

Since the late sixties, many a user of Zermelo's *Nachlass* also paid a visit to his widow, enjoying her warm hospitality and open-minded cheerfulness. They listened to what she had to tell about her late husband, and they heard of the mutual care which had shaped their relationship. One or the other participant of the regular Oberwolfach conferences on set theory also combined the meeting with a visit to Bernshof. When she learnt about plans for a biography, she instantly offered her help, provided information and material about Zermelo's family and, in particular, conveyed a picture of his personality beyond what can be found in documents. On 3 September 2002 she celebrated her 100th birthday still in good health. But shortly afterwards her vitality started to fade away. On 15 December 2003, more than fifty years after her husband, she died in her apartment on beloved Bernshof. Before her death she had made her farewells to those who had been close to her during her last years.

5

A Final Word

Looking back on Zermelo's life one gets the impression that its characteristic features condense in a variety of broad scenes partly extending into controversial extremes: His scientific interests range from theoretical physics via applied mathematics to number theory and the foundations of mathematics; among his intellectual interests, engineering practice comes along with music theory, philosophy, and classical poetry. Descriptions of his personality use terms such as "irascible" and "strange," but also "helpful" and "generous."

Due to the fact that his important publications are linked to scientific controversies where he chooses a polemical form of discussion, showing nearly no willingness to compromise, the picture as present in literature tends towards the "irascible" side. Sharpness could also be a feature in non-scientific affairs. However, as a rule the controversial character was tied to objective facts or strong personal convictions. Furthermore, he did not stop committed criticism in cases where he had to fear serious consequences for himself.

Zermelo's controversial nature was accompanied by the impression of being inadequately recognized. He often reveals an intensive wish for a greater resonance of his work. The feeling of insufficient recognition becomes understandable when taking into account that his university career was slowed down by illness and, after having finally led to a permanent professorship, ended when he was at the age of forty-five.

Private judgements reveal the other side of his personality. Because of his broad interests and his enthusiasm in following them, he was considered a stimulating conversationalist. Paul Bernays "experienced much intellectual stimulation from him – also beyond what [he] learnt scientifically from him," Helmuth Gericke described Zermelo as an "original personality with many witty ideas,"[1] and Bronisław Knaster remembered his "aphorisms" and his "wise and astute remarks about the world and the people." In a report accompanying the application of the Freiburg Faculty of Science and Mathematics in

---

[1]Letter from Gericke to the editor of [vanR76] of 13 December 1970; MLF.

1946 to re-appoint Zermelo to an honorary professorship,[2] Wilhelm Süss gives the following characterization that closes up with the impression conveyed in the conversations with his widow:

> Yet Zermelo is by no means only the withdrawn scholar who knows nothing else but his science. Only those who have got into closer contact with him, certainly know about the enthusiasm and the good understanding by which he has occupied himself, for instance, with the classical world. Only very few people will know for example his translations of Homer. Where he has confidence, he shows a childlike and pure nature and – despite an often hard destiny – a generous heart.

During his time in Göttingen, Zermelo, being described as nervous and solitary ([Rei70], 97), was nevertheless a cooperative member of Hilbert's group. His scientific contacts sometimes developed into real friendships such as those with Constantin Carathéodory, Gerhard Hessenberg, and Erhard Schmidt.[3] Documentation of his role as a colleague in the post-Göttingen era is rather lacking. Apparently he was considered a competent partner. Some liked to discuss mathematical matters with him, like Albert Einstein in Zurich, others seem to have ignored him to a great extent. In Freiburg he developed a closer scientific and personal relationship only to the most talented assistants, Reinhold Baer and Arnold Scholz, inspiring both and arousing their interest in foundational questions.

Zermelo liked to attend courses given by colleagues in mathematics or other disciplines such as physics and philosophy. Concerning his own courses, there is evidence that he prepared them carefully but was not a lecturer who inspired the average student. As Mrs. Zermelo remembered from the time around 1934, her husband liked to invite his students to Bernshof where he was together with them in a relaxed atmosphere.

Zermelo's most influential scientific work falls into his first period of activity. Not large in number, the corresponding papers led to new conceptual developments. By his doctoral dissertation of 1894 he promoted field-theoretic aspects in the calculus of variations. His controversy with Ludwig Boltzmann in 1896/1897 led to a reconsideration of the preconditions of the kinetic theory of heat. His 1913 paper about a set-theoretic application to the theory of chess is considered the beginning of game theory. Even more important, his papers from 1904 and 1908 became the starting point of modern axiomatic set theory. His work on the axiom of choice was ahead of the conceptions of set theory at that time.

The papers in applied mathematics which he published around 1930, during his second period of activity, found a positive echo. But in contrast to the fruits of his early research, the foundational papers of this second period –

---

[2] UAF, B 15/736.

[3] According to Sanford Segal he "seems to have inspired people to friendship" ([Seg03], 468).

the interplay of large cardinals and inner models of set theory in the cumulative hierarchy and the development of an infinitary logic – failed to have an immediate influence or even had no resonance at all. There are various reasons: The development of set theory and logic in the thirties was based on formal systems of logic. As proved by Kurt Gödel and Thoralf Skolem, the restriction to such formal systems leads to limitations on the characterizability of mathematical structures and the completeness of mathematical reasoning, leaving mathematicians with an unavoidable weakness of their methods. Instead of regarding these results as contributing to a methodological clarification, Zermelo looked upon them as a danger for mathematics itself. He did not strive for their meticulous technical understanding and provided his alternatives without the standards of precision which Gödel and Skolem had set and which promoted the development of mathematical logic into a discipline of its own inside mathematics. According to a characterization by John W. Dawson ([Daw97], 75–77), Zermelo's fight was a "reactionary" one. In the time following he was therefore considered a logician of the past and suffered the fate of scientific isolation. Taking into consideration that it was he who had created axiomatic set theory and, with his highly debatable notion of definiteness, had initiated the road to a first-order version of set theory and the triumphal march of first-order logic, this development is regrettable. About half a century later, some logicians were ready to share reservations against a dominant role of first-order logic. Jon Barwise argued against the "first-order thesis" according to which logic in essence is first-order logic ([Bar85], 3–6):

> The reasons for the widespread, often uncritical, acceptance of the first-order thesis are numerous. Partly it grew out of interest in and hopes for Hilbert's program. Partly it was spawned by the great successs in the formalization of parts of mathematics in first-order theories like Zermelo-Fraenkel set theory. And partly, it grew out of a pervasive nominalism in the philosophy of science in the mid-twentieth century. [...] The first-order thesis [...] confuses the subject matter of logic with one of its tools. First-order language is just an artificial language structured to help investigate logic, much as a telescope is a tool constructed to help study heavenly bodies. From the perspective of the mathematician in the street, the first-order thesis is like the claim that astronomy is the study of the telescope.

Whether such opinions are justified or not, that they have been held by well-known logicians attests to a certain uneasiness with logical systems if measured against mathematical needs. In this sense, Zermelo might have felt in principle confirmed by them.

Summing up, we see an intellectually extraordinarily broad personality who with brilliant incisiveness and creativity contributed to mathematical and physical science. Not a friend of systematic development and exploration, he would have needed a group to fully unfold and discuss the consequences of his ideas. In Göttingen he had enjoyed such a surrounding. However, he himself did not found what could be called a school. There are several reasons, each of them a sufficient one. His life did not readily offer an opportunity to build up a school: illness and his early retirement from a first professorship, continuing illness and the loss of a second teaching position under deplorable circumstances. Furthermore, he was not a man with a penchant for cooperation, but a man with strong reservations against "cliques" and with an urge for scientific freedom which despised all opportunism, a man of solitude who liked controversial debate and only in individual cases opened himself to others. Modifying the picture of the magic wand which Reinhold Baer used when congratulating him on his 80th birthday, one might say that he did not leave us with a well-cultivated garden, but with precious plants, among them the plant of axiomatic set theory which in the meantime and thanks to the care of those who were given it has unfolded to impressive beauty.

Zermelo's *Curriculum Vitae*

1871

27 July: Zermelo is born in Berlin (p. 1).

1878

3 June: Death of Zermelo's mother Auguste Zermelo née Zieger (p. 2).

1889

24 January: Death of Zermelo's father Theodor Zermelo (p. 4).

March: Zermelo finishes school. Remarks in his leaving certificate show that he suffers from physical fatigue (p. 6).

Summer semester – summer semester 1890: Zermelo studies mathematics and physics at Berlin University (p. 7).

1890

Winter semester 1890/91: Studies at the University of Halle-Wittenberg, among others with Georg Cantor and Edmund Husserl (p. 7).

1891

Summer semester 1891: Studies at Freiburg University (p. 7).

Winter semester 1891/92 – summer semester 1897: Studies again at Berlin University (p. 8).

1894

23 March: Zermelo applies to begin the Ph. D. procedure (p. 8).

6 October: Zermelo obtains his Ph. D. degree. His dissertation *Untersuchungen zur Variations-Rechnung* was supervised by Hermann Amandus Schwarz (p. 8, p. 10).

December – September 1897: Zermelo is an assistant to Max Planck at the Institute for Theoretical Physics of Berlin University (p. 9).

1895

December: Zermelo completes his first paper ([Zer96a]) opposing Ludwig Boltzmann's statistical theory of heat (p. 18).

1896

Summer: Zermelo applies for an assistantship at the Deutsche Seewarte in Hamburg (p. 9).

September: Zermelo completes his second paper ([Zer96b]) opposing Boltzmann (p. 21). p

1897

2 February: Zermelo passes his state exams *pro facultate docendi* (p. 9).

19 July: Zermelo asks Felix Klein for support for his *Habilitation* (p. 27).

Winter semester 1897/98: Zermelo continues his studies at Göttingen University (p. 9, p. 27).

1899

February: David Hilbert presents Zermelo's first paper in applied mathematics ([Zer99a]) to the Königliche Gesellschaft der Wissenschaften zu Göttingen; it treats differential equations with inequalities (p. 28). Opening of the *Habilitation* procedure with the *Habilitation* thesis *Hydrodynamische Untersuchungen über die Wirbelbewegungen in einer Kugelfläche* whose first part is published as [Zer02b] (p. 29). The second part of the thesis remains unpublished; it contains a solution of the 3-vortex problem on the sphere (p. 31).

4 March: Zermelo gives his *Habilitation* address ([Zer00a]) which opposes Boltzmann's statistical mechanics (p. 23, p. 31).

Around 1900

Beginning of the cooperation with Hilbert on the foundations of mathematics. Zermelo formulates the Zermelo-Russell paradox (p. 45).

1900

Wintersemester 1900/01: Zermelo gives a course on set theory. According to [Moo02a] it is the first course of lectures ever given which is entirely devoted to set theory (p. 48).

1901

9 March: David Hilbert presents Zermelo's result on the addition of cardinals ([Zer02a]) to the Königliche Gesellschaft der Wissenschaften zu Göttingen. The proof uses the axiom of choice (p. 49).

1902

12 May: Zermelo gives a talk on Frege's foundation of arithmetic before the Göttingen Mathematical Society. It shows that the set-theoretic paradoxes were now taken seriously by the Hilbert group (p. 78).

Summer semester – winter semester 1906/07: Zermelo receives a *Privatdozenten* grant (p. 33).

Zermelo publishes his first paper on the calculus of variations ([Zer02c]). It treats shortest lines of bounded steepness with or without bounded torsion (p. 14).

1903

June: Zermelo is under consideration for an extraordinary professorship of mathematics at Breslau University. He is shortlisted in the second position after Gerhard Kowalewski, Franz London, and Josef Wellstein who are shortlisted *aequo loco* in the first position (p. 34).

December: Zermelo completes his second paper on the calculus of variations ([Zer03]). It gives two simple proofs of a result of Paul du Bois-Reymond on the range of the method of Lagrange (p. 14).

1904

Beginning of a life-long friendship with Constantin Carathéodory (p. 32).

Together with Hans Hahn, Zermelo writes a contribution on the calculus of variations for the *Encyklopädie der mathematischen Wissenschaften* ([HahZ04]) (p. 15).

August: Third International Congress of Mathematicians at Heidelberg. Julius König gives a flawed refutation of Cantor's continuum hypothesis. Zermelo's role in detecting the error is still being discussed (p. 50).

24 September: Zermelo informs Hilbert about his proof of the well-ordering theorem and the essential role of the axiom of choice. The letter is published as the well-ordering paper [Zer04] (p. 47, p. 53).

15 November: During a meeting of the Göttingen Mathematical Society, Zermelo defends his well-ordering proof against criticism of Julius König, Felix Bernstein, and Arthur Schoenflies (p. 59).

1905

January: Zermelo falls seriously ill. In order to recover he spends spring and early summer in Italy (p. 105).

German translation ([Gib05]) of [Gib02] (p. 24).

May: Work on the theory of finite sets which finally results in the papers [Zer09a] and [Zer09b] (p. 61).

21 December: Zermelo receives the title "Professor." The application had been filed by Hilbert in December 1904 (p. 34).

1906

Work on a book on the calculus of variations together with Carathéodory (p. 33, p. 35). Final criticism of Boltzmann's statistical interpretation of the second law of thermodynamics in a review ([Zer06a]) of [Gib02] (p. 25).

Early that year: Zermelo catches pleurisy (p. 105).

Summer semester: Zermelo lectures on "Mengenlehre und Zahlbegriff." He formulates an axiom system of set theory which comes close to the Zermelo axiom system published by him in 1908 (p. 83).

June: Medical doctors diagnose tuberculosis of the lungs (p. 105).

Summer: Zermelo spends a longer time at the seaside (p. 105, p. 107).

Autumn: Zermelo is under discussion for a full professorship of mathematics at Würzburg University. The professorship is given to the extraordinary Würzburg professor Georg Rost (p. 106). According to Hermann Minkowski Zermelo's difficulties in obtaining a paid professorship are rooted in his "nervous haste" (p. 107).

Death of Zermelo's sister Lena (p. 2).

Winter 1906/07 – winter 1907/08: Several extended stays in Swiss health resorts for lung diseases (p. 105).

1907

March: Zermelo applies for a professorship at the Academy of Agriculture in Poppelsdorf without success (p. 93).

14 July and 30 July: During a stay in the Swiss alps Zermelo completes his papers on a new proof of the well-ordering theorem and on the axiomatization of set theory ([Zer08a] and [Zer08b]), respectively (p. 76).

20 August: Following an application by the Göttingen Seminar of Mathematics and Physics, the ministry commissions Zermelo to also represent in his lectures mathematical logic and related topics, thus installing the first official lectureship for mathematical logic in Germany (p. 95).

December – January 1908: Correspondence with Leonard Nelson, showing Zermelo's critical attitude toward Felix Bernstein (p. 72).

1908

April: Fourth International Congress of Mathematicians in Rome. Zermelo presents his work on finite sets ([Zer09b]) (p. 64). He gets acquainted with Bertrand Russell (p. 71). Together with Gerhard Hessenberg and Hugo Dingler he conceives plans for establishing a quarterly journal for the foundations of mathematics. The project fails because of diverging views between the group and the Teubner publishing house (p. 102).

Summer semester: Zermelo gives a course on mathematical logic in fulfilment of his lectureship for mathematical logic and related topics (p. 96).

Death of Zermelo's sister Marie (p. 2).

1909

July: Zermelo is under consideration for an extraordinary professorship of mathematics at Würzburg University and shortlisted in the first position. The professorship is given to Emil Hilb shortlisted in the second position (p. 107).

1910

24 January: The board of directors of the Göttingen Seminar of Mathematics and Physics applies to the minister to appoint Zermelo an extraordinary professor (p. 111). Zermelo is under consideration for a full professorship of higher mathematics at Zürich University and shortlisted in the first position (p. 114).

24 February: The *Regierungsrat* of the Canton of Zurich approves Zermelo's choice (p. 117).

15 April: Zermelo is appointed a full professor at Zurich University, preliminarily for a period of six years (p. 117).

1911

28 January: Zermelo applies for time off for the coming summer semester because of an outbreak of his tuberculosis (p. 119).

February and March: Together with a partner, Zermelo applies for several patents concerning, for example, a regulator for controlling the revolutions of a machine (p. 118).

Zermelo is awarded the interest from the Wolfskehl Prize, Hilbert being chairman of the Wolfskehl committee of the Gesellschaft der Wissenschaften zu Göttingen (p. 111).

Summer semester – winter semester 1911/12: Leave for a cure because of tuberculosis (p. 119).

1912

Beginning of the cooperation with Paul Bernays who completes his *Habilitation* with Zermelo in 1913 and stays at Zurich University as an assistant to Zermelo and later as a *Privatdozent* until 1919 (p. 125).

August: Fifth International Congress of Mathematicians in Cambridge. Following an invitation by Bertrand Russell, Zermelo gives two talks, one on axiomatic and genetic methods in the foundation of mathematical disciplines and one on the game of chess (p. 71, p. 130, p. 153). The second one results in his chess paper ([Zer13]) which may be considered the first paper in game theory (p. 130).

Zermelo conceives plans for an edition of his collected papers (p. 120).

1913

Spring: Zermelo is discussed for a full professorship in mathematics at the Technical University of Breslau. He is shortlisted in the first position. The professorship is given to Max Dehn, shortlisted in the second position together with Issai Schur (p. 117).

December: Zermelo completes his paper on maximal proper subrings of the field of the real numbers and the complex numbers, respectively ([Zer14]); it makes essential use of the axiom of choice (p. 133).

1914

Early that year: Regular discussions with Albert Einstein (p. 127).

March: Operation of the thorax by Ferdinand Sauerbruch, the pioneer of thorax surgery (p. 120).

Around 1915

Zermelo develops a theory of ordinal numbers where the ordinals are defined as by John von Neumann in 1923 (p. 133).

1915

Spring: A new serious outbreak of tuberculosis forces Zermelo to take a one-year leave (p. 120).

July: Waldemar Alexandrow completes his Ph. D. thesis ([Ale15]). It is the only thesis guided by Zermelo alone (p. 129). Kurt Grelling's thesis ([Grel10]), which extends Zermelo's theory of finite sets, was supervised by David Hilbert, but guided by Zermelo (p. 65).

1916

Februar: As Zermelo has not yet recovered, he is urged to agree to retire. He does so on 5 April (p. 121).

15 April: Zermelo retires from his professorship (p. 121).

31 October: Zermelo is awarded the annual Alfred Ackermann-Teubner Prize of Leipzig University for the Promotion of the Mathematical Sciences. Later prize winners include, for example, Emil Artin and Emmy Noether (p. 121).

November – February 1917: Zermelo stays in Göttingen (p. 122). On 7 November 1916 he gives a talk on his theory of ordinal numbers before the Göttingen Mathematical Society (p. 122, p. 134).

1917

March – October 1919: Zermelo stays in various health resorts in the Swiss alps (p. 122).

October: Death of Zermelo's sister Anna (p. 232).

1919

July: First draft of the tournament paper ([Zer28]) wherein Zermelo develops a procedure for evaluating the result of a tournament by using a maximum likelihood method (p. 129).

November – March 1921: Zermelo stays at Locarno, Switzerland (p. 122).

1921

Spring: Zermelo stays in Southern Tyrol (p. 122, p. 135). Correspondence with Abraham A. Fraenkel (p. 135).

10 May: In a letter to Fraenkel Zermelo formulates a second-order version of the axiom of replacement, at the same time criticizing it because of its non-definite character (p. 137).

17 July (?): Zermelo formulates his "infinity theses" where he describes the aims of his research in infinitary languages and infinitary logic as performed in the early 1930s (p. 204).

22 September: Fraenkel announces his axiom of replacement in a talk delivered at the annual meeting of the Deutsche Mathematiker-Vereinigung. Zermelo agrees in principle, but maintains criticism because of a deficiency of definiteness (p. 138).

1 October: Zermelo settles in Freiburg, Germany (p. 139).

1923

Winter semester 1923/24: Zermelo attends Edmund Husserl's course "Erste Philosophie" (p. 139).

– 1929: Cooperation with Marvin Farber on the development of a semantically based logic system (p. 154), in 1927 leading to plans for a monograph on logic (p. 157).

1924

Summer: Zermelo loses interest in Husserl's phenomenology (p. 147).

1926

Zermelo starts a translation of parts of Homer's Odyssey, thereby aiming at "liveliness as immediate as possible" (p. 144).

22 April: Zermelo is appointed a "full honorary professor" at the Mathematical Institute of Freiburg University (p. 143).

Summer semester – winter semester 1934/35: Zermelo gives regular courses in various fields of mathematics (p. 143).

– 1932: Work on the edition of Cantor's collected papers ([Can32]) with support from the mathematicians Reinhold Baer and Arnold Scholz and the philosopher Oskar Becker. The participation of Abraham A. Fraenkel leads to a mutual estrangement (p. 158).

1927

June: Zermelo completes his paper on measurability ([Zer27a]), where he presents results which he had obtained when guiding Alexandrow's thesis ([Ale15]) (p. 148).

1928

August: Zermelo completes his tournament paper ([Zer28]) (p. 149).

1929

– 1931: Zermelo receives a grant from the Notgemeinschaft der Deutschen Wissenschaft (Deutsche Forschungsgemeinschaft) for a project on the nature and the foundations of pure and applied mathematics and the significance of the infinite in mathematics (p. 163).

May and June: Zermelo spends several weeks in Poland, giving talks in Cracow and Lvov and a series of talks in Warsaw. In the latter ones he develops his view of the nature of mathematics, arguing strongly against intuitionism and formalism (p. 167).

July: Zermelo completes his definiteness paper ([Zer29a]) wherein he meets criticism of his notion of definiteness as put forward by Abraham A. Fraenkel, Thoralf Skolem, Hermann Weyl, and others (p. 174, p. 179).

September: At the annual meeting of the Deutsche Mathematiker-Vereinigung in Prague Zermelo gives a talk on the solution of what is now called the "Zermelo Navigation Problem" ([Zer30c]). An extension of the result is published as [Zer31a] (p. 150).

Arnold Scholz becomes an assistant at the Mathematical Institute of Freiburg University, staying there for five years. Until his death on 1 February 1942, he will be Zermelo's closest friend and scientific partner (p. 221).

1930

Spring: Zermelo completes the *Grenzzahlen* paper ([Zer30a]) where he formulates the second-order Zermelo-Fraenkel axiom system and conceives a convincing picture of the cumulative hierarchy (p. 187).

– 1932: Zermelo's controversy with Skolem and Gödel about finitary mathematics, in particular about Skolem's first-order approach to set theory (p. 200) and Gödel's first incompleteness theorem (p. 212).

1931

Development of infinitary languages and an infinitary logic as a response to Skolem and Gödel (p. 202, p. 206).

15 September: Zermelo presents his work on infinitary languages and infinitary logic at the annual meeting of the Deutsche Mathematiker-Vereinigung. The talk results in the polemical Bad Elster paper ([Zer32a]) and the uncontroversial version [Zer32b] (p. 215).

September/October: Correspondence with Gödel about the proof of Gödel's first incompleteness theorem and Zermelo's infinitistic point of view (p. 216).

18 December: Zermelo is elected a corresponding member of the Gesellschaft der Wissenschaften zu Göttingen. The academy follows a proposal of Richard Courant (p. 177).

– 1935: Continuation of research on infinitary languages and infinitary logic, resulting in the *Satzsysteme* paper ([Zer35]) (p. 206). Isolation from the mainstream of mathematical logic (p. 177). Work on large cardinals (p. 190, p. 193) and on a monograph on set theory (p. 185, p. 245).

1932

June: Zermelo devises an electrodynamic clutch for motorcars (p. 145).

July: Zermelo is invited to contribute a paper to a special issue of *Zeitschrift für angewandte Mathematik und Mechanik* in honour of its founder and editor Richard von Mises. He complies with the invitation with his paper [Zer33a] (p. 152).

1933

16 February: Zermelo is elected an extraordinary member of the Heidelberger Akademie der Wissenschaften. The academy follows a proposal by Heinrich Liebmann and Artur Rosenthal (p. 178).

1935

2 March: Zermelo resigns from the honorary professorhip when denounced because of refusing the Hitler salute (p. 228).

– 1940: Smaller scientific projects in various fields of mathematics, further work on a book on set theory and work on a collection of mathematical "miniatures" representing several of his own results (p. 243).

1937

4 October: Zermelo gives a flawed refutation of the existence of countable models of set theory (p. 203, p. 244).

1941

19 July: Arnold Scholz organizes a colloquium in Göttingen on the occasion of Zermelo's 70th birthday; Zermelo gives three talks which correspond to three items of his collection of mathematical "miniatures." Further speakers include Konrad Knopp and Bartel van der Waerden (p. 247).

1944

14 October: Marriage to Gertrud Seekamp (p. 243).

Zermelo suffers from a glaucoma that can no longer be treated adequately and will finally lead to total blindness (p. 250).

1946

23 July: Zermelo is reappointed an honorary professor. Because of age and increasing blindness, he is unable to lecture (p. 251).

– 1950: In order to escape financial need, Zermelo plans to move back to Switzerland. His applications fail; the Swiss authorities argue that his Swiss pension does not suffice to provide for his wife, too (p. 251, p. 253).

1949

Spring: Zermelo tries to arrange an edition of his collected works (p. 254).

1952

6 February: Death of Zermelo's sister Elisabeth.

1953

21 May: Zermelo dies in Freiburg (p. 256).

1959

12 April: Death of Zermelo's sister Margarete.

2003

15 December: Gertrud Zermelo, then 101 years old, dies in Freiburg (p. 258).

7

Appendix
Selected Original Versions

7.1 Berlin

OV 1.01 Zermelo to his father, 28 July 1888 (p. 4):

Lieber Vater!
Durch Deine treuen Glückwünsche zum gestrigen Tage und durch Deine Geschenke, mit denen Du mich diesmal wie auch sonst so sehr reichlich bedachtest, hast Du mir große Freude bereitet, und für beides spreche ich Dir jetzt vor allen Dingen meinen innigsten Dank aus. Vollkommen überrascht war ich durch Dein so sehr freundliches Anerbieten der Reise, das ich mir doch gewiß nicht besonders verdient habe. Ich kann nicht allein in meinem Interesse sondern auch vor allem in dem Deinen nur hoffen, daß Dein Plan wirklich zur Ausführung kommt, da es eben von Deinem Gesundheitszustande abhängen soll. Daß ich mich sehr gern Deinen "Bedingungen" fügen würde, ist ja selbstverständlich. [...]
    Dein gehorsamer Sohn Ernst.

OV 1.02 Max Planck about his assistant Zermelo, 7 July 1896 (p. 9):

Ich [war] mit seinen Leistungen, in denen er seine besondere mathematische Begabung stets auf das Gewissenhafteste verwerthet hat, jederzeit außerordentlich zufrieden.

OV 1.03 Hermann Amandus Schwarz about Zermelo's Ph.D. thesis, 5 July 1894 (p. 12):

Es gelingt dem Verf[asser] die Hauptuntersuchungen des Herrn Weierstraß [...] nach meinem Urtheile richtig zu verallgemeinern und er hat auf diese Weise eine m. E. sehr werthvolle Vervollständigung zu unserem bisherigen Wissen in diesem Theile der Variationsrechnung gewonnen. Wenn mich nicht alles trügt, so werden alle künftigen Forscher auf diesem schwierigen Gebiet an die Ergebnisse dieser Arbeit und an die Art und Weise ihrer Herleitung anknüpfen müssen.

OV 1.04 Felix Klein about Zermelo's papers against Ludwig Boltzmann, 13 February 1899 (p. 24):

Ich möchte betonen, dass die beiden gegen Boltzmann gerichteten Arbeiten des Cand[idaten] zwar recht scharfsinnig sind und dass man namentlich auch den Muth anerkennen muß, mit welchem sich ihr jugendlicher Verfasser einem Manne wie Boltzmann entgegenstellt, dass aber schliesslich doch wohl Boltzmann Recht behalten hat, dessen tiefgehende und originale Betrachtungsweise allerdings schwer zu verstehen war.

OV 1.05 Woldemar Voigt about Zermelo's controversy with Ludwig Boltzmann, mid February 1899 (p. 24):

Ich möchte aber hinzufügen, dass schon die eindringliche Beschäftigung mit der überaus interessanten und von der Mehrzahl der Physiker kaum wirklich berücksichtigten B[oltzmann]'schen Arbeiten eine Art von Leistung darstellt, sowie daß die von Herrn Dr. Zermelo geübte Kritik unzweifelhaft zur Klärung des schwierigen Gegenstandes beigetragen hat.

7.2 Göttingen

OV 2.01 Zermelo to Felix Klein, 19 July 1897 (p. 27):

Hochgeehrter Herr Professor,
Indem ich Ihnen die beifolgenden Abzüge der bisher von mir veröffentlichten Arbeiten zu freundlicher Berücksichtigung empfehle, gestatte ich mir, mich ganz ergebenst an Sie zu wenden in folgender Angelegenheit.

Seit meiner Promotion Oktober 1894 bin ich jetzt fast 3 Jahre Assistent am hiesigen "Institut für theoretische Physik" für Herrn Prof. Planck, gedenke aber mit dem Beginn des Wintersemesters diese Stelle aufzugeben, um mich in einer kleineren Stadt durch wissenschaftliche Arbeiten zu einer eventuellen späteren Habilitation in theoretischer Physik, Mechanik etc an irgend einer Universität oder technischen Hochschule weiter vorzubereiten, gleichzeitig aber, wenn sich Gelegenheit findet, meine praktisch-physikalische Ausbildung zu vervollständigen.

Da mir nun bekannt ist, welch reiche Anregung und Förderung Sie, Herr Professor, jüngeren Mathematikern zuteil werden lassen und welch reges Interesse Sie immer auch mathematisch-physikalischen Problemen entgegenbringen, so würde ich es mir zur besonderen Ehre rechnen, mich hierbei Ihres geschätzten wissenschaftlichen Rates bedienen zu können, durch den Sie mich zu verbindlichstem Danke verpflichten würden.

Wenn Sie es gütigst gestatten, so bin ich gern bereit, mich Ihnen demnächst persönlich vorzustellen, und ich wäre Ihnen äußerst dankbar, wenn Sie die Güte hätten, mir mitzuteilen, zu welcher Zeit, etwa am nächsten Sonntag oder Montag, Ihnen event[uell] mein Besuch angenehm sein würde.

Mit dem Ausdruck meiner vorzüglichsten Hochachtung gestatte ich mir, mich Ihnen zu empfehlen als
    Ihr ganz ergebener
    Dr. E. Zermelo.

OV 2.02 David Hilbert about Zermelo's *Habilitation* thesis, 9 February 1899 (p. 29):

Die von Herrn Dr. E. Zermelo eingereichte Arbeit "Hydrodynamische Untersuchungen über die Wirbelbewegungen auf einer Kugelfläche" ist eine rein mathematische, wenngleich sie an das physikalisch-meteorologische Problem der Cyclonen-Bewegung auf der Erdoberfläche anknüpft.

Das *erste* Kapitel der Abhandlung beschäftigt sich mit der Flüssigkeitsbewegung auf einer beliebigen Fläche im Raume. Der Verfasser entwickelt auf Grund der Lagrange'schen Differentialgleichungen 2ter Art durch Einführung zweier unabhängiger Gauss'scher Coordinaten die Form, welche die allgemeinen Gesetze der Flüssigkeitsbewegung für den vorliegenden Fall annehmen. Mit Hilfe des Helmholtz' schen Theorems von der Constanz der Wirbelmomente stellt der Verfasser im Falle einer incompressiblen Flüssigkeit eine partielle Differentialgleichung 3ter Ordnung für die Stromfunktion ψ auf, durch die der zeitliche Verlauf der Strömung vollständig bestimmt ist.

Das *zweite* Kapitel wendet diese allgemeine Theorie auf die Kugelfläche an. Es wird für die partielle Differentialgleichung $D\psi = 2\rho$, wo D ein gewisser Differentialparameter und ρ die Wirbeldichte bedeutet, eine solche specielle Lösung als Grundlösung eingeführt, die einen einzigen singulären sogenannten "Strudelpunkt" und auf der ganzen übrigen Kugel constante Wirbeldichte darstellt. Durch Benutzung dieser Grundlösung an Stelle der gewöhnlichen Green'schen Function in der Ebene wird das allgemeine Integral obiger partieller Differentialgleichung aufgestellt: das sogenannte sphärische Potential – entsprechend dem gewöhnlichen logarithmischen Potential in der Ebene, wobei die Wirbeldichte ρ die Rolle der Massendichte übernimmt. Es folgt das Studium der Eigenschaften des sphärischen Potentials. Denkt man sich auf der Kugel eine Masse ausgebreitet, deren Dichte von der Wirbeldichte sich überall um eine bestimmte von 0 verschiedene Constante unterscheidet, so bleibt der Schwerpunkt dieser Massenvertheilung auf der Kugel ein fester Punkt im Raum: der sogenannte "repräsentirende Schwerpunkt". Zum Schlusse dieses Kapitels werden besondere Klassen von continuirlichen stationären und rotirend-stationären Bewegungen auf der Kugel angegeben, die durch Kugelfunctionen darstellbar sind.

Im *dritten* Kapitel führt der Verfasser die Berechnung der Geschwindigkeit eines Strudelpunktes aus und stellt die Gleichungen der Bewegung für den Fall auf, dass ausser der continuirlichen Wirbelbewegung noch eine endliche Anzahl von Strudelpunkten vorhanden ist. Insbesondere wird eine solche Strömung ein Strudelsystem genannt, wenn die continuirliche Wirbeldichte auf der ganzen Kugel constant ist. Ein Strudelsystem ist durch die Anzahl, die Configuration und die Momente der Strudelpunkte völlig bestimmt; die Aufgabe besteht darin[,] die Änderung anzugeben, welche die Configuration während der Zeit erfährt. Ein besonderes Interesse beansprucht der Fall des Gleichgewichts, der sich dadurch charakterisirt, dass das sogenannte Selbstpotential der Strudelmomente ein Maximum oder Minimum ist. Die Bedingung wird auf verschiedene Weise geometrisch gedeutet, wobei sich die regulären Polyeder als besondere Gleichgewichtsfiguren herausstellen.

Im *letzten* Kapitel wird das Problem dreier Strudel eingehend behandelt, was im wesentlichen auf die Untersuchung der Gestaltsaenderung des von demselben gebildeten Dreiecks hinausläuft. Es werden die Differentialgleichungen der Bewegung aufgestellt und diskutirt, namentlich der Fall gleicher Strudelmomente, wo die

Integration durch elliptische Funktionen möglich ist. Dabei ist es wesentlich, dass als neue zu bestimmende Funktionen die Quadrate der Seiten desjenigen ebenen Dreieckes eingeführt werden, welches man durch gradlinige Verbindung der Strudelpunkte erhält.

Die Arbeit ist eine sorgfältige und gründliche Untersuchung und hat zu neuen und bemerkenswerthen Resultaten geführt; sie legt Zeugnis ab nicht nur für die Kenntnisse und Fähigkeiten, sondern auch für den wissenschaftlichen Sinn des Verfassers und sein ideales Streben. Ich stimme daher für eine Zulassung zur Habilitation.

Hilbert.

OV 2.03 From David Hilbert's assessment of Zermelo on the occasion of Zermelo's *Habilitation*, 9 February 1899 (p. 30):

Dr. Zermelo ist mir aus persönlichem Umgange wohl bekannt; derselbe hat während seines Hierseins stets das regste wissenschaftliche Interesse bekundet und insbesondere in der mathematischen Gesellschaft durch Vorträge und wissenschaftliche Bemerkungen bewiesen, dass er sich wohl für den akademischen Beruf eignet.

OV 2.04 Zermelo to Max Dehn concerning the publication of his *Habilitation* thesis, 25 September 1900 (p. 31):

Dagegen hapert es noch sehr mit [...] Colleg-Vorbereitung, ganz zu schweigen von der unglückseligen Wirbel-Arbeit, deren Vollendung und Drucklegung noch immer aussteht. Wie das wohl noch werden wird?

OV 2.05 From David Hilbert's assessment of Zermelo concerning a *Privatdozenten* grant, 20 March 1901 (p. 34):

Herr Dr. Zermelo ist ein begabter Gelehrter mit scharfem Urteil und raschem Auffassungsvermögen; er hat ein lebendiges Interesse und offenes Verständnis für die Fragen unserer Wissenschaft und zudem besitzt er umfassende theoretische Kenntnisse im Gebiete der mathematischen Physik. Ich stehe mit ihm [...] in stetem wissenschaftlichen Verkehr.

OV 2.06 David Hilbert's recommendation of Zermelo, ibid. (p. 34):

Da Herr Zermelo ein schätzenswertes Mitglied unseres Lehrkörpers geworden ist, [empfehle ich] denselben auf das wärmste zur Berücksichtigung bei Verleihung eines Privatdocentenstipendiums.

OV 2.07 From David Hilbert's assessment of Zermelo concerning a professorship at Breslau University, 31 May 1903 (p. 35):

Was nun weitere Namen betrifft, so will ich gleich mit demjenigen beginnen, den ich an Stelle der Breslauer Fa[kultät] als den eigentlichen Kandidaten bezeichnen würde, das ist Zermelo.

Zerm[elo] ist ein moderner Math[ematiker], der in seltener Weise Vielfältigkeit mit Tiefe verbindet. Er kennt die moderne Variationsrechnung (bereitet darüber ein umfassendes Lehrbuch vor.) In der Variationsrechnung aber erblicke ich für die Zukunft einen der wichtigsten Zweige der Math[ematik]. Die Ausbildung d[ie]s[es]

Zweiges und die Fruchtbarmachung desselben für die Nachbargebiete wird, hoffe ich[,] in Zukunft dazu beitragen, der Math[ematik] das Ansehen wieder zurückzugeben, welches sie vor 100 Jahren besaß, und welches ihr in letzter Zeit verloren zu gehen droht. [...]

Sie werden nicht glauben, daß ich Zerm[elo] wegloben will. Bevor Mink[owski] herkam und ehe Blumenthal herangewachsen war, war Zer[melo] mein hauptsächl[icher] math[ematischer] Verkehr und ich habe viel von ihm gelernt, z. B. die Weierstraßsche Variationsrechnung, ich würde ihn hier also am meisten vermissen.

OV 2.08 David Hilbert in his lecture course on logical principles of mathematical thinking, summer semester 1905 (p. 44):

Es ist in der Entwicklungsgeschichte der Wissenschaft wohl immer so gewesen, dass man ohne viele Scrupel eine Disciplin zu bearbeiten begann und soweit vordrang wie möglich, dass man dabei aber, oft erst nach langer Zeit, auf Schwierigkeiten stieß, durch die man gezwungen wurde, umzukehren und sich auf die Grundlagen der Disciplin zu besinnen. Das Gebäude der Wissenschaft wird nicht aufgerichtet wie ein Wohnhaus, wo zuerst die Grundmauern fest fundamentiert werden und man dann erst zum Auf- und Ausbau der Wohnräume schreitet; die Wissenschaft zieht es vor, sich möglichst schnell wohnliche Räume zu verschaffen, in denen sie schalten kann, und erst nachträglich, wenn es sich zeigt, dass hier und da die locker gefügten Fundamente den Ausbau der Wohnräume nicht zu tragen vermögen, geht sie daran, dieselben zu stützen und zu befestigen. Das ist kein Mangel, sondern die richtige und gesunde Entwicklung.

OV 2.09 Zermelo to Heinrich Scholz on the discovery of the Zermelo-Russell paradox, 10 April 1936 (p. 46):

Über die mengentheoretischen Antinomien wurde um 1900 herum im Hilbert'schen Kreise viel diskutiert, und damals habe ich auch der Antinomie von der größten Mächtigkeit die später nach Russell benannte präzise Form (von der "Menge aller Mengen, die sich nicht selbst enthalten") gegeben. Beim Erscheinen des Russellschen Werkes (Principles of Mathematics 1903?) war uns das schon geläufig.

OV 2.10 Zermelo to Max Dehn, 27 October 1904 (p. 52, p. 59):

Lieber D[ehn]!
Besten Dank für freundliche Zustimmung! Eine solche erhielt ich auch vorgestern von der "jüngeren Berliner Schule", d. h. also wohl von Landau u[nd] Schur. Wenn Du aber wüsstest, wie ich von den Spezialisten der Mengenlehre, von Bernstein, Schoenflies u[nd] König mit Einwänden, vorläufig brieflich, bombardirt werde! Da wird sich noch eine hübsche Polemik in den Annalen entwickeln; aber ich fürchte sie nicht. Von dem Schicksal des König'schen Vortrages weisst Du also noch nichts? Harmloses Gemüt! Die ganzen Ferien war eigentlich von nichts anderem die Rede. Sowohl Hilbert als mir gegenüber hat K[önig] seinen damaligen Beweis feierlich revozirt, ebenso auch B[ernstein] seinen Potenzen-Satz. K[önig] kann sich gratuliren, dass in Heidelberg die Bibliothek so früh geschlossen wurde; er hätte sich sonst ev[entuell] noch in persona blamirt. So musste ich erst meine Heimkehr nach G[öttingen] abwarten, um nachzusehen, wo es dann sofort offenbar war; doch hatte das K[önig] mit[t]lerweile bei der Ausarbeitung schon selbst gefunden. Versuche also, bitte, die

Dissertation S. 50(?) [two unreadable letters] auf den von K[önig] behandelten Fall $\mathfrak{s} = \mathfrak{m}_1 + \mathfrak{m}_2 + \ldots$ anzuwenden, um den Fehler, der allein von K[önig] benutzt wird, sofort zu sehen. Also eine richtige "Ruine Bernstein", sagt Blument[h]al. Aber K[önig] möchte noch retten, was zu retten ist, und [...] phantasirt von der Menge W *aller* Ordnungstypen, die in mancher Menge enthalten sein *könne*, sodass mein $L_\gamma = W$ und dann meine Folgerung $L_\gamma = M$ unrichtig wäre. Sekundirt wird er darin merkwürdigerweise von Bernstein u[nd] Schoenflies, von letzterem freilich ganz verworren u[nd] missverständlich. Die W-Gläubigen leugnen bereits die Def[inition] der Teilmenge u[nd] womöglich noch den Satz vom Widerspruch!

Mit herzl[ichem] Gruss Dein E. Zermelo.

OV 2.11 Zermelo to David Hilbert on the concepts of number and finite set, 29 June 1905 (p. 61):

Es ist mir [...] gelungen, das Problem, das ich das "Dedekind'sche Problem des Zahlbegriffs" nennen möchte, vollständig und übersichtlich zu lösen und damit die wenig anziehende und schwer verständliche Dedekind'sche Deduktion überflüssig zu machen.

OV 2.12 Zermelo to David Hilbert on the finiteness of Dedekind finite sets, 29 June 1905 (p. 62):

Um [...] zu beweisen, *dass jede "endliche" Menge* (die keinem ihrer Abschnitte aequivalent ist) *einem* (echten) *Abschnitte der "Zahlenreihe"* (Typus ω) *aequivalent ist,* denke man sie sich *wohlgeordnet.* Enthielte dabei irgend ein Abschnitt oder die ganze Menge *kein letztes Element,* so könnte man jedes seiner Elemente auf das unmittelbar folgende abbilden, wobei das *erste* Element natürlich *nicht* verwendet wird, und die Menge wäre auf einen Teil von sich selbst abgebildet gegen die Annahme. Eine "Folge", die nebst allen Abschnitten ein letztes Element besitzt, ist aber ähnlich einem Abschnitt der "Zahlenreihe".

OV 2.13 Zermelo to David Hilbert on the role of the axiom of choice in the theory of finite sets, 29 June 1905 (p. 62):

Dass ich somit zum Beweis den "Wohlordnungssatz" verwende, das "Prinzip der Auswahl" also als gültig voraussetzen muss, ist nicht zufällig und kein Übelstand. Denn dieses Prinzip ist in der Tat eine *notwendige Voraussetzung* des Theorems, das sich *ohne* "Auswahl" *überhaupt nicht* beweisen ließe. Auch *Cantor* beweist den Satz, daß "jede unendliche Menge eine Teilmenge v[on] d[er] Mächtigkeit $\aleph_0$ besitze", indem er erst ein Element heraus zieht, dann ein anderes und sodann das berüchtigte "u.s.w." verwendet. Bei richtiger Präzisierung kommt dies genau wieder auf meinen Wohlordnungssatz hinaus. Man kann auch *nicht* geltend machen, daß bei *endlichen* Mengen nur eine endliche Anzahl von Malen "ausgewählt" zu werden brauchte, was nach der Ansicht jener "Empiristen" noch erlaubt sein soll. Denn von der betrachteten Menge wissen wir *nur* die *negative* Eigenschaft, keinem ihrer Teile äquivalent zu sein; die "successive Auswahl" im empirischen Sinne erfordert aber den endlichen *Ordnungstypus*, dessen Vorhandensein nicht vorausgesetzt sondern eben *bewiesen* werden soll.

OV 2.14 Zermelo to David Hilbert, as above (p. 63):

Die Theorie der endlichen Mengen ist unmöglich ohne das "Prinzip der Auswahl", und der "Wohlordnungssatz" ist die wahre Grundlage für die ganze Theorie der Anzahl. Dies wäre eine wertvolle Waffe gegen meine "empiristisch-skeptischen" Gegner (Borel, Enriques, Peano u. a.), während die "dogmatischen" Gegner (Bernstein, Schoenflies u[nd] Genossen) in dieser Frage allgemein nicht ernst genommen werden, wie ich mich im Gespräch mit Enriques (in Florenz) überzeugen konnte. Der letztere bezweifelt folgerichtig so ziemlich die gesamte Mengenlehre, einschließlich des Mächtigkeitsbegriffes, sodass sich mit ihm eigentlich nicht mehr streiten läßt.

OV 2.15 Gerhard Hessenberg to Leonard Nelson, 7 September 1907 (p. 68):

Zermelo hat einen neuen Beweis des Wohlordnungssatzes heraus. Ich hatte gestern die Korrektur in Händen. Der Beweis ist famos und sehr einfach. Ich hatte aber selbst vergeblich auf diesem Weg vorzudringen versucht. Außerdem hat er 8 Seiten Polemik mit Poincaré, Bernstein, Jourdain, Peano, Hardy, Schoenflies etc, die sehr witzig ist und die Logizisten scharf mitnimmt. Er betont überall den synthetischen Charakter der Mathematik und wirft Poincaré vor, daß er gerade die Mengenlehre mit dem Logizismus verwechsele.

OV 2.16 The Göttingen Seminar of Mathematics and Physics to the Prussian Minister of Cultural Affairs, 24 January 1910 (p. 68):

Dass [Zermelo] noch keinerlei Berufung an eine andere Hochschule erhielt, liegt nicht zum wenigsten daran, dass sich an sein Hauptergebnis, den sog. Wohlordnungssatz, [...] einige (nach Ansicht der Sachverständigen unter uns) unberechtigte Kritiken anschlossen, welche die kaum erst für die Notwendigkeit und Nützlichkeit der Mengenlehre gewonnene öffentliche Meinung in eine Verwirrung brachten, aus welcher die in der Form wohldurchdachte und massvolle, sachlich uns völlig überzeugend erscheinende Antikritik Zermelos hoffentlich bald die Geister erlösen wird.

OV 2.17 Zermelo to David Hilbert, 25 March 1907 (p. 78):

Meine Mengenlehre werde ich Ihrem Wunsche entsprechend baldmöglichst fertig stellen, obwohl ich eigentlich noch den Beweis der Widerspruchslosigkeit hatte hinzufügen wollen.

OV 2.18 Zermelo's proof of the Schröder-Bernstein equivalence theorem, postcard to David Hilbert of 28 June 1905 (p. 89):

Sehr geehrter Herr Geh[eimrat]!
Als ein weiteres Resultat meiner mengentheoretischen Studien teile ich noch folgendes mit als *neuen Beweis des Äquivalenzsatzes* (auf Grund der Dedekindschen "Kettentheorie"). Es handelt sich um den Satz:
"Ist $S = (P, Q, S') = (P, S_1)$, wo $S' \simeq S$ und P, Q, S' keine gem[einsamen] El[emente] enthalten, so ist immer auch: $S_1 = (Q, S') \simeq S$."
Es sei nämlich (Q, V) gemeinsamer Bestandteil aller Teilsysteme von S, welche Q und bei der betreffenden Abbildung von S auf S' ihre eigenen Bilder enthalten (auf e[inen] Teil von sich abgebildet werden). Dann hat auch (Q, V) d[ie] gleiche Eigenschaft, und V enthält überhaupt nur Bilder von Elementen von (Q, V). Denn jedes

andere Element könnte man fortlassen und behielte ein *kleineres* System (Q, V_1), das gleichfalls auf sich selbst abgebildet würde. Es ist also V Bestandteil von S', also etwa $S' = (V, R)$. Das Bild von (Q, V) ist nun V selbst, da Q überhaupt keine Bilder enthält and V keine anderen Elemente. Somit ist $(Q, V) \simeq V$ und $S_1 = (Q, S') = (Q, V, R) \simeq (V, R) = S'$. [Daher] $S_1 \simeq S'$, q.e.d. Hieraus folgt nun in bekannter Weise der "Äquivalenzsatz".

Beste Grüße von E. Zermelo

OV 2.19 Zermelo's appointment to a lectureship for mathematical logic, 20 August 1907 (p. 95):

Einem Antrage der Direktoren des Mathematisch-physikalischen Seminars der dortigen Universität entsprechend beauftrage ich Sie, vom nächsten Semester ab in Ihren Vorlesungen auch die mathematische Logik und verwandte Gebiete zu vertreten.

OV 2.20 David Hilbert in his lecture course on logical principles of mathematical thinking, summer semester 1905 (p. 96):

Die Paradoxieen zeigen zur Genüge, dass eine Prüfung und Neuaufführung der Grundlagen der Mathematik und Logik unbedingt nötig ist. Man erkennt ja sofort, dass die Widersprüche auf der Zusammenfassung gewisser Allheiten zu einer Menge beruhten, die nicht erlaubt zu sein scheinen; aber damit ist noch nichts gewonnen, denn jedes Denken beruht gerade auf solchen Zusammenfassungen, und das Problem bleibt, hier das Erlaubte von dem Unerlaubten zu sondern.

OV 2.21 Zermelo on the analytic character of mathematics in his course on mathematical logic, summer semester 1908 (p. 99):

Man hat gesagt, die Mathematik sei doch nicht oder wenigstens nicht ausschließlich Selbstzweck, man wolle sie doch auf die Wirklichkeit anwenden, wie könne man das aber, wenn die Mathematik aus Definitionen und daraus abgeleiteten analytischen Sätzen bestünde, von denen man gar nicht wüßte, ob sie in der Wirklichkeit gelten. Darauf ist zu sagen, daß man sich freilich erst überzeugen muß, ob die Axiome einer Theorie in dem Gebiet der Wirklichkeit gelten, auf das man die Theorie anwenden will, und diese Feststellung erfordert jedenfalls ein außerlogisches Verfahren.

OV 2.22 Zermelo to Leonard Nelson, 14 March 1908 (p. 102):

Eine *periodische* Zeitschrift direkt für "Philosophie der Mathematik" oder so ähnlich würde m. Er. sehr viel größeres Entgegenkommen in beteiligten Kreisen zu erwarten haben. Wie denkt Hessenberg darüber? Meinerseits würde ich einem solchen zeitgemäßen Unternehmen meine *volle* Unterstützung widmen, während ich Bedenken hätte, mich auf irgendein spez[ielles] philosophisches System festzulegen.

OV 2.23 Gerhard Hessenberg to Leonard Nelson after Hermann Minkowki's death, 27 January 1909 (p. 104):

Das beste wäre, man liesse jetzt endlich Zermelo anstellen. Die Vernachlässigung eines unserer hervorragendsten Köpfe ist bereits ein öffentliches Ärgernis und eine böse Schattenseite des an und für sich ja recht erfolgreichen Regiments Felix des Einzigen. Noch jeder Ausländer, dem diese Sache ein Novum ist, ist empört, wenn er erfährt, daß Zermelo noch immer Privatdozent ist. – In Betracht kommen, wie ich fürchte *vor* Zermelo, Landau und Blumenthal. Letzterem mis[s]gönne ich es direkt; Landau hätte sich einen solchen Erfolg redlich verdient, aber Berlin würde wieder um einen seiner sehr spärlichen guten Lehrer ärmer ohne Aussicht auf Ersatz.

OV 2.24 Hermann Minkowki's assessment of Zermelo concerning a professorship in Würzburg, 12 October 1906 (p. 106):

Zermelo ist wirklich ein Mathematiker von ersten Qualitäten, von vielseitigstem Wissen, von rascher durchdringender Auffassung, seltener kritischer Begabung. Letzthin beschäftigte er sich viel mit der sogenannten Mengenlehre [...]. Auf diesem an die Philosophie streifenden Gebiete ist Zermelo eine Autorität geworden [...]. Endlich ist er [...] sehr bewandert und interessiert in mathematischer Physik [...]. Es ist mehrfach hervorgetreten, dass er mit grosser Gewandtheit Probleme zu meistern verstand, die ihm, noch in embryonalem Zustande, aus mathematischen Wehen hiesiger angewandter Kollegen merklich wurden. [...] Dass er trotz seiner Bedeutung noch nicht zu äusserlicher Anerkennung durchgedrungen ist, ist eine Ungerechtigkeit, die ein ständiges Thema in den Unterhaltungen zwischen Hilbert und mir bildet. Descoudres bezeichnete Zermelo einmal als Prototyp eines tragischen Helden. Sein auffälliger Mangel an Glück beruht zum grossen Teil auf seiner äusseren Erscheinung, der nervösen Hast, die bei ihm in Reden und Tun hervortritt und erst in allerletzter Zeit einem mehr ruhigen, abgeklärten Wesen Platz macht. Er ist durch die Klarheit seines Verstandes ein Pädagoge ersten Ranges für die reiferen Studenten, denen es auf ein Eindringen in die Tiefen der Wissenschaft ankommt; diese wie alle jüngeren Dozenten hier, mit denen er näher befreundet ist, schätzen ihn ausserordentlich, aber er ist kein Pädagoge für die Anfänger. [...] Leider liess Zermelo's Gesundheitszustand letzthin zu wünschen übrig. Im Winter hatte er sich durch Unvorsichtigkeit eine Brustfellentzündung zugezogen, musste längere Zeit pausieren, neulich kam er sehr erfrischt von der See zurück, doch rät ihm der Arzt, den Winter im Süden zu verbringen; ich hörte noch nicht, wie er sich entschlossen hat.

OV 2.25 The Senate of Würzburg University about Zermelo, 14 July 1909 (p. 108):

Er ist ein Mathematiker von ersten Qualitäten, von vielseitigem Wissen, rascher durchdringender Auffassung und seltener kritischer Begabung. Auf den Gebieten der Mengenlehre, Axiomatik, Logik[,] überhaupt in den Prinzipienfragen der Mathematik, ist er die erste Autorität; auf dem Gebiete der Variationsrechnung hat er durch seine Arbeiten geradezu epochemachend gewirkt. An der Universität Göttingen hat Zermelo eine vielseitige erfolgreiche Lehrtätigkeit entfaltet; er gilt besonders bei den reiferen Studenten, denen es auf ein Eindringen in die Tiefen der Wissenschaft ankommt, als ein vorzüglicher Lehrer.

OV 2.26 Karl Schwarzschild about Zermelo in a letter to Arnold Sommerfeld, 19 April 1909 (p. 109):

Die Frage, ob Zermelo einigermaßen verständlich vorträgt, ist sehr glatt zu beantworten, da er manchmal sogar ganz ausgezeichnet vorträgt. Er sieht blühend aus, hat sein Kolleg durchgehalten und, soviel ich weiß, nicht über seine Gesundheit geklagt. Ich vermute, daß er das kritische Alter für Schwindsuchtskandidaten glücklich überstanden hat.

OV 2.27 David Hilbert's assessment of Zermelo for a professorship in Würzburg, 5 May 1909 (p. 109):

Lieber Herr Kollege,
Göttinger Mathematiker empfehle ich unaufgefordert überhaupt nicht mehr, da ich die Erfahrung gemacht habe, dass ich den Herrn dadurch nicht nütze. Insbesondere habe ich längst aufgehört, zu glauben, dass eine Fakultät die Klugheit besitzen könnte, Zermelo zu berufen. Seit längerer Zeit habe ich ihm daher hier einen Lehrauftrag für math[ematische] Logik erwirkt und die Berufung Landau's benutzt, um das Ministerium zu veranlassen seine Renumeration wesentlich zu erhöhen. Zum Ordinarius an Minkowski's Stelle konnten wir ihn nicht machen, da Landau für uns der gegebene Mann war – aus vielen Gründen. Z[ermelo] ist glücklicherweise dick und fett geworden und seine Lunge ist völlig ausgeheilt. Er befindet sich in jeder Hinsicht sehr wohl. Er liest immer mit viel Erfolg: in diesem Semester Differential u[nd] Integralrechnung vor 90 Zuhörern und Uebungen dazu zusammen mit Toeplitz, in denen diese beiden Kollegen ebenfalls gegen 100 Teilnehmer haben. Auf dem Gebiet der Mengenlehre, Axiomatik, Logik, Prinzipienfragen der Mathematik ist er die erste Autorität, die sich noch letztens in sehr eklatanter Weise gegen Poincaré bewährt hat. [...]
   Mit den besten Grüßen
   Ihr ergebenster
   Hilbert.

7.3 Zurich

OV 3.01 The Dean of the Philosophical Faculty (Section for Science and Mathematics) of Zurich University to the Council of Education of the Canton Zurich, 22 January 1904 (p. 115):

In allen seinen Publikationen zeigt sich Zermelo als ein *moderner Mathematiker, der in seltener Weise Vielseitigkeit mit Tiefe verbindet. Während er ein gründlicher Kenner der mathematischen Physik ist,* was mir übereinstimmend von Männern wie Voigt, Carl Runge, Sommerfeld, Nernst bestätigt wird, ist er zugleich *die Autorität in der mathematischen Logik. Er vereinigt damit in sich das Verständnis der am weitesten von einander entfernt liegenden Teile in dem gewaltigen Gebiete des mathematischen Wissens.* Zermelo übt hier [in Göttingen] seit 11 Jahren *die denkbar vielseitigste Vorlesungspraxis* aus. Er hat wiederholt auch elementare Vorlesungen wie *analytische Geometrie, Differential- und Integralrechnung* vor grösserem Publikum (70–80 Zuhörer), das ihm mehrere Semester hindurch in unverminderter Zahl treu geblieben ist, mit gutem Erfolg gelesen. Und wenn er *über Kapitel der höheren Mathematik vor reiferen Zuhörern* vorträgt, gestalten sich seine Vorlesungen zu originellen Leistungen. Er vertritt in ihnen den modernen Standpunkt des betreffenden Wissenszweiges; sie gehören zu den besten, die hier gelesen werden. Eine stattliche Zahl von heutigen Privatdozenten und Professoren haben bei ihnen Belehrung geschöpft. Noch eben rühmt mir Kollege Landau, der zur Zeit die Vorlesung Zermelos über die logischen Grundlagen der Mathematik regelmässig hört, wie klar und fesselnd sein Vortrag ist. Er war vor Jahren krank. Er ist jedoch als völlig gesund anzusehen [...] *Er würde zweifellos mit grösster Freudigkeit dem Ruf nach Zürich Folge leisten und sein Amt, dessen bin ich überzeugt, mit äusserster Pflichttreue verwalten.*

OV 3.02 Ibid. (p. 115):

Herr Kollege Schmidt [...] erklärt, die Angaben Prof. Hilberts über die wissenschaftliche Bedeutung der Arbeiten Zermelos, soweit sie rein mathematisch sind, Wort für Wort unterschreiben zu können. [...] Er erwähnt ferner, daß Prof. Zermelo trotz seines undeutsch klingenden Namens Deutscher ist und rühmt an ihm, auf Grund langjährigen, persönlichen Verkehrs mit ihm in Göttingen, *daß er nicht nur als Charakter ein ganz vortrefflicher Mensch ist,* sondern auch *den Vorzug strenger Sachlichkeit, Aufrichtigkeit und Unbestechlichkeit im Urteil* besitzt.

OV 3.03 The Senate of the Technical University of Breslau to the Prussian Minister for Spiritual and Educational Matters, 21 February 1913 (p. 117):

Zermelo ist einer der auch im Auslande bekanntesten deutschen Mathematiker; seine wissenschaftlichen Arbeiten sind ausgezeichnet [...] und haben teilweise geradezu Aufsehen erregt, so eine Arbeit über kinetische Gastheorie und seine Untersuchungen zur Mengenlehre, vor allem der Beweis des berühmten Wohlordnungssatzes. Obwohl er sich hier in den vielleicht abstraktesten Gebieten der mathematischen Logik bewegt, zeigen doch seine Assistententätigkeit bei Max Planck und die hieraus entstandenen Arbeiten zur theoretischen Physik, dass Zermelo gerade die für eine technische Hochschule bedeutungsvollsten Anwendungen der Mathematik vollkommen beherrscht. Die Lebendigkeit seines Geistes gestattet ihm überhaupt, sich den verschiedensten Ansprüchen schnell und leicht anzupassen.

OV 3.04 Aurel Stodola about Zermelo to Ernst Reichel, 15 March 1913 (p. 118):

Ihre Befürchtungen betreffend seinen Gesundheitszustand sind nicht unbegründet, er mußte in der Tat vor kurzem ein ganzes Semester Urlaub nehmen und weilte in Davos. Man sprach aber davon, dass es kein Lungenleiden, sondern eher ein Nervenleiden war. Im übrigen eine liebenswürdige, aber etwas eigenartige Persönlichkeit; halb Sonderling, ziemlich unstet, vielleicht doch eher für Universitäten, als für Technische Hochschulen geeignet. Es würde mich sehr wundern, wenn *technische* Probleme sein Interesse erregen würden.

Dies das Wenige und Ungewisse, was ich Ihnen mitteilen kann.

OV 3.05 Address of the Zurich University Commission concerning Zermelo's continuing inability to lecture (p. 121):

[...] daß es äußerste Zeit sei, für das Sommersemester 1916 [...] alle Anordnungen für einen ungestörten, vollwertigen Unterrichtsbetrieb in den mathematischen Disziplinen zu treffen.

OV 3.06 Zermelo to the Zurich *Regierungsrat* concerning his retirement, 5 April 1916 (p. 121):

Hochgeehrter Herr Regierungsrat!
[...] Wie ich schon dem Herrn Erziehungssekretär seiner Zeit mitteilte, hatte ich bisher immer noch Bedenken getragen, wegen einer voraussichtlich nur vorübergehenden

Arbeitsunfähigkeit meine ganze Berufstätigkeit vorzeitig aufzugeben, und hatte deshalb um einen Aufschub der Frage bis zum nächsten Semester gebeten. [...] Nachdem nun aber die Oberbehörde, wie ich Ihrer Zuschrift entnehme, auf Grund eines ärztlichen Gutachtens davon abgesehen hat, entsprechend dem Vorschlag meiner Fakultät meine Amtsperiode provisorisch um ein weiteres Semester zu verlängern, vielmehr im Interesse eines geordneten Unterrichts eine unmittelbare Entscheidung wünscht, bin ich nunmehr bereit, auf die von der Regierung vorgeschlagene Regelung einzugehen und hiermit meine Versetzung in den Ruhestand aus Gesundheitsrücksichten nachzusuchen. In der Bemessung des mir in Aussicht gestellten Ruhegehaltes darf ich wohl ein freundliches Entgegenkommen der Behörde dankbar anerkennen, doch ist es meine Absicht, von diesem Angebot nur solange Gebrauch zu machen, bis es mir nach Wiedererlangung meiner Arbeitsfähigkeit gelingen sollte, einen meinen Kräften entsprechenden neuen Wirkungskreis zu finden.

Indem ich hoffe, daß die nunmehr bevorstehende Neubesetzung der Lehrstelle zum Gedeihen der Wissenschaft wie unserer Hochschule beitragen werde, zeichne ich mit vorzüglichster Hochachtung ganz ergebenst Dr. Ernst Zermelo.

OV 3.07 Zermelo to the Zurich *Regierungsrat* concerning the insufficiency of his pension, 6 December 1923 (p. 124):

Das ist etwa so viel, wie ich vor 30 Jahren als junger Student zu einfachster Lebensführung benötigte, und entspricht in keiner Weise den heutigen Teuerungsverhältnissen. Schon jetzt bin ich gezwungen, meine Einnahmen zu überschreiten, und werde in kurzem, wenn meine geringen Ersparnisse vom Vorjahr verbraucht sind, nicht mehr in der Lage sein, meiner geschwächten Gesundheit eine ausreichende Ernährung zu sichern.

OV 3.08 Decree of the Zurich *Regierungsrat* concerning Zermelo's pension, 15 December 1923 (p. 124):

Wenn auch Prof. Zermelo nur wenige Jahre an der Universität Zürich als Dozent gewirkt hat, so sollte, vom rein menschlichen Standpunkt betrachtet, das Ruhegehalt doch so angesetzt werden, dass Prof. Zermelo bei bescheidenen Lebensansprüchen daraus leben kann.

OV 3.09 Albert Einstein to Zermelo, 4 October 1912 (p. 127):

Ich bin abends immer auf meiner Wohnung [...] u[nd] freue mich sehr, wenn Sie mich aufsuchen.

OV 3.10 Zermelo to the Educational Council of the Canton of Zurich, 27 March 1911 (p. 128):

Ich darf hier wohl daran erinnern, dass ich mich seit meiner Berufung eifrig bemüht habe, vorzüglich qualifizierte junge Mathematiker für unsere Universität zu gewinnen, dass es mir aber schon bei zweien misslungen ist, weil sich ihnen anderswo günstigere Aussichten boten.

OV 3.11 Zermelo to his Faculty concerning difficulties in improving the situation of mathematics, 10 June 1911 (p. 129):

Denn da es mir vor allen Dingen darauf ankommen muß, überhaupt wieder arbeitsfähig zu werden, so würde ich mich nicht an eine Stellung klammern, die mir die eigenen Kollegen nach Möglichkeit zu erschweren suchen.

OV 3.12 Dénes König to Zermelo on the chess paper, 13 February 1927 (p. 132):

[Von Neumann] hat mich darauf aufmerksam gemacht, dass der Satz über die Beschränktheit der Zugzahlen mit Hilfe meines Lemmas einfach bewiesen werden kann. Damals kannte er Ihren Cambridger Vortrag nicht, erst ich machte ihn darauf aufmerksam. Jetzt arbeitet er an einer Arbeit über Spiele, die in den Math[ematischen] Ann[alen] erscheinen wird.

OV 3.13 Paul Bernays to Zermelo concerning Zermelo's work on the von Neumann ordinal numbers, 20 December 1920 (p. 134):

Auch Ihre Theorie der Wohlordnungszahlen fand bei der Vorlesung Erwähnung, nämlich im Zusammenhang mit der Paradoxie von Burali-Forti, welche dadurch eine prägnantere Fassung erhält. Ich nehme an, dass Sie einverstanden damit sind, dass ich Ihre diesbezüglichen Gedanken Hilbert mitgeteilt habe. Jedenfalls geschah die Erwähnung in der Vorlesung unter ausdrücklicher Angabe Ihrer Autorschaft, sodass Sie Ihrerseits für die umfassendste mathematische Axiomatik als Urheber genannt wurden.

OV 3.14 Zermelo to Abraham A. Fraenkel about the empty set, 9 May 1921 (p. 135):

Nebenbei wird mir selbst die Berechtigung dieser "Nullmenge" immer zweifelhafter. Könnte sie nicht entbehrt werden bei geeigneter Einschränkung des "Aussonderungsaxioms"? Tatsächlich dient sie doch nur zur *formalen* Vereinfachung.

OV 3.15 Zermelo to Abraham A. Fraenkel about the axiom of choice, 31 March 1921 (p. 135):

Als "Auswahlaxiom" bezeichne ich [das Axiom VI] nur uneigentlich der üblichen Sprechweise zu Liebe. Von einer *wirklichen* "Auswahl" ist in meiner Theorie nicht die Rede. Der Satz "Man kann das Axiom etc." auf S. 266 wäre auch nur als eine Anmerkung, welche die Theorie nicht berührt, aufzufassen. Für mich ist das Axiom VI ein reines *Existenz-Axiom*[,] nicht mehr und nicht weniger. [...] Die "gleichzeitige Auswahl" der Elemente und Zusammenfassung zu einer Menge [...] gilt mir nur als eine Vorstellungsweise, sich den Sinn und die (psychologische) Notwendigkeit meines Axioms *anschaulich* zu machen.

OV 3.16 Abraham A. Fraenkel to Zermelo about a gap in Zermelo's axiom system, 6 May 1921 (p. 136):

Hochgeehrter Herr Professor,
auf die 2 Briefe, die ich Mitte April an Sie von Amsterdam aus richtete – namentlich auf den zweiten, der das Beispiel zur Unabhängigkeit des Auswahlaxioms, wie ich

es verstand, enthielt – bin ich noch ohne Ihre Antwort; ich wäre Ihnen für *baldige* solche außerordentlich zu Dank verpflichtet, weil ich gern mit dem Abschluße meiner Note auf Ihre Meinungsäußerung warten möchte. Ist diese nochmalige Bitte um Ihre Antwort der Hauptzweck dieser Zeilen, so benutze ich doch gerne die Gelegenheit, um noch zwei weitere Punkte zur Sprache zu bringen.

1) Es sei Z_0 eine unendliche Menge (z. B. die von Ihnen so bezeichnete) und $\mathfrak{U}(Z_0) = Z_1$, $\mathfrak{U}(Z_1) = Z_2$, usw.. Wie folgt dann aus Ihrer Theorie (Grundl[agen] d[er] M[engenlehre] I), daß $\{Z_0, Z_1, Z_2, \ldots\}$ eine Menge ist, daß also z. B. die Vereinigungsmenge existiert? Würde Ihre Theorie zu einem solchen Beweis nicht genügen, so wäre offenbar z. B. die Existenz von Mengen von der Kardinalzahl $\aleph_\omega$ nicht beweisbar.

2) [...]

Sollte Ihre Zeit nicht mehr erlauben, so wäre ich doch wenigstens für je ein kurzes Wort zu dem aus Amsterdam angegebenen Beispiel und zur 1. Frage des vorliegenden Briefes sehr dankbar. Ich glaube im übrigen Ihre Freundlichkeit in nächster Zeit nicht mehr zu bemühen. Mit den besten Empfehlungen

Ihr ganz ergebener

A. Fraenkel

OV 3.17 Zermelo to Abraham A. Fraenkel on the axiom of replacement, 9 May 1921 (p. 137):

Sehr geehrter Herr College!

Gestern Abend hatte ich gerade meinen Brief zur Beantwortung der Ihrigen vom 11. und 14. v[om] M[onat] geschrieben, als Ihr letzter Brief vom 6. V. eintraf, auf den ich nun doch auch noch etwas eingehen muß.

Ihre Bemerkung bezüglich $Z^* = Z_0 + Z_1 + Z_2 + \cdots$ scheint berechtigt und war mir bei der Abfassung meiner Gr[und]l[agen] entgangen. In der Tat wird da wohl noch ein Axiom nötig sein, aber welches? Man könnte es so versuchen: Sind die Dinge $A, B, C, \ldots$ durch ein-eindeutige Beziehung den Dingen $a, b, c, \ldots$ zugeordnet, welche die Elemente einer Menge m bilden, so sind auch die Dinge $A, B, C, \ldots$ Elemente einer Menge M. Dann brauchte man nur die $Z_0, Z_1, Z_2 \cdots$ den Elementen der Menge Z_0 zuzuordnen, um die Menge $\Theta = \left\{ Z_0, Z_1, Z_2, \ldots \right\}$ und $Z^* = \mathfrak{S}\Theta$ zu gewinnen. Aber das gefällt mir wenig, da mir dieser abstrakte Zuordnungsbegriff zu wenig "definit" vorkommt. Ich habe ihn doch gerade durch meine "Theorie der Äquivalenz" zu ersetzen gesucht. Also: diese Schwierigkeit ist noch ungelöst, und jedenfalls bin ich Ihnen dankbar, daß Sie mich darauf aufmerksam gemacht haben. Nebenbei: Bei der Annahme eines solchen "Zuordnungs-Axioms" wäre der von Ihnen versuchte Unabhängigkeitsbeweis undurchführbar. Denn der Bereich $\mathfrak{B}$, welcher doch alle *Elemente* m_i, m_i' enthält, müßte dann auch alle *Mengen*

$$\{m_1, m_2, m_3, \ldots\}$$

enthalten und der Beweis verfehlte sein Ziel. Also: mit dem Zuordnungs-Axiom geht es wohl nicht. Aber mit welchem anderen? Sobald mir etwas Aussichtsvolles einfällt, werde ich es Ihnen mitteilen, wie mich auch umgekehrt darauf bezügliche Bemerkungen Ihrerseits lebhaft interessieren werden. [...]

Mit Hochachtung

Ihr Zermelo

OV 3.18 Zermelo to Abraham A. Fraenkel on the axiom of choice and the axiom of replacement, 8 July 1921 (p. 137):

Ihr "Beweis" des Auswahlaxioms vermittels des "Ersetzungsaxioms" ist mir [...] doch einigermaßen bedenklich. Bei der "Ersetzung" müßte doch eine *eineindeutige* Zuordnung der Elemente [...] zu den Mengen M gefordert werden, die gerade hier eben nicht möglich ist.

Die Schwierigkeit besteht eben in einer hinreichenden Präzisierung solcher "Ersetzung" oder "Zuordnung", die hier allerdings in anderer Weise als in meiner "Theorie der Äquivalenz" überwunden werden müßte.

7.4 Freiburg

OV 4.01 Zermelo to his sisters Elisabeth and Margarete concerning a honorary professorship at Freiburg University, 27 March 1927 (p. 142):

Aber es wäre mir doch sehr wertvoll, *überhaupt* wieder akademisch tätig zu sein und mit der Universität kollegial zu verkehren.

OV 4.02 Lothar Heffter and Alfred Loewy in the application for Zermelo's honorary professorship, 19 February 1926 (p. 142):

Professor Zermelo ist ein hervorragender Mathematiker, der namentlich durch seine Arbeiten zur Mengenlehre internationalen Ruhm genießt. Seine Aufnahme in den Lehrkörper unserer Hochschule wäre daher für den mathematischen Unterricht und damit für die naturwissenschaftlich-mathematische Fakultät eine ausserordentlich wertvolle Bereicherung. Einen Mann von seinem Lebensalter und von seiner Bedeutung als Privatdozent der Fakultät anzugliedern, scheint uns indessen nicht angängig. Wir beantragen deshalb seine Ernennung zum ordentlichen Honorarprofessor.

OV 4.03 A sample of Zermelo's translation of parts of Homer's Odyssey (p. 144):

Fünfter Gesang
KALYPSO. Das Floss

Als Eos sich erhob vom Rosenlager
Den Göttern und den Menschen Licht zu bringen,
Da sassen sie versammelt zur Beratung
Die Götter alle und als mächtigster
Der Donnrer Zeus. Zu ihnen sprach Athene,
Der Leiden des Odysseus eingedenk,
der immer noch im Haus der Nymphe weilte:
"Hört Vater Zeus und all ihr andern Götter!
Bald wird doch keiner mehr von all den Fürsten,
die jetzt das Szepter führen auf der Erde,
gerecht und milde seines Amtes walten,
Nein hart und grausam werden sie hinfort
Nur Unrecht tun und wilde Frevel üben,
Wenn keiner des Odysseus mehr gedenkt

Von seinen Völkern, die er liebreich pflegte,
Wie je ein Vater seine Kinder nur!
Und er, in bittrem Harme sich verzehrend,
Er weilt noch immer auf Kalypsos Eiland,
Die ihn gefangen hält: denn ohne Schiffe
Und ohne Fahrgenossen, die ihn rudern,
Kann er die liebe Heimat nie erreichen.
Und nun beschloss man gar, ihm seinen Sohn
Zu töten, wenn er heimkehrt von der Reise,
Der jetzt nach Pylos zog und Lakedämon,
Um Kunde von dem Vater zu erlangen!"
Antwortend sprach zu ihr der Göttervater:
"Mein liebes Kind, wie soll ich das verstehn?
Ist's dir um den Odysseus so zu tun,
So hast Du selbst für ihn den Plan ersonnen,
Wie er heimkehrend Rache nehmen soll!
So führe denn zunächst, du kannst es ja,
Telemachos zurück in seine Heimat,
Dass er von dir behütet sicher sei
Und jene auf dem Schiff das Nachsehn haben!"
So sprach er, und zu seinem lieben Sohn
Dem Hermes dann sich wendend, sagte er:
"Nun, Hermes, da du doch mein Bote bist,
So künde du sogleich der schönen Nymphe
Von mir den unabänderlichen Ratschluss,
Dass jetzt Odysseus nach der Heimat kehre,
Von Göttern nicht, von Menschen nicht geleitet,
Auf wohlgefügtem Floss und dass er dann
In zwanzig Tagen leidensreicher Fahrt
Nach Scheria der fruchtbaren gelange,
Wo die Phäaken nah den Göttern hausen.
Die werden dort wie einen Gott ihn ehren
Und ihn zu Schiff nach seiner Heimat senden,
Mit Erz und Gold und Kleidern reich beschenkt,
Wie er sie nicht vor Troja selbst gehabt,
Als er mit aller Beute heimwärts zog.
Denn ihm ist's vorbestimmt, zum Vaterlande
Und zu den Seinen endlich heimzukehren!"
So sprach er, und der Götterbote tat's:
Flugs band er sich die goldenen Sandalen,
Die ihn mit Windeseile übers Meer
Und weites Land hintrugen, an die Füsse
Und nahm den Stab, womit er nach Gefallen
Der Menschen Augen bald in Schlaf versenken,
Bald aus dem Schlummer wieder wecken kann.

OV 4.04 Zermelo's description of his electrodynamic clutch, June 1932 (p. 146):

Im Folgenden handelt es sich um ein Verfahren, durch eine elastisch wirkende Kuppelung zwischen der Motor- und der Antriebswelle des Wagens mit kontinuierlich

veränderlicher und automatisch regulierbarer Untersetzung die Einschaltung verschiedener "Gänge" durch Zahngetriebe entbehrlich zu machen und dadurch ein stoßfreies Anfahren und Abbremsen des Wagens sowie einen gleichmäßig ruhigen Gang auf kupiertem Terrain, insbesondere auch auf Gebirgsstraßen zu ermöglichen.

OV 4.05 Zermelo to Marvin Farber about the philosophy at Freiburg University, 24 August 1928 (p. 147):

Ich habe hier leider niemand, mit dem ich [über Logik] sprechen könnte. Auch Prof. Becker hat kein Interesse: die "Intuitionisten" verachten die Logik! Und wenn erst Heidegger kommt: die "Sorge des Sorgers" und der "Wurf der Geworfenheit"! – Solche "Lebens"- bzw. "Todes"-Philosophie hat mit Logik und Mathematik schon gar nichts zu tun.

OV 4.06 Zermelo to Marvin Farber concerning a professorship in the United States, 21 September 1924 (p. 147):

Nur über die Möglichkeit, *selbst* hinzukommen, mache ich mir keine Illusionen [...]. Derartige Berufungen erhalten immer nur solche Gelehrte, die schon in ihrer Heimat hinreichen anerkannt sind und wohlplaziert sind wie Sommerfeld, Weyl u. a. Aber für einen, der einmal "kaltgestellt" ist, hat man auch im Ausland nichts übrig. Das ist eben die "Solidarität" der internationalen Bonzo-kratie!

OV 4.07 Zermelo to Adolf Kneser, then chairman of the Deutsche Mathematiker-Vereinigung, when announcing his talk on the navigation problem, 7 September 1929 (p. 148):

Auch erinnere ich mich noch deutlich unseres Gesprächs in Breslau, wo sie mir eine Wiederaufnahme der Variationsrechnung anstelle der modischen Grundlagen-Forschung ans Herz zu legen suchten. Sie sehen also, daß bei mir noch nicht Hopfen und Malz der "klassischen" Mathematik verloren ist und daß meine alte, wenn auch bisher meist unglückliche Liebe zu den "Anwendungen" im Stillen weiter geglommen ist.

OV 4.08 Zermelo to Marvin Farber on the project of a common monograph in logic, 23 April 1928 (p. 157):

Auch möchte ich zunächst einmal eine Probe Ihrer Logik-Darstellung gesehen haben, um mir klar zu werden, ob eine *gemeinsame* Weiterarbeit hier wirklich angängig und erfolgversprechend sein würde. Wenn Sie mir bis zum August etwas schicken, so könnte ich mich in den Ferien damit beschäftigen[,] und wenn Sie dann im Sommer 1929 event[uell] [...] herkämen, so wäre schon der Grund für fruchtbare Weiterarbeit gelegt.

OV 4.09 Zermelo to Abraham A. Fraenkel concerning Fraenkel's biography of Georg Cantor, 30 October 1927 (p. 159):

Leider ergab aber schon die erste Durchsicht, daß der Text Ihrer Darstellung noch erheblicher Änderungen, namentlich in stilistischer Beziehung bedarf, um für den vorliegenden Zweck brauchbar und druckfertig zu sein [...] Der Name des Helden, der dargestellten Person, sollte in einer Biographie überall ausgeschrieben werden, sofern er nicht durch eine Umschreibung ersetzt werden kann; das erfordert schon die Achtung vor der Person [...]; "der Naturfreund C." (Seite 10) macht einen geradezu grotesken Eindruck. Es ist, wie wenn in einer festlichen Gesellschaft ein Gast (oder gar der Gastgeber selbst!) mit einer Garderobenmarke herumliefe [...] Auf Seite 41 erscheint eine "Fusion" zwischen Mathematik und Philosophie: hier scheinen aktuelle Vorgänge des Bankwesens mitgespielt zu haben. Es war wohl eine "innere Verwandschaft" gemeint. Die Wissenschaft möchten wir lieber von "Fusion" wie von "Konfusion" bewahrt wissen [...]. Intime Familienbriefe, die dem Biographen vertrauensvoll zur Verfügung gestellt sind, können ihm selbst überaus wertvoll sein, indem sie ihm das Bild der darzustellenden Persönlichkeit ergänzen. Aber bei ihrem Abdruck kann man gar nicht zurückhaltend genug sein; überall, wo er entbehrlich ist, wirkt er auch taktlos [...]. Die Vorgänge des Jahres 1904 im Anschluß an den Heidelberger Kongress sollten vollständig im Zusammenhang oder gar nicht berichtet werden [...]. Auch weltanschauliche Bekenntnisse des Biographen scheinen mir hier nicht am Platze zu sein. Wer z. B. der Meinung anhängt, daß logische Durchdringung auch in der Wissenschaft von Übel und durch eine mystische "Intuition" zu ersetzen sei, hat heutzutage reichlich Gelegenheit, dieser modischen Überzeugung in Wort und Schrift Ausdruck zu geben; aber es braucht doch nicht gerade in der Lebensbeschreibung eines Mannes zu geschehen, der als Mathematiker ein Leben lang um logische Klarheit gerungen hat. Die Frage, ob z. B. der "logistische" Einfluß Dedekinds auf Cantor nützlich oder schädlich gewesen sei (Seite 10), wollen wir doch lieber der Zukunft überlassen, die noch über so manche der heute geräuschvoll diskutierten Streitfragen zur Tagesordnung übergehen wird [...].

Soviel für heute, obwohl ich noch manches Weitere in prinzipieller Beziehung zu sagen hätte. Ich hoffe aber, daß Sie meine Gesichtspunkte im Großen und Ganzen billigen und sich schließlich auch selbst von der Notwendigkeit überzeugen werden, daß Ihr Manuskript vor der Drucklegung noch einer wesentlichen Umarbeitung in Form und Inhalt bedarf. Es fragt sich nur, ob Sie diese Überarbeitung *selbst* übernehmen oder *mir* überlasssen wollen.

OV 4.10 Zermelo to Abraham A. Fraenkel on the same matter, 27 November 1927 (p. 160):

Es freut mich, daß Sie trotz allem bereit sind, die erforderlichen Änderungen Ihres Manuskripts vorzunehmen und also wohl meine Ausstellungen wenigstens teilweise als berechtigt anerkennen. Und das war es ja, was ich wünschte; sonst hätte ich mir die Mühe der ausführlichen Begründung doch nicht zu machen brauchen. Natürlich wäre es für mich weit bequemer gewesen, Ihre Arbeit ohne weiteres anzunehmen – und wozu dann erst lesen? – aber das schien mir eben mit der übernommenen Pflicht des Herausgebers unvereinbar. Und Sie sollten bemerkt haben, daß ich die Arbeit sorgfältig und gründlich durchgesehen habe und für meine Kritik keine anderen als sachliche Gründe im Interesse des Werkes maßgebend waren. Was nun den "Ton" meiner Ausführungen betrifft, durch den Sie sich augenscheinlich gekränkt fühlen, so weiß ich nicht, was Sie damit meinen. Etwa die scherzhafte Form, [...] die "Fusion" zu persiflieren? Ist das denn nicht eigentlich die gegebene und naturgemäße Form, einem unglücklich gewählten Ausdruck gegenüber an die sich von selbst anknüpfenden Assoziationen zu erinnern? Es gibt zwar Personen, die *jeden* Scherz als Beleidigung empfinden, weil ihnen eben der Sinn für Humor vollständig abgeht; aber niemand möchte doch gern zu diesen Unglücklichen gerechnet werden

[...] Aber nichts für ungut. Hauptsache ist, daß Sie bereit sind, die notwendigen Änderungen vorzunehmen.

OV 4.11 Abraham A. Fraenkel to Zermelo on the same matter, 12 December 1929 (p. 160):

Von den sachlichen u[nd] vielfach auch den stilistischen Abänderungsvorschlägen, die Sie machen, werde ich aber auf die Mehrzahl nicht eingehen können, weil ich weder meine wiss[enschaftlichen] Anschauungen noch meinen Stil zu verleugnen wünsche.

OV 4.12 Zermelo to Richard Courant about his financial situation, 4 February 1932 (p. 163):

Meine Cantor-Ausgabe hoffe ich bis zum März fertig zu stellen und wäre dann zur Übernahme irgend einer anderen literarisch-wissenschaftlichen Arbeit bereit und erbötig, falls sich Gelegenheit dazu finden sollte. Meine Einkommensverhältnisse machen mir solchen Nebenverdienst zur Notwendigkeit. [...] In Freiburg bzw. in Karlsruhe habe ich keinerlei Aussicht auf einen Lehrauftrag oder dergleichen. [...] Sollte sich also irgendwo die Gelegenheit zu Übersetzungen oder dergleichen bieten, so wäre ich dankbar für jede entsprechende Mitteilung. In einer Zeit allgemeiner Arbeitslosigkeit darf man sich freilich in dieser Beziehung keinen Illusionen hingeben. Aber man darf ebensowenig irgend eine sich bietende Gelegenheit versäumen. [...] In erster Linie denke ich dabei an Hessenbergs "Grundbegriffe der Mengenlehre", die [...] jetzt vergriffen sind, aber eine Neuauflage gewiß verdienten. Sollte ich nicht darüber an den Verlag schreiben, sofern er noch existiert? [...] Eile hat es ja wohl nicht, aber ich möchte doch nicht, daß mir darin ein anderer zuvorkommt.

OV 4.13 Zermelo to the *Notgemeinschaft* about his financial situation, 3 December 1930 (p. 164):

Ohne die weitere Unterstützung der Notgemeinschaft wäre ich also im nächsten Jahr gänzlich auf mein völlig unzureichendes Ruhegehalt angewiesen, während ich andererseits nicht wie jüngere Forscher von einer staatlichen Anstellung eine Besserung meiner wirtschaftlichen Lage zu erwarten hätte. Unter Würdigung dieser besonderen Umstände bitte ich daher die Notgemeinschaft zu erwägen, ob nicht eine Weiterbewilligung des Stipendiums möglich wäre.

OV 4.14 Zermelo to Lothar Heffter, 13 March 1929 (p. 166):

Für den mathematischen Gesamtunterricht kann ich als nichtbeamteter Dozent ohne Lehrauftrag in keiner Weise verantwortlich gemacht werden, stehe aber jedem Kollegen, der sich über ihn betreffende Fragen mit mir zu besprechen wünscht, jederzeit gern zur Verfügung, allerdings im Geiste der Kollegialität, auf dem Boden der Gleichheit und Gegenseitigkeit.

OV 4.15 The topics of Zermelo's Warsaw talks (p. 168):

T1 Was ist Mathematik? Die Mathematik als die Logik des Unendlichen.
T2 Axiomensysteme und logisch vollständige Systeme als Grundlage der allgemeinen Axiomatik.
T3 Über disjunktive Systeme und den Satz vom ausgeschlossenen Dritten.

T4 Über unendliche Bereiche und die Bedeutung des Unendlichen für die gesamte Mathematik.

T5 Über die Widerspruchslosigkeit der Arithmetik und die Möglichkeit eines formalen Beweises.

T6 Über die Axiomatik der Mengenlehre.

T7 Über die Möglichkeit einer independenten Mengendefinition.

T8 Theorie der "Grundfolgen" als Ersatz der "Ordnungszahlen".

T9 Über einige Grundfragen der Mathematik.

OV 4.16 The titles of Zermelo's notes for the Warsaw talks (p. 168):

W1 Was ist Mathematik?

W2 Disjunktive Systeme und der Satz vom ausgeschlossenen Dritten.

W3 Endliche und unendliche Bereiche.

W4 Wie rechtfertigt sich die Annahme des Unendlichen?

W5 Kann die Widerspruchslosigkeit der Arithmetik "bewiesen" werden?

W6 Über Mengen, Klassen und Bereiche. Versuch einer Definition des Mengen-Begriffs.

OV 4.17 On the principle of the excluded middle, Warsaw notes W2 (p. 171):

In der Tat ist es auch immer diese (auf hypothetische Modelle) eingeschränkte Anwendungsform des allgemeinen logischen Prinzipes, die in der Beweisführung der klassischen Mathematik aller Disziplinen eine so wesentliche Rolle spielt und m. Er. auch gar nicht entbehrt werden kann. Eine Mathematik ohne den (richtig verstandenen) Satz vom ausgeschlossenen Dritten, wie sie die "Intuitionisten" fordern und selbst glauben bieten zu können, wäre überhaupt keine Mathematik mehr [...] Die "Realisierbarkeit" durch Modelle ist eben die Grundvoraussetzung aller mathematischen Theorien, und ohne sie verliert auch die Frage nach der "Widerspruchsfreiheit" eines Axiomen-Systemes seine eigentliche Bedeutung. Denn die Axiome selbst tun einander nichts, bevor sie nicht auf ein und dasselbe (gegebene oder hypostasierte) Modell angewendet werden. "Widerspruchslosigkeit" hat erst einen Sinn, wenn ein geschlossener Kreis logischer Operationen und Prinzipien für die möglichen Schluß-folgerungen zugrunde gelegt wird. Zu diesen Prinzipien gehört aber ganz wesentlich eben der Satz vom ausgeschlossenen Dritten.

OV 4.18 On the hypothesis of the infinite, Warsaw notes W4 (p. 171):

Könnte nicht gerade diese scheinbar so fruchtbare Hypothese des Unendlichen geradezu Widersprüche in die Mathematik hineingebracht und damit das eigentliche Wesen dieser auf ihre Folgerichtigkeit so stolzen Wissenschaft von Grund auf zerstört haben?

OV 4.19 On consistency proofs, Warsaw notes W4 (p. 172):

Ein solcher Nachweis müßte sich, sofern er möglich ist, gründen auf eine durchgehende und vollständige Formalisierung der ganzen für die Mathematik in Betracht kommenden Logik. Jede "Unvollständigkeit" der zugrunde gelegten "Beweis-Theorie", jede etwa vergessene Schlußmöglichkeit würde den ganzen Beweis in Frage stellen.

Da nun aber eine derartige "Vollständigkeit" augenscheinlich niemals verbürgt werden kann, so entfällt damit m. Er. auch jede Möglichkeit, die Widerspruchslosigkeit formal zu beweisen.

OV 4.20 On consistency proofs versus the idea of infinite domains, Warsaw notes W5 (p. 172):

Überhaupt geht es nun einmal nicht an, den Formalismus wieder auf den Formalismus zu stützen; irgend einmal muß doch wirklich gedacht, muß etwas gesetzt, etwas angenommen werden. Und die einfachste Annahme, die gemacht werden kann und die zur Begründung der Arithmetik (wie auch der gesamten klassischen Mathematik) ausreicht, ist eben jene Idee der "unendlichen Bereiche", die sich dem logisch-mathematischen Denken geradezu zwangsmäßig aufdrängt und auf die auch tatsächlich unsere ganze Wissenschaft, so wie sie sich historisch entwickelt hat, aufgebaut ist.

OV 4.21 On the hypothesis of the infinite, Warsaw notes W4 (p. 172):

Rechtfertigen läßt sich eine solche Annahme lediglich durch ihren Erfolg, durch die Tatsache, daß sie (und sie allein!) die Schöpfung und Entwickelung der ganzen bisherigen Arithmetik, die eben wesentlich eine Wissenschaft des Unendlichen ist, ermöglicht hat.

OV 4.22 Zermelo to Marvin Farber on David Hilbert's and Wilhelm Ackermann's *Grundzüge der theoretischen Logik*, 24 August 1928 (p. 173):

Hilberts nun erschienene Logik ist mehr als dürftig, und auch von seiner immer angekündigten "Grundlegung der Mathematik" erwarte ich schon nichts Überwältigendes mehr.

OV 4.23 Zermelo to Richard Courant after having been elected a corresponding member of the Göttingen academy, 4 February 1932 (p. 179):

In der Tat habe ich mich über die Göttinger Ernennung sehr gefreut, umso mehr, als ich ja doch, wie Sie wissen, mit Anerkennung nicht gerade verwöhnt bin. In der Annahme, daß ich diese Ernennung in erster Linie meinem ersten und einzigen Lehrer in der Wissenschaft, Hilbert zu verdanken hätte, habe ich gleich zu Weihnachten an ihn geschrieben, ohne freilich bisher eine Antwort erhalten zu haben. [...]
  Zu Hilberts Geburtstag bin ich nicht nach Göttingen gekommen, da die Entfernung allzu groß ist und ich auch nicht wissen konnte, ob meine Anwesenheit wirklich erwünscht gewesen wäre. Aber vielleicht findet sich später die Gelegenheit, meine "angestammte Universität" wiederzusehen.

OV 4.24 Zermelo to Abraham A. Fraenkel on urelements, 20 January 1924 (p. 190):

Bedenklich scheint mir die Forderung, daß jedes Element einer Menge *selbst* eine Menge sein soll. Formell geht es wohl und vereinfacht die Formulierung. Aber wie steht es dann mit der *Anwendung* der Mengenlehre auf Geometrie und Physik?

OV 4.25 Zermelo to the *Notgemeinschaft* about a continuation of the *Grenzzahlen* theme, 3 December 1930 (p. 193):

Andere [fortsetzende Untersuchungen] dagegen, die ich gleichfalls für aussichtsvoll halte, bereiten noch erhebliche Schwierigkeiten und werden voraussichtlich noch längere intensive Forschungsarbeit erfordern, ehe sie zur Veröffentlichung reif sind.

OV 4.26 Zermelo to Abraham A. Fraenkel about Fraenkel's version of definiteness, 4 December 1921 (p. 201):

Ist Ihre Einschränkung [...] nicht doch vielleicht zu *eng*? Damit würden ja doch wohl immer nur *abzählbar* viele Untermengen gewonnen werden, und käme man dadurch nicht geradezu auf das Weyl-Brouwersche (abzählbare!) Pseudo-Continuum? Diese Frage scheint mir noch eingehenderer Erörterung wert zu sein. Aber vielleicht fließt die Nichtabzählbarkeit aus der Kombination dieses mit *anderen* Axiomen.

OV 4.27 Zermelo to an unknown addressee about his cumulative hierarchy, 25 May 1930 (p. 203):

Indem ich so den Mengenbegriff "relativiere", glaube ich andererseits den Skolemschen "Relativismus", der die ganze Mengenlehre in einem *abzählbaren* Modell darstellen möchte, widerlegen zu können. Es können eben nicht alle Mengen konstruktiv gegeben sein [...] und eine auf diese Annahme gegründete Theorie wäre überhaupt keine Mengenlehre mehr.

OV 4.28 Zermelo's infinity theses (p. 205):

(1) Jeder echte mathematische Satz hat "infinitären" Charakter, d. h. er bezieht sich auf einen *unendlichen* Bereich und ist als eine Zusammenfassung von unendlich vielen "Elementarsätzen" aufzufassen.

(2) Das Unendliche ist uns in der Wirklichkeit weder physisch noch psychisch gegeben, es muß als "Idee" im Platonischen Sinne erfaßt und "gesetzt" werden.

(3) Da aus finitären Sätzen niemals infinitäre abgeleitet werden können, so müssen auch die "Axiome" jeder mathematischen Theorie infinitär sein, und die "Widerspruchslosigkeit" einer solchen Theorie kann nicht anders "bewiesen" werden als durch Aufweisung eines entsprechenden widerspruchsfreien Systems von unendlich vielen Elementarsätzen.

(4) Die herkömmliche "Aristotelische" Logik ist ihrer Natur nach finitär und daher ungeeignet zur Begründung der mathematischen Wissenschaft. Es ergibt sich daraus die Notwendigkeit einer erweiterten "infinitären" oder "Platonischen" Logik, die auf einer Art infinitärer "Anschauung" beruht – wie z.B. in der Frage des "Auswahlaxioms" –, aber paradoxerweise gerade von den "Intuitionisten" aus Gewohnheitsgründen abgelehnt wird.

(5) Jeder mathematische Satz ist aufzufassen als eine Zusammenfassung von (unendlich vielen) Elementarsätzen, den "Grundrelationen", durch Konjunktion, Disjunktion und Negation, und jede Ableitung eines Satzes aus anderen Sätzen, insbesondere jeder "Beweis" ist nichts anderes als eine "Umgruppierung" der zugrunde liegenden Elementarsätze.

OV 4.29 Zermelo on the role of the axiom of foundation, around 1932 (p. 210):

[Das Fundierungsaxiom] soll die in der Mengenlehre wirklich brauchbaren Mengenbereiche vollständig charakterisieren [. . .] und die natürlichste und sicherste Grundlage für eine allgemeine ïnfinitistischeünd damit wahrhaft mathematische Syllogistik
der Satzsysteme bilden.

OV 4.30 Zermelo to an unknown addressee about his planned talk at the Bad
Elster conference, summer 1931 (p. 210):

Bleiben Sie wirklich die ganzen Ferien in Ihrer Sommerfrische oder kommen Sie
nicht doch lieber zur Tagung nach Bad Elster? Mir würde sehr daran liegen, für
meinen "metamathematischen" Vortrag unter den Zuhörern einen oder den anderen
zu haben, der meine Fundamenta-Arbeit gelesen (und verstanden!) hat. Denn ich
bin überzeugt, mit meiner "infinitistischen Logik" auf allen Seiten Widerspruch zu
finden: weder die "Intuitionisten" (natürlich, sie sind ja Feinde der Logik überhaupt)
noch die Formalisten, noch die Russellianer werden sie gelten lassen wollen. Ich
rechne nur noch auf die junge Generation, die allen solchen Dingen vorurteilsloser
gegenübersteht. Auch die Gegner des "Auswahlprinzips" und der Wohlordnung sind
ja nicht widerlegt, aber beides hat sich allmählich durch seine natürliche Kraft bei
der neuen Generation durchgesetzt.

OV 4.31 Gregory Moore on the Gödel controversy, 2002 (p. 212):

Der alte und erfahrene Zermelo, dessen Ansichten zur Logik größtenteils Jahrzehnte
vorher geformt wurden, [wurde] mit dem brillianten jungen Gödel konfrontiert.

OV 4.32 Heidegger's decree of 3 November 1933 concerning the opening of lectures
with the Hitler salute (p. 221):

Die Studentenschaft grüßt zu *Beginn* des Unterrichts nicht mehr durch Trampeln,
sondern durch Aufstehen und Erheben des rechten Armes. Die Dozenten grüßen
wieder vom Katheder aus in derselben Weise. Das Ende der Unterrichtsstunden
bleibt wie bisher.

OV 4.33 Wilhelm Magnus to Wilhelm Süss about the letter of Gustav Doetsch
under 4.34 (p. 222):

[Ich füge] einen Brief, genauer die Abschrift eines Briefes, bei, den Doetsch 1934 an
Tornier geschrieben hat. [. . .] Ich hätte nicht mit Arnold Scholz so gut bekannt sein
müssen um den dringenden Wunsch zu empfinden, daß es den Schreibern derartiger
Briefe in Zukunft erschwert werden möge, andere in einer derartigen Weise zu verleumden. Scholz hat mir immer versichert, dass er das Leben in Freiburg nicht lange
mehr hätte aushalten können; ich sehe nun, dass er in einem Masse recht gehabt
hat, das ich mir damals nicht richtig vorstellen konnte!

OV 4.34 Gustav Doetsch to Erhard Tornier about Arnold Scholz, 31 July 1934
(p. 222):

Ich verlange von niemandem, daß er Nat[ional]-Soz[ialist] ist, ich bin selbst nicht
Parteimitglied. Ich würde aber im Einklang mit den Intentionen der Regierung nur

jemanden fördern, der zum mindesten eine positive Einstellung zum heutigen Staat hat. Davon kann bei Sch[olz] keine Rede sein. Er ist im Gegenteil hier zusammen mit seinem einzigen Freund, den er hat, nämlich Zermelo, als der typische Meckerer bekannt, der nichts lieber sähe, als wenn das nat[ional]-soz[ialistische] Regiment, das im Gegensatz zu den früheren solchen Charakteren wie er es ist, zu Leibe zu gehen droht, so bald wie möglich wieder verschwände. – Daß ein solcher Waschlappen wie Sch[olz] einen politischen Einfluß ausüben könnte, ist allerdings ausgeschlossen. [...] Wenn aber solche Leute wie Sch[olz] sehen, daß es ihnen trotz aller Ankündigungen der Regierung auch heute noch sehr gut geht, [Footnote: Vor einem halben Jahr hat er sich ein Stipendium der Notgemeinsch[aft] besorgt, die Sache wurde hinter meinem Rücken eingeleitet.] so darf man sich nicht wundern, wenn sie ihr Unwesen ruhig und nur umso frecher weitertreiben. [...]

Sie werden jetzt vielleicht begreifen, daß ich baß erstaunt war, daß dieser Mann ausgerechnet zu einer Zeit, wo solchen Drohnenexistenzen schärfster Kampf angesagt wird, mit der Auszeichnung bedacht werden soll, nach auswärts für einen Lehrauftrag geholt zu werden.

OV 4.35 Zermelo to Heinrich Liebmann, 7 September 1933 (p. 223):

Im Übrigen hoffe ich, daß die Heidelberger Universität die Stürme des neuen Reiches ohne wesentlichen Schaden überstanden hat. Hier ist allerdings Loevy [!] endgültig emeritiert worden – was ich nach Lage der Dinge für kein Unglück halte; freiwillig wäre er niemals zurückgetreten. Aber den Nachfolger Heffters müssen wir leider behalten; über diesen Kollegen gibt es im Grunde nur eine Meinung. Nur ein Wunder kann hier noch helfen. Hoffen wir das beste!

OV 4.36 Eugen Schlotter's letter of denunciation, 18 January 1935 (p. 125, p. 225):

Ich fühle mich verpflichtet, über einen Vorfall Meldung zu erstatten, der bei der Saarkundgebung der Universität am 15. 1. 35 die Empörung meiner ganzen Umgebung hervorgerufen hat. An dieser Kundgebung hat Prof. Zermelo in einer Weise teilgenommen, die unbedingt als Schädigung des Ansehens der Freiburger Dozentenschaft anzusehen ist. Vorausschicken muss ich einige Bemerkungen, die geeignet sind, das Auftreten dieses Herrn zu charakterisieren.

Prof. Zermelo hat sich vor einem Menschenalter bei der Begründung der Mengenlehre einen unsterblichen Namen in der Wissenschaft gemacht. Seither hat er aber wissenschaftlich nie mehr etwas geleistet. Er ruhte auf seinen Lorbeeren aus und wurde zum Säufer. Ein wohlwollender Gönner verschaffte ihm aufgrund seines berühmten Namens eine Professur an der Universität Zürich. Als ihm diese Stelle sicher war, ging Zermelo auf Reisen, ohne sich um seine Verpflichtungen an der Universität zu kümmern. Er verlor seine Professur und lebte von dem ihm gewährten reichlichen Ruhegehalt. Mit unübertrefflicher Bosheit vergalt er die Gastfreundschaft seiner Wohltäter. Einmal trug er sich ins Fremdenbuch eines schweizer Hotels mit folgenden Worten ein: "E. Zermelo, schweizer Professor, aber *nicht* Helvetier!" Der Erfolg war eine erhebliche Kürzung seines Ruhegehaltes. – Er kam nach Freiburg und wurde aufgrund seines Namens Honorarprofessor. Nach der nationalen Erhebung fiel er öfters durch schwerste Beleidigungen des Führers und der Einrichtungen des Dritten Reiches auf. Bei Studenten und Dozenten genoss er bisher Narrenfreiheit. Als ich ihn vor einiger Zeit aufforderte, der Freiburger Junglehrerschaftbeizutreten und die seinem Einkommen entsprechende Aufnahmegebühr zu zahlen, erklärte er mir, dass er nicht einsehen könne, weshalb er das tun müsse. Sein Geld bezöge er aus der Schweiz. Er beabsichtige sowieso, im nächsten und übernächsten Semester nicht in Freiburg zu sein. Im übrigen warte er ab, bis er von einer massgebenderen Stelle (als ich es ihm zu sein schien) zum Eintritt aufgefordert würde. [...]

Bei der Saarkundgebung am 15.1.35 glaubte Professor Zermelo, sich ganz im Vordergrund der Dozentenschaft zeigen zu müssen, wohl weniger in der Absicht, seine nationale Gesinnung zu zeigen, sondern wahrscheinlich, um ja alles am besten sehen zu können. Als das Deutschlandlied angestimmt wurde, merkte er, wie alle Anwesenden die Hand erhoben. Scheu sah er sich nach allen Seiten um, und als er gar keinen anderen Ausweg sah, hielt er das Händchen bis auf Hüfthöhe vor und bewegte die Lippen, als ob er ebenfalls das Deutschlandlied sänge. Ich bin aber überzeugt, dass er nicht einmal den Text des Deutschlandliedes kennt, denn seine immer seltener werdenden Lippenbewegungen passten gar nicht zu diesem Text. Ueber diese Beobachtung war ich wie meine ganze Umgebung empört.

Als dann das Horst-Wessellied gesungen wurde, fühlte er sich sichtlich unbehaglich. Das Händchen wollte zunächst überhaupt nicht hoch. Noch scheuer blickte er um sich, bis er beim dritten Vers keine andere Möglichkeit sah, als das Händchen wieder mit sichtlichem Widerstreben in Hüfthöhe vorzuhalten. Dabei machte er den Eindruck, als müsse er Gift schlucken, verzweifelte Lippenbewegungen waren immer seltener zu beobachten. Er erschien erlöst, als die Zeremonie zu Ende war. Der Anblick dieser furchtbaren Jammergestalt verdarb uns die ganze Freude. Jeder studentische Beobachter musste doch den denkbar schlechtesten Eindruck mitnehmen von einer Dozentenschaft, die noch solche Elemente bei sich duldet. Wenn schon jemand nicht aus nationaler Ueberzeugung zu einer derartigen Kundgebung geht, aber zu neugierig ist, um nicht fernzubleiben, dann soll er sich wenigstens im Hintergrund aufhalten, wo er durch sein feindseliges Verhalten kein Aergernis erregt.

Ich bitte die Freiburger Dozentenschaft, zu untersuchen, ob Prof. Zermelo (noch) würdig ist, ehrenhalber Professor der Universität Freiburg zu sein und es ihm ermöglicht werden soll, im nächsten und übernächsten Semester irgendwo anders das Ansehen der deutschen Dozentenschaft zu schädigen.

OV 4.37 From the letter of denunciation of the Freiburg student union, 12 February 1935 (p. 226):

Herr Zermelo ist seit der nationalsozialistischen Erhebung schon öfters durch schwerste Beleidigungen des Führers und durch Nichtbeachtung der Einrichtungen des Dritten Reiches aufgefallen; so weigert er sich z.B., mit dem deutschen Gruss zu grüssen; dass er überhaupt noch an der Freiburger Universität existieren kann, liegt lediglich daran, dass er als ungefährlich bis jetzt geduldet wurde und nur noch auf wissenschaftlichem Gebiet ernstgenommen wird!

Diese Duldung darf jedoch nicht so weit gehen, dass er ungestraft das Ansehen der Freiburger Dozenten- und Studentenschaft schädigen darf. Ein Fall, der allgemeine Empörung hervorrief, war sein Auftreten bei der Saarkundgebung der Universität. [...]

Um zu verhindern, dass Prof. Zermelo noch länger die Belange und das Ansehen der Dozenten- und Studentenschaft öffentlich schädigen kann, ersuchen wir das Ministerium, nach Prüfung der Sachlage für eine Entfernung des Herrn Prof. hon. Dr. E. Zermelo aus dem Lehrkörper Sorge zu tragen.

OV 4.38 Gustav Doetsch's statement, 25 February 1935 (p. 227):

Zermelo verweigert mir und Anderen im Institut den Hitler-Gruß und zwar nicht
aus Zufall, sondern absichtlich, denn es geschah nicht nur hin und wieder und
zufälligerweise, sondern *regelmäßig*; selbst auf ein an ihn allein gerichtetes lautes
"Heil Hitler, Herr Zermelo", als er wieder den üblichen Gruß ignoriert hatte, antwortete er nur mit einem gequetschten "Heil". Aber auch an den folgenden Tagen
drückte er sich regelmäßig wiederum vor der Beantwortung des Hitlergrußes und
tat so, als ob er mein Hereinkommen und meinen Gruß überhaupt nicht bemerkt
hätte. Dies Ereignis vollzog sich etwa Anfang Februar ds. Js. Auch mein Assistent
Schlotter und mein Kollege Süss haben mir das Gleiche bestätigt. Obwohl Zermelo
von reiferen Studenten nicht ernst genommen wird, halte ich ein solches politisches
Gebaren für die jüngeren doch für gefährlich.

OV 4.39 Wilhelm Süss' statement, 28 February 1935 (p. 228):

Ich bin seit dem 31. 10. 34 in Freiburg und stehe mit Zermelo korrekt. Mir ist seitdem aufgefallen, daß Z[ermelo] den deutschen Gruß nicht erwidert. Weitere negative
Äußerungen gegen Führer und Bewegung sind in meiner Gegenwart nicht gefallen.

OV 4.40 Lothar Heffter's statement, 28 February 1935 (p. 228):

Über die Verweigerung des Hitlergrußes durch Zermelo habe ich keine eigene Erfahrung, da ich ihm nur selten begegne; auch Äußerungen Zermelos gegen Führer und
das neue Reich sind mir nicht bekannt. Ich halte Zermelo für einen außerordentlich genialen Mathematiker, bei dem die Grenze zwischen Genie und Abnormität
vielleicht etwas flüssig ist.

OV 4.41 Zermelo's statement, 1 March 1935 (p. 228):

Ich habe nicht die Absicht gehabt, durch Verweigerung des Hitlergrußes gegen eine
Gesetzesbestimmung zu verstoßen und habe auch bei Beginn meiner Vorlesungen
die vorgeschriebene Handbewegung gemacht; ich habe mich nur gegen Übertreibung
wehren wollen, wenn ich nicht immer den Gruß vorschriftsmäßig beantwortet habe.
    Auf Vorhalt: Ich bin vielleicht im Gasthaus gelegentlich mit kritischen Äuße-
rungen gegen die heutige Politik unvorsichtig gewesen, habe sie aber immer nur im
privaten Kreise vorgebracht, jedenfalls nicht im Dozentenzimmer.

OV 4.42 Zermelo's letter of renunciation, 2 March 1935 (p. 228):

Sehr geehrter Herr Dekan!
Hiermit teile ich Ihnen mit, daß ich nach einer Besprechung mit dem stellvertretenden Rektor mich entschlossen habe, auf eine weitere Lehrtätigkeit an der hiesigen
Universität zu verzichten und bitte Sie, mich meinen bisherigen Herren Kollegen
bestens zu empfehlen.
    In vorzüglichster Hochachtung
    Dr. E. Zermelo
    bisher ord[entlicher] Honorarprofessor an der Universität.

OV 4.43 The Rector's report to the ministry, 4 March 1935 (p. 230):

Aufgrund der Berichte gegen Professor Zermelo vom 18. 1. und 12. 2. 35 habe ich am
14. 2. gegen den Beschuldigten eine Untersuchung eingeleitet.

Die Vernehmung der Zeugen, soweit sie bis jetzt möglich war, ergibt einwandfrei, daß Z[ermelo] den deutschen Gruß nicht erwidert und dadurch Ärgernis in Dozenten- und Studentenkreisen erregt. Es scheint auch festzustehen, daß er dies willentlich und prinzipiell tut.

Über die Behauptung Schlotters, daß Z[ermelo] öfters durch schwerste Beleidigungen des Führers und der Einrichtungen des Dritten Reiches aufgefallen sei, konnte bis jetzt nichts Greifbares ermittelt werden; es scheint sich nach Zermelos Andeutungen um abfällige Kritiken in Gasthäusern zu handeln.

Für die Beurteilung dieses Falles muß die Persönlichkeit Zermelos berücksichtigt werden. Geheimrat Heffter charakterisiert ihn als genialen Mathematiker, bei dem die Grenze zwischen Genie und Abnormität flüssig sei. Als Forscher hat er Weltruf, als Lehrer hat er anscheinend stark nachgelassen. Von anderer Seite wird er als seltsamer Querkopf bezeichnet, der nicht aus bösem Willen, sondern aus Prinzip opponiert und sich über alles ärgert. Äußerlich sind an ihm starke Zerfallserscheinungen bemerkbar, dem Alkohol scheint er ergeben zu sein. Somit dürfte sein weiteres Verbleiben im Lehrkörper nicht wünschenswert sein.

Ich beantrage, mit Rücksicht auf die gegen ihn erhobenen begründeten Anklagen ihn aufzufordern, auf seine weitere Lehrtätigkeit zu verzichten. Anderfalls wäre gegen ihn ein Verfahren auf Entziehung der venia legendi einzuleiten.

OV 4.44 Zermelo's request for a confirmation of his letter of renunciation, 13 March 1935 (p. 230):

Ew. Magnifizenz
teile ich ergebenst mit, daß ich am 2. März dem Dekanat meiner Fakultät meinen Verzicht offiziell angezeigt habe, ohne bisher irgendeine Bezugnahme oder Empfangsbestätigung erhalten zu haben. Aus dem Umstande, daß mir neuerdings wieder amtliche Verfügungen die Universität betreffend übermittelt wurden, glaube ich aber schließen zu müssen, daß von meinem Austritt aus dem Lehrkörper amtlich noch nicht Kenntnis genommen ist. Möglicherweise ist mein Schreiben bei zeitweiliger Abwesenheit des Dekans liegen geblieben. Ich lege aber Wert darauf, festzustellen, daß ich unmittelbar nach meiner Besprechung mit dem stellvertretenden Rektor die erforderlichen Schritte getan habe, um mein Verhältnis zur Universität klarzustellen.
In geziemender Hochachtung
Dr. E. Zermelo.

OV 4.45 Zermelo to his sisters Elisabeth and Margarete, 19 April 1935 (p. 232):

Meine lieben Schwestern,
besten Dank für Eure frdl. Karte vom 28. III., durch die ich nach langer Pause endlich einmal wieder Nachricht von Euch erhalte. Es tut mir sehr leid, daß Eure Gesundheit wieder so viel zu wünschen läßt und Ihr so viel zu leiden hattet. Endlich erfahre ich Eure neue Adresse, aber ich hörte gern Näheres, wie Ihr jetzt wohnt. Mir selbst ging es wenigstens gesundheitlich weiter befriedigend. Aber ich bin hier recht einsam, und das neue Jahr begann ich unter trüben Vorahnungen und ich sollte Recht behalten. Meine hiesige Dozententätigkeit, für die ich ja keine Besoldung bezog, die mir aber früher noch manche Freude machte, habe ich jetzt unter dem Druck der Verhältnisse, zwischen intrigierenden Kollegen und denunzierenden Studenten, nachdem ich schon für das Sommersemester keine Vorlesung mehr angekündigt hatte, jetzt endgültig aufgegeben. Das Semester hatte kaum geschlossen, am 1. März, wo man sonst überall

die Saar-Rückkehr feierte, wurde ich vor den stellvertretenden Rektor zitiert, der mir Folgendes eröffnete: es sei eine Untersuchung gegen mich im Gange wegen "Verweigerung" (d. h. nicht vorschriftsmäßige Erwiderung) des Hitler-Grußes. Ich antwortete, daß ich mich in und außerhalb meiner Lehrtätigkeit bemüht hätte, allen gesetzlichen Vorschriften des neuen Staates peinlich nachzukommen, da ich mit Sokrates u. a. der Meinung sei, daß der Bürger jeder Regierung Gehorsam schuldig sei, auch wenn er ihre Maßregeln nicht gutheißen könne. Dagegen hätte ich als Nicht-Beamter, der auch keine staatliche Besoldung bezöge, mich durchaus nicht verpflichtet gefühlt, unabläßig eine nicht vorhandene Partei-Gesinnung zum Ausdruck zu bringen. Die Politik des Geßler-Hutes scheine sich doch, wie man bei Schiller nachlesen könne, nicht gerade bewährt zu haben. Und so hätte ich denn die Hitler-Grüße in einer Form beantwortet, wie ich auch sonst Grüße zu beantworten pflegte. Allfällige kritische Äußerungen über Regierungsmaßregeln hätte ich immer nur ganz privatim und niemals, soweit ich wisse, innerhalb der Universitätsräume verlauten lassen. Ich glaubte also in sachlicher Beziehung meinen Standpunkt voll vertreten zu können. Wenn mir aber die Universität zu verstehen geben wolle, daß sie auf meine (freiwillige und unbesoldete) Lehrtätigkeit keinen Wert mehr lege, so sei ich bereit, freiwillig auf eine Tätigkeit zu verzichten, die mir unter solchen Umständen keine Freude mehr machen könne. Damit schien der Stellvertreter auch zufrieden zu sein und wies mich an den Dekan, an den ich meine Verzichterklärung zu richten hätte; auch machte er keinen Versuch einer gütlichen Beilegung. Auf mein an den Dekan gerichtetes und mit einer "Empfehlung an die bisherigen Kollegen" schließendes Schreiben habe ich aber *niemals* eine Antwort oder Empfangsbestätigung erhalten, so daß ich 14 Tage später beim Rektorat anfragen mußte, ob denn meine Verzichterklärung nicht zur amtlichen Kenntnis gekommen sei. Nach weiteren 14 Tagen amtliche Mitteilung des Kultusministeriums in Karlsruhe, daß ich aus dem Lehrkörper ausgeschieden sei. Alles ohne Anrede oder die geringste Höflichkeitsformel, nachdem ich doch sieben Jahre ohne Entlohnung für die Universität tätig gewesen war. Die Interessen und das Ansehen der Hochschule scheinen für ihre offiziellen Vertreter nicht so wichtig gewesen zu sein wie das Bedürfnis, sich nach oben zu empfehlen. Auch von den mir befreundeten Fach- und anderen Kollegen hat niemand bisher den Versuch gemacht, sich mit mir über die Sache auszusprechen, ich gelte eben als verdächtig, und da will jeder sich fernhalten. Ich weiß aber auch, woher der Wind weht und wer mir diese Suppe eingebrockt hat: ein mißgünstiger Kollege, der es verstanden hat, den Mangel an wissenschaftlicher Begabung durch besonders üble Charaktereigenschaften zu kompensieren und sich eine Stellung zu erschleichen, die ihm nicht zukam. Aber was hat er nun davon? Werde ich jetzt günstiger über ihn urteilen, wenn ich meine Erlebnisse anderen Fachgenossen erzähle? Und was hat die Partei davon? Hat man wirklich geglaubt, daß ich staatsgefährliche Vorlesungen über Zahlentheorie oder Mengenlehre halten könnte? Bedauerlich ist nur, daß heute, wo die Selbstverwaltung der Universitäten aufgehoben ist, alle Intriganten freie Bahn haben, wenn sie nur geschickt genug den Mantel nach dem Winde zu drehen wissen. Ich selbst brauche die Vorlesungstätigkeit nicht, da ich meine Forschungsarbeit auch ohne das fortsetzen kann. Ob ich freilich in Freiburg bleibe, steht noch dahin. Vielleicht gehe ich nach Darmstadt, das eine zentrale Lage mit mildem Klima vereinigt und durch Bibliotheken Arbeitsmöglichkeiten bietet; auch habe ich ja die befreundete Familie in Jugenheim. Das wird sich erst im Laufe des Sommers entscheiden. Im Auslande käme außer der Schweiz, die ich nicht liebe, höchstens noch Österreich in Betracht, während Italien sich in geräuschvollen Kriegsdrohungen gefällt.

Nun aber genug, vielleicht seht Ihr aus diesem einen noch recht humorvollen Falle, wie es hier zugeht.

Mit besten Grüßen
Euer Bruder Ernst.
Fröhliche Ostern!

OV 4.46 Gustav Doetsch to the Rector of Freiburg University, 1 June 1936 (p. 235):

Der frühere Honorarprofessor an unserer Universität, Dr. Ernst Zermelo, wurde vor mehr als Jahresfrist veranlaßt, seine Honorarprofessur niederzulegen, anderenfalls ein Disziplinarverfahren gegen ihn eröffnet worden wäre wegen seiner gegnerischen Einstellung zum Dritten Reich. Nichtsdestoweniger läßt Zermelo sich in dem Mitglieder-Verzeichnis der Deutschen Mathematiker-Vereinigung noch immer als Honorarprofessor an der Universität Freiburg führen. Da dieses Verzeichnis allgemein mit Recht als authentisches Verzeichnis der deutschen mathematischen Hochschullehrer angesehen wird, gilt Zermelo noch immer als Mitglied unserer Universität [...].

Es scheint mir daher angebracht, wenn Zermelo von seiten des Rektorats offiziell darauf aufmerksam gemacht wird, daß es dem Rektorat zu Ohren gekommen sei, daß er sich noch immer als Honorarprofessor an der Univ[ersität] Freiburg im Mitglieder-Verzeichnis der Deutschen Mathematiker-Vereinigung führen lasse und daß das unstatthaft sei.

Ich halte es durchaus für möglich, daß Zermelo es geradezu darauf anlegt, noch weiter als Mitglied unserer Universität zu gelten; denn er drängt sich auch heute noch in die Sitzungen der in der Universität tagenden Freiburger Mathematischen Gesellschaft ein, ohne dazu eingeladen zu sein.

Heil Hitler!
G. Doetsch

OV 4.47 Zermelo's ten commandments indicting the Nazi regime (p. 236):

Neuer Keilschriftfund in Ninive

Die zehn Gebote Baals auf zwei Tontafeln

1) Ich bin Baal, Dein Gott. Du sollst keine andern Götter haben.

2) Du sollst Bildnisse machen von Mir, soviel Du kannst, und soll kein Ort sein ohne Baals-Bild auf Erden noch unter der Erde.

3) Du sollst den Namen Deines Gottes immer und überall im Munde führen, denn Baal wird den nicht ungestraft lassen, der seinen Namen verschweigt.

4) Du sollst den Feiertag heiligen und Feste feiern Tag für Tag. Die Arbeit aber überlaß den andern.

5) Du sollst den Priestern Baals unbedingten Gehorsam leisten, sei es auch wider Vater und Mutter.

6) Du sollst töten jeden, der Baal im Wege steht, Männer, Frauen und Kinder, auf daß die Ungläubigen ausgerottet werden auf Erden.

7) Du sollst nicht ehelos bleiben, sondern Nachkommen in die Welt setzen wie Sand am Meer.

8) Du sollst stehlen zur Ehre Baals, was Du kannst, allen denen, die ihn nicht anbeten.

9) Du sollst auch falsches Zeugnis ablegen wider die Ungläubigen und lügen und verleumden, was das Zeug hält: etwas bleibt immer hängen.

10) Du sollst begehren Ämter und Würden, die es gibt und die es nicht gibt, und keinem Ungläubigen lassen, was sein ist.

OV 4.48 Zermelo to Heinrich Scholz on the edition of Frege's collected papers, 10 April 1936 (p. 239):

Im Übrigen wäre ich selbst zu jeder gewünschten Mitarbeit an der beabsichtigten Frege-Ausgabe bereit, da ich nun doch schon eine gewisse Übung in solcher redaktionellen Tätigkeit gewonnen zu haben glaube. Jedenfalls kann ich es nur sehr begrüßen, daß dieser so lange verkannte Logiker der Mathematik durch die Neuausgabe nachträglich noch zu seinem Rechte kommt.

OV 4.49 A sample of Zermelo's little poems on everyday events, Easter 1937 (p. 240):

Das Osterei

Benjamin, das jüngste Hühnchen,
Ist so zierlich, klein und schwach,
Auf dem großen Hühnerhofe
Hat es nichts als Ungemach.
Ach das arme kleine Hühnchen,
Wenn es nur zu scharren wagt,
Wird es von den andern Hühnern
Futterneidisch weggejagt.
Hungrig flieht es zu den Menschen,
Schlüpft durchs Küchenfenster ein,
Läßt sich dort mit Hund und Katze
Brüderlich das Futter streun.
Und zum Dank der guten Köchin,
Die so treulich es gepflegt,
Hat es dann am Ostermorgen
Ihr ein Ei ins Bett gelegt.

OV 4.50 Heinrich Scholz to Zermelo, 3 October 1942 (p. 246):

Ich werde Ihnen nicht sagen müssen, dass ich an allem, was Sie mir zugewendet haben, mit dem grössten Interesse teilgenommen habe. Dass Sie an einem Lehrbuch der Mengenlehre arbeiten, erfahre ich durch Sie zum ersten Mal. Dieses Lehrbuch wird uns im höchsten Grade interessieren, und ich würde Ihnen sehr dankbar sein, wenn Sie mir unverbindlich sagen könnten, wann es voraussichtlich fertig sein wird. Dasselbe trifft zu auf die Arbeit an den endlichen Mengen. Ich nehme an, dass sie in einer mathematischen Zeitschrift erscheinen wird, und ich möchte Sie schon jetzt sehr bitten, dass Sie mir von dieser Arbeit wenigstens 2 Abzüge für mein Seminar zuwenden.

OV 4.51 The themes of Zermelo's planned *Mathematical Miniatures*, October 1940 (p. 246):

(1) Fliege und Spinne: Kürzeste Linie auf einer Körperoberfläche.

(2) Die Perlen des Maharadja: Verteilungsprobleme.

(3) Der Rattenfänger: Verfolgung in der Ebene.

(4) Straßenbau im Gebirge: Kürzeste Linien von begrenzter Steilheit.

(5) Navigation in der Luft: Ausnutzung des Windes bei der Luftfahrt.

(6) Schiffahrt auf einem See: Kürzeste Linien innerhalb einer Begrenzung.

(7) Der Springbrunnen: Wurfparabeln aus gemeinsamer Quelle.

(8) Wie zerbricht man ein Stück Zucker? Bruchlinien auf einem Rechteck.

(9) Gummiball und Lampenschirm: Faltung und Knickung biegsamer Flächen.

(10) Schachturnier: Bewertung seiner Teilnehmer: Maximumsproblem der Wahrscheinlichkeitsrechnung.

(11) Stationäre Strömungen in einer kompressiblen oder inkompressiblen Flüssigkeit.

(12) Die ewige Wiederkehr nach Nietzsche und Poincaré: Ein Satz der Dynamik.

(13) Der Schweppermann oder das Schubkastenprinzip.

(14) Tanzstunde und Hotelbetrieb bei uns und in Infinitalien: Ein-eindeutige Zuordnung endlicher und unendlicher Gesamtheiten.

(15) Lebensordnung und Verfassung im Lande Infinitalien.

(16) Holzabfuhr und Marschkolonnen.

OV 4.52 Heinrich Scholz to Zermelo on the occasion of Arnold Scholz's death, 3 October 1942 (p. 247):

Dass Sie Herrn Arnold Scholz sehr vermissen, bewegt mich in einem ungewöhnlichen Sinne. Ich habe ihn zwar nur ein einziges Mal hier in Münster gesehen; aber es war der Mühe wert, und das zahlentheoret[ische] Büchlein, das er bald darauf in der Göschensamml[un]g herausgebracht hat, ist auch in der Grundlegung von einer ganz ungewöhnlichen Sorgfalt und Genauigkeit. Ich habe bisher geglaubt, dass wir Grundlagenforscher im engsten Sinne die Einzigen sind, die ihn betrauern. Es tut mir wohl, von Ihnen zu hören, dass Sie von unserer Meinung sind. Ich habe nicht begriffen, warum er so im Verborgenen geblieben ist.

OV 4.53 Zermelo's answer to Paul Bernays' congratulations on his 80th birthday, 1 October 1941 (p. 102, p. 218, p. 249):

Ich freue mich sehr darüber, daß immer noch einige meiner früheren Kollegen und Mitarbeiter sich meiner erinnern, während ich schon so manche meiner Freunde durch den Tod verloren habe. Man wird eben immer einsamer, ist aber umso dankbarer für jedes freundliche Gedenken. Wenn ich auch immer noch wissenschaftlich interessiert und beschäftigt bin, so vermisse ich doch allzu sehr jeden wissenschaftlichen Gedankenaustausch, der mir früher, namentlich in meiner Göttinger Zeit so reichlich zuteil geworden war. [...] Über die Auswirkung meiner eigentlichen Lebensarbeit, soweit sie die "Grundlagen" und die Mengenlehre betrifft, mache ich mir freilich keine Illusionen mehr. Wo mein Name noch genannt wird, geschieht es immer *nur* in Verbindung mit dem "Auswahlprinzip", auf das ich *niemals* Prioritätsansprüche gestellt habe. So war das auch bei dem Grundlagen-Kongreß in Zürich, zu dem ich nicht eingeladen wurde, dessen Bericht Sie mir aber freundlicherweise zugesandt haben. In *keinem* der Vorträge oder Diskussionen, soweit ich mich bisher überzeugte, wurde eine meiner seit 1904 erschienenen Arbeiten (namentlich die beiden Annalen-Noten von 1908 und die beiden in den "Fundamenta" erschienenen im letzten Jahrzehnt) angeführt oder berücksichtigt, während die fragwürdigen Verdienste eines Skolem oder Gödel reichlich breitgetreten wurden. Dabei erinnere ich mich, daß schon bei der Mathematiker-Tagung in Bad Elster mein Vortrag über Satz-Systeme durch eine Intrige der von Hahn und Gödel vertretenen Wiener Schule von der Diskussion ausgeschlossen wurde, und habe seitdem die Lust verloren, über Grundlagen vorzutragen. So geht es augenscheinlich jedem, der keine "Schule" oder Klique hinter sich hat. Aber vielleicht kommt noch eine Zeit, wo auch meine Arbeiten wieder entdeckt und gelesen werden.

OV 4.54 Zermelo's request for a reinstatement as an honorary professor, 23 January 1946 (p. 250):

Als früherer Universitätsprofessor an der Universität Zürich für Mathematik gestatte ich Ihrer Magnifizenz Folgendes geziemend mitzuteilen. Nachdem ich im Jahre 1916 durch Krankheit genötigt war, meine Lehrtätigkeit vorläufig aufzugeben und in den dauernden Ruhestand versetzt wurde, habe ich mich bald darauf als Reichsdeutscher in der hiesigen Universitätsstadt zu ständigem Wohnsitz niedergelassen und als Ordentlicher *Honorarprofessor* eine Reihe von Jahren regelmäßige Vorlesungen aus dem Gebiete der reinen und angewandten Mathematik gehalten, bis ich unter der Hitler-Regierung durch parteipolitische Intrigen genötigt wurde, diese Tätigkeit aufzugeben. Unter den jetzt veränderten Umständen wäre ich aber gern bereit, sie meinen Kräften entsprechend in bescheidenem Umfange wieder aufzunehmen, und bitte daher die Universität im Einverständnis mit der Naturwissenschaftlichen Fakultät um geneigte Wiederernennung zum Honorarprofessor.

  In geziemender Hochachtung
  Dr. Ernst Zermelo
  Universitätsprofessor im Ruhestand.

OV 4.55 A list of possible talks which Zermelo set up on 17 October 1946 (p. 251):

(1) Über die Bedeutung des allgemeinen Wohlordnungssatzes für die Theorie der endlichen und der abzählbaren Mengen.
(2) Über die logische Form der mathematischen Sätze und Beweise.
(3) Über die Knickung der Flächen und die Entstehung der Mondkrater.
(4) Über den zweiten Hauptsatz der Wärmelehre und die Kinetische Gastheorie.
(5) Über ein mechanisches Modell zur Darstellung der Tonleitern und Harmonien.

OV 4.56 Zermelo to an unknown addressee on his efforts to settle in Switzerland, 18 January 1947 (p. 252):

Sehr geehrter lieber Kollege!
Besten Dank für Ihre freundliche Karte vom [gap] l. Leider sind Ihre Bemühungen wieder vergeblich gewesen. Von dem hiesigen Schweizerischen Konsulat habe ich nach 6-monatlichem Warten zwar keinen direkten Bescheid, aber eine auf mein Gesuch um Einreise-Genehmigung bezügliche Mitteilung erhalten, welche weitere Bemühungen wohl als aussichtslos erscheinen läßt und die ich hier im Ausschnitt beilege. Da ich keine Verwandten in der Schweiz habe und auch keine Einladung von einer Universität oder Behörde und noch weniger die Bereitwilligkeit einer solchen

Stelle habe, die Kosten meines dortigen Aufenthaltes zu tragen oder zu garantieren, so kann mein Gesuch wohl als endgültig abgelehnt betrachtet werden. Mein mir bei meinem Ausscheiden vom Amt ausdrücklich als "lebenslänglich" bewilligtes "Ruhegehalt" darf ich wohl ebenfalls in den bekannten "Schornstein" schreiben, nachdem alle meine bezüglichen Anfragen unbeantwortet geblieben sind. So kommt es, wenn man eine noch so verlockende Berufung in ein noch so neutrales Ausland angenommen hat. Man ist dann eben völlig rechtlos, und es hilft nichts.

Mit besten Grüßen und vielem Dank

Ihr Kollege E. Zermelo.

OV 4.57 Zermelo to his wife Gertrud from Zurich, 25 October 1947 (p. 252):

Überall gibt es hier gut zu essen und zu trinken und Straßenbahnen fahren in allen Richtungen. Nur schade, daß ich nicht ganz hier bleiben kann.

OV 4.58 Zermelo to the publisher Johann Ambrosius Barth, trying to arrange an edition of his collected papers, 20 January 1950 (p. 254):

Da ich für meine eignen Arbeiten math[ematischen] Inhalts bisher noch keinen Herausgeber gefunden habe, hat der Verlag Keiper mir den Ihrigen empfohlen, der mir schon durch die Herausgabe der Gibbs'schen Publikationen bekannt war, und ich bitte Sie daher um die freundliche Mitteilung, ob Sie wohl in der Lage wären, diese Herausgabe zu übernehmen, nachdem ich jetzt aus Gesundheitsgründen nicht mehr in der Lage bin, neue wissenschaftliche Arbeiten erscheinen zu lassen.

OV 4.59 Wilhelm Süss to Zermelo on the occasion of Zermelo's 80th birthday, 27 July 1951 (p. 255):

Ihr Name ist in aller Welt unter den Mathematikern auf lange Zeit durch die grundlegenden Arbeiten zur Mengenlehre bekannt. Das Schicksal u[nd] die Bosheit der Menschen haben später manchen schwarzen Schatten über den strahlenden Ruhm ihrer Jugendzeit geworfen. Wer aber wie wir das Glück hatte, Sie persönlich kennen zu lernen, weiß auch davon, wie Ihr gütiges Herz auch alle Mißhelligkeiten zu ertragen verstanden hat.

Das hiesige mathematische Institut durfte sich viele Jahre Ihrer Mitwirkung erfreuen, bis dann eine Denunziation Ihnen die Lösung von der Universität für lange Jahre aufzwang. Sie haben Böses hier nicht mit Haß vergolten, sondern sich später mit einer formalen Rehabilitation begnügt, ein Zeichen mehr für die gütige Wesensart, die Ihnen eignet.

OV 4.60 Paul Bernays' letter of condolence to Gertrud Zermelo, 17 July 1953 (p. 256):

In der Zeit meines ersten Züricher Aufenthaltes – er hatte damals die Züricher Mathematik-Professur inne, und ich war hier Privatdozent – hatten wir oft geselliges Beisammensein und Gedankenaustausch. Er hatte viel Witz und Esprit, und ich habe – auch ausser demjenigen, was ich wissenschaftlich von ihm lernte – viel geistige Anregung von ihm empfangen. Stets auch erfreute ich mich seines persönlichen Wohlwollens.

Seine wissenschaftlichen Leistungen, besonders im Gebiet der Mengenlehre, kommen in der heutigen Grundlagenforschung stark zur Geltung; sie werden gewiss für die Dauer ihre Bedeutung behalten und seinen Namen in der Wissenschaft lebendig erhalten.

Sie, liebe Frau Zermelo, können das befriedigende Bewusstsein haben, dass Sie Ihrem Mann in seinen späten Lebensjahren viele Mühen des Daseins abgenommen und ihn vor Verlassenheit und trauriger Einsamkeit bewahrt haben. Dafür werden auch alle, die der markanten und originellen Forscherpersönlichkeit Ihres Mannes würdigend gedenken, Ihnen Dank wissen.

References

[Ale15] Alexandrow, Waldemar: *Elementare Grundlagen für die Theorie des Maßes*, Ph. D. Thesis, University of Zurich 1915.

[AreRT92] Aref, Hassan, Rott, Nicholas, and Thomann, Hans: "Gröbli's Solution of the Three Vortex Problem," *Annual Reviews of Fluid Mechanics* **24** (1992), 1–20.

[AreM06] Aref, Hassan and Meleshko, Vyacheslav V.: "A Bibliography of Vortex Dynamics 1858–1956," manuscript, Virginia Polytechnic Institute and State University 2006. To appear in *Advances in Applied Mathematics*.

[Bae28] Baer, Reinhold: "Über ein Vollständigkeitsaxiom der Mengenlehre," *Mathematische Zeitschrift* **27** (1928), 536–539.

[Bae29] Baer, Reinhold: "Zur Axiomatik der Kardinalzahlarithmetik," *Mathematische Zeitschrift* **29** (1929), 381–396.

[Bae52] Baer, Reinhold: "Endlichkeitskriterien für Kommutatorgruppen," *Mathematische Annalen* **124** (1952), 161–177.

[BaeL31] Baer, Reinhold and Levi, Friedrich: "Stetige Funktionen in topologischen Räumen. Ernst Zermelo zum 60. Geburtstag am 27. Juli 1931 gewidmet," *Mathematische Zeitschrift* **34** (1931), 110–130.

[BaoCS04] Bao, David, Robles, Colleen, and Shen, Zhongmin: "Zermelo Navigation on Riemannian Manifolds," *Journal of Differential Geometry* **66** (2004), 391–449.

[Bar67] Barwise, Jon: *Infinitary Logic and Admissible Sets*, Ph. D. Thesis, Stanford University 1967.

[Bar75] Barwise, Jon: *Admissible Sets and Structures. An Approach to Definability Theory*, Springer-Verlag: Berlin, Heidelberg, and New York 1975.

[Bar85] Barwise, Jon: "Model-Theoretic Logics: Background and Aims," in [BarF85], 3–23.

[Bar89] Barwise, Jon: *The Situation in Logic*, Center for the Study of Language and Information: Stanford 1989.

[Bar77] Barwise, Jon (ed.): *Handbook of Mathematical Logic*, North-Holland Publishing Company: Amsterdam, New York, Oxford 1977.

[BarF85] Barwise, Jon and Feferman, Solomon (eds.): *Model-Theoretic Logics*, Springer-Verlag: New York et al. 1985.

[BecDW87] Becker, Heinrich, Dahms, Hans-Joachim, and Wegeler, Cornelia (eds.): *Die Universität Göttingen unter dem Nationalsozialismus. Das verdrängte Kapitel ihrer 250jährigen Geschichte*, Saur: München et al. 1987.

[Beck27] Becker, Oskar: "Mathematische Existenz. Zur Logik und Ontologie mathematischer Phänomene," *Jahrbuch für Philosophie und phänomenologische Forschung* **8** (1927), 1–370. Also separately published. 2nd edition, Max Niemeyer: Tübingen 1973.

[Beh69] Behrend, Hannelore: "Der Wiederkehreinwand gegen Boltzmanns H-Theorem und der Begriff der Irreversibilität," *NTM Schriftenreihe für Geschichte der Naturwissenschaften, Technik und Medizin* **6** (1969), No. 2, 27–36.

[Bel33] Bell, Eric Temple: Review of [Can32], *Bulletin of the American Mathematical Society* **39** (1933), 17.

[Ber12] Bernays, Paul: *Über die Darstellung von positiven, ganzen Zahlen durch die primitiven, binären, quadratischen Formen einer nicht-quadratischen Diskriminante*, Ph.D. Thesis, University of Göttingen 1912.

[Ber13] Bernays, Paul: "Zur elementaren Theorie der Landauschen Funktion $\phi(\alpha)$," *Vierteljahrsschrift der Naturforschenden Gesellschaft in Zürich* **58** (1913), Habilitation Thesis University of Zurich.

[Ber26] Bernays, Paul: "Axiomatische Untersuchung des Aussagenkalküls der 'Principia Mathematica'," *Mathematische Zeitschrift* **25** (1926), 305–320.

[Ber32] Bernays, Paul: "Methode des Nachweises von Widerspruchsfreiheit und ihre Grenzen," in Walter Saxer (ed.): *Verhandlungen des Internationalen Mathematiker-Kongresses Zürich 1932, Zweiter Band: Sektionsvorträge*, Orell Füssli Verlag: Zurich and Berlin 1932, 342–343.

[Ber35] Bernays, Paul: "Hilberts Untersuchungen über die Grundlagen der Geometrie," in [Hil35], 196–216.

[Ber41] Bernays, Paul: "A System of Axiomatic Set Theory II," *The Journal of Symbolic Logic* **6** (1941), 1–119.

[Ber55] Bernays, Paul: "Betrachtungen über das Vollständigkeitsaxiom und verwandte Axiome," *Mathematische Zeitschrift* **63** (1955), 219–229.

[Bern01] Bernstein, Felix: *Untersuchungen aus der Mengenlehre*, Halle a.S., Ph.D. Thesis Göttingen 1901.

[Bern05a] Bernstein, Felix: "Über die Reihe der transfiniten Ordnungszahlen," *Mathematische Annalen* **60** (1905), 187–193.

[Bern05b] Bernstein, Felix: "Zum Kontinuumproblem," *Mathematische Annalen* **60** (1905), 463–464.

[Bern05c] Bernstein, Felix: "Die Theorie der reellen Zahlen," *Jahresbericht der Deutschen Mathematiker-Vereinigung* **15** (1905), 447–449.

[Bet59] Beth, Evert Willem: *The Foundations of Mathematics. A Study in the Philosophy of Science*, North-Holland: Amsterdam 1959.

[Bett96] Bettazzi, Rodolpho: "Gruppi finiti ed infiniti di enti," *Atti della Accademia delle Scienze di Torino, Classe di Scienze Fisiche, Matematiche, e Naturale* **31** (1896), 506–512.

[Bie33] Bieberbach, Ludwig: Review of [Can32], *Jahresbericht der Deutschen Mathematiker-Vereinigung* **42** (1933), 2nd section, 92–93.

[Bier88] Biermann, Kurt-R.: *Die Mathematik und ihre Dozenten an der Berliner Universität 1810–1933. Stationen auf dem Wege eines mathematischen Zentrums von Weltgeltung*, Akademie-Verlag: Berlin 1988.

[Bil51] Bilharz, Herbert: "Über die Gaußsche Methode zur angenäherten Berechnung bestimmter Integrale. Ernst Zermelo zum 80. Geburtstag gewidmet," *Mathematische Nachrichten* **6** (1951), 171–192.

[Ble60] Blencke, Erna: "Leonard Nelsons Leben und Wirken im Spiegel der Briefe an seine Eltern, 1891–1915," in Hellmut Becker, Willi Eichler, and Gustav Heckmann (eds.): *Erziehung und Politik. Minna Specht zu ihrem 80. Geburtstag,* Verlag Öffentliches Leben: Frankfurt a. M., 9–72.

[Bli46] Bliss, Gilbert A.: *Lectures On the Calculus of Variations,* The University of Chicago Press: Chicago 1946, 2nd impression 1947.

[Blu35] Blumenthal, Otto: "Lebensgeschichte," in [Hil35], 388–429.

[Bog77] Bogomolov, V. A.: "Dynamics of Vorticity at a Sphere," *Fluid Dynamics* **12** (1977), 863–870.

[Bog79] Bogomolov, V. A.: "Two-Dimensional Fluid Dynamics on a Sphere," *Izvestiya of Atmospherec and Oceanic Physics* **15** (1979), 18–22.

[Bol68] Boltzmann, Ludwig: "Studien über das Gleichgewicht der lebendigen Kraft zwischen bewegten materiellen Punkten," *Sitzungsberichte der Kaiserlichen Akademie der Wissenschaften in Wien, Mathematisch-Naturwissenschaftliche Classe* **58** (1868), 517–560. Reprinted in [Bol09], Vol. 1, 49–96

[Bol72] Boltzmann, Ludwig: "Weitere Studien über das Gleichgewicht unter Gasmolekülen," *Sitzungsberichte der Kaiserlichen Akademie der Wissenschaften in Wien, Mathematisch-Naturwissenschaftliche Classe* **66** (1872), 275–370. Reprinted in [Bol09], Vol. 1, 316–402 and in [Bru70], 115–225; English translation [Bru66], 88–175.

[Bol77] Boltzmann, Ludwig: "Über die Beziehung zwischen dem zweiten Hauptsatz der mechanischen Wärmetheorie und der Wahrscheinlichkeitsrechnung resp. den Sätzen über das Wärmegleichgewicht," *Sitzungsberichte der Kaiserlichen Akademie der Wissenschaften in Wien, Mathematisch-Naturwissenschaftliche Classe* **76** (1877), 373–435. Reprinted in [Bol09], Vol. 2, 164–223.

[Bol96] Boltzmann, Ludwig: "Entgegnung auf die wärmetheoretischen Betrachtungen des Hrn. E. Zermelo," *Annalen der Physik und Chemie* n. s. **57** (1896), 773–784. Reprinted in [Bol09], Vol. 3, 567–578 and in [Bru70], 276–289; English translation [Bru66], 218–228.

[Bol97a] Boltzmann, Ludwig: "Zu Hrn. Zermelo's Abhandlung 'Über die mechanische Erklärung irreversibler Vorgänge'," *Annalen der Physik und Chemie* n. s. **60** (1897), 392–398. Reprinted in [Bol09], Vol. 3, 579–586 and in [Bru70], 301–309; English translation [Bru66], 238–245.

[Bol97b] Boltzmann, Ludwig: *Vorlesungen über die Principe der Mechanik,* Barth: Leipzig 1897.

[Bol05a] Boltzmann, Ludwig: *Populäre Schriften,* Barth: Leipzig 1905.

[Bol05b] Boltzmann, Ludwig: "Reise eines deutschen Professors ins Eldorado," in [Bol05a], 403–435.

[Bol09] Boltzmann, Ludwig: *Wissenschaftliche Abhandlungen,* Vols. 1–3, Leipzig: J. A. Barth 1909.

[BolN07] Boltzmann, Ludwig and Nabl, Joseph: "Kinetische Theorie der Materie," in *Encyklopädie der mathematischen Wissenschaften mit Einschluss ihrer Anwendungen,* Vol. 5: *Physik,* Part 1. 8, 493–557, Teubner: Leipzig, volume published 1903–1921, article published 1907.

[Bolz04] Bolza, Oskar: *Lectures on the Calculus of Variations,* University of Chicago Press: Chicago 1904.

[Bolz09] Bolza, Oskar: *Vorlesungen über Variationsrechnung,* Teubner: Leipzig 1909.

[Boo54]   Boole, George: *An Investigation of the Laws of Thought on which are Founded the Mathematical Theories of Logic and Probabilities*, Walton & Maberly: London 1854.

[Bor98]   Borel, Émile: *Leçons sur la théorie des fonctions*, Gauthier-Villars: Paris 1898, ⁴1950.

[Bor05]   Borel, Émile: "Quelques remarques sur les principes de la théorie des ensembles," *Mathematische Annalen* **60** (1905), 194–195.

[Bou39]   Bourbaki, Nicholas: *Eléments de mathématique*, Part 1: *Les structures fondamentales de l'analyse*, Book I: *Théorie des ensembles*, Hermann: Paris 1939.

[Bou60]   Bourbaki, Nicholas: *Éléments d'histoire de mathématiques*, Hermann: Paris 1960.

[BraT52]   Bradley, Ralph Allen, Terry, Milton E.: "The Rank Analysis of Incomplete Block Designs. 1. The Method of Paired Comparisons," *Biometrika* **39** (1952), 324–345.

[Bre98]   Brémaud, Pierre: *Markov Chains*, Springer-Verlag: Berlin et al. 1998.

[Bru66]   Brush, Stephen G.: *Kinetic Theory II. Irreversible Processes*, Pergamon Press: Oxford 1966.

[Bru70]   Brush, Stephen G.: *Kinetische Energie II. Irreversible Prozesse. Einführung und Originaltexte*, Vieweg: Braunschweig 1970 (*WTB-Texte*; 67).

[Bru78]   Brush, Stephen G.: *The Temperature of History. Phases of Science and Culture in the Nineteenth Century*, Franklin: New York 1978.

[Buh66]   Buhl, Günter: "Die algebraische Logik im Urteil der deutschen Philosophie des 19. Jahrhunderts," *Kant-Studien* **57** (1966), 360–372.

[Bul02]   Buldt, Bernd et al. (eds.): *Kurt Gödel. Wahrheit & Beweisbarkeit*, Vol. 2: *Kompendium zum Werk*, öbv & hpt: Vienna 2002.

[Bur97]   Burali-Forti, Cesare: "Una questione sui numeri transfiniti," *Rendiconti del circolo matematico di Palermo* **11** (1897), 154–164.

[Can71a]   Cantor, Georg: "Notiz zu dem Aufsatze: Beweis, daß eine für jeden reellen Wert von x durch eine trigonometrische Reihe gegebene Funktion $f(x)$ sich nur auf eine einzige Weise in dieser Form darstellen läßt," *Journal für die reine und angewandte Mathematik* **73** (1871), 294–296. Reprinted in [Can32], 84–86.

[Can71b]   Cantor, Georg: "Über trigonometrische Reihen," *Mathematische Annalen* **4** (1871), 139–143. Reprinted in [Can32], 87–91.

[Can72]   Cantor, Georg: "Über die Ausdehnung eines Satzes aus der Theorie der trigonometrischen Reihen," *Mathematische Annalen* **5** (1872), 123–132. Reprinted in [Can32], 92–102.

[Can73]   Cantor, Georg: "Historische Notizen über die Wahrscheinlichkeitsrechnung," *Sitzungsberichte der Naturforschenden Gesellschaft zu Halle* (1873), 34–42. Reprinted in [Can32], 357–367.

[Can78]   Cantor, Georg: "Ein Beitrag zur Mannigfaltigkeitslehre," *Journal für die reine und angewandte Mathematik* **84** (1878), 242–258. Reprinted in [Can32], 119–133.

[Can82]   Cantor, Georg: "Ueber unendliche, lineare Punktmannichfaltigkeiten [3.]," *Mathematische Annalen* **20** (1882), 113–121. Reprinted in [Can32], 149–157.

[Can83]   Cantor, Georg: "Ueber unendliche, lineare Punktmannichfaltigkeiten [5.]," *Mathematische Annalen* **21** (1883), 545–591. Reprinted in [Can32], 165–204. English translation in [Ewa96], 878–920.

[Can87]   Cantor, Georg: "Mitteilungen zur Lehre vom Transfiniten I," *Zeitschrift für Philosophie und philosophische Kritik* **91** (1887), 81–125. Reprinted in [Can32], 378–419.

[Can90/91]   Cantor, Georg: "Über eine elementare Frage der Mannigfaltigkeitslehre," *Jahresbericht der Deutschen Mathematiker-Vereinigung* **1** (1890/91; printed 1892), 75–78. Reprinted in Cantor1932, 278–280; English translation in [Ewa96], 920–922.

[Can95]   Cantor, Georg: "Beiträge zur Begründung der transfiniten Mengenlehre (Erster Artikel)," *Mathematische Annalen* **46** (1895), 481–512. Reprinted in [Can32], 282–311. English translation [Can41].

[Can97]   Cantor, Georg: "Beiträge zur Begründung der transfiniten Mengenlehre (Zweiter Artikel)," *Mathematische Annalen* **49** (1897), 207–246. Reprinted [Can32], 312–351.

[Can99]   Cantor, Georg: Letter to R. Dedekind, 3 August 1899. Reprinted in [Can1991], 407–411; English translation in [Ewa96], 931–935.

[Can32]   Cantor, Georg: *Gesammelte Abhandlungen mathematischen und philosophischen Inhalts. Mit erläuternden Anmerkungen sowie mit Ergänzungen aus dem Briefwechsel Cantor-Dedekind*, edited by Ernst Zermelo, Springer-Verlag: Berlin 1932. Reprinted Olms: Hildesheim 1962, 1980; Springer-Verlag: Berlin et al. 1980.

[Can41]   Cantor, Georg: *Contributions to the Founding of the Theory of Transfinite Numbers*, translated and provided with an introduction and notes by Philip E. B. Jourdain, The Open Court: La Salle, Illinois 1941.

[Can1991]   Cantor, Georg: *Briefe*, edited by Herbert Meschkowski and Winfried Nilson, Springer-Verlag: Berlin, Heidelberg, New York et al. 1991.

[Car04]   Carathéodory, Constantin: *Über die diskontinuierlichen Lösungen in der Variationsrechnung*, Doctoral thesis, University of Göttingen; reprinted in [Car54], Vol. I (1954), 3–79.

[Car10]   Carathéodory, Constantin: Reviews of Oskar Bolza, *Vorlesungen über Variationsrechnung,* and Jaques Hadamard, *Leçons sur le calcul des variations recueillies par M. Fréchet, Archiv der Mathematik und Physik* **16** (1910), 221–224.

[Car18]   Carathéodory, Constantin: *Vorlesungen über reelle Funktionen*, B. G. Teubner: Leipzig 1918.

[Car19]   Carathéodory, Constantin: "Über den Wiederkehrsatz von Poincaré," *Sitzungsberichte der Preußischen Akademie der Wissenschaften zu Berlin, Mathematisch-physikalische Klasse* **1919** (1919), 580–584. Reprinted in [Car54], Vol. IV (1956), 296–301.

[Car35]   Carathéodory, Constantin: *Variationsrechnung und partielle Differentialgleichungen erster Ordnung*, B. G. Teubner: Leipzig 1935.

[Car54]   Carathéodory, Constantin: *Gesammelte Mathematische Schriften*, Vol. I–V, C. H. Beck: Munich 1954–1957.

[Carn36]   Carnap, Rudolf: "Truth in Mathematics and Logic" (abstract), *The Journal of Symbolic Logic* **1** (1936), 59.

[Cav38a]   Cavaillès, Jean: *Remarques sur la formation de la théorie abstraite des ensembles*, Hermann: Paris 1938.

[Cav38b]   Cavaillès, Jean: *Méthode axiomatique et formalisme. Essai sur le problème du fondement des mathématiques*, Hermann: Paris 1938.

[Coh66]   Cohen, Paul J.: *Set Theory and the Continuum Hypothesis*, The Benjamin/Cummings Publishing Company: Reading, MA, 1966.

[Cou04] Couturat, Louis: "[IIme Congrès de Philosophie, Genève. Comptes rendu critiques] II. Logique et Philosophie des Sciences. Séances générales," *Revue de Métaphysique et de Morale* **12** (1904), 1037–1077.

[DahH88] Dahms, Hans-Joachim and Halfmann, Frank: "Die Universität Göttingen in der Revolution 1918/19," in *1918. Die Revolution in Südhannover. Begleitheft zur Dokumentation des Museumsverbundes Südniedersachsen*, Städtisches Museum: Göttingen 1988, 59–82.

[Dau79] Dauben, Joseph W.: *Georg Cantor. His Mathematics and Philosophy of the Infinite*, Harvard University Press: Cambridge, Mass. and London 1979. Reprinted Princeton University Press: Princeton 1990.

[Dav88] David, Herbert A.: *The Method of Paired Comparisons*, 2nd edition, Oxford University Press: Oxford 1988, first edition Charles Griffin & Company Ltd.: London 1963.

[Daw85] Dawson, John W., Jr.: "Completing the Gödel-Zermelo Correspondence," *Historia Mathematica* **12** (1985), 66–70.

[Daw97] Dawson, John W., Jr.: *Logical Dilemmas. The Life and Work of Kurt Gödel*, A. K. Peters: Wellesley, Mass 1997.

[Ded72] Dedekind, Richard: *Stetigkeit und irrationale Zahlen*, Vieweg: Braunschweig 1872. Reprinted in [Ded1932], 315–334; English translation in [Ewa96], 765–779.

[Ded88] Dedekind, Richard: *Was sind und was sollen die Zahlen?*, Vieweg: Braunschweig, ^{2}1893, ^{3}1911, ^{8}1960. Reprinted in [Ded1932], 335–391; English translation in [Ewa96], 787–833.

[Ded1932] Dedekind, Richard: *Gesammelte Werke*, Vol. III, edited by R. Fricke, E. Noether and Ø. Ore, Vieweg: Braunschweig 1932.

[Dei05] Deiser, Oliver: "Der Multiplikationssatz der Mengenlehre," *Jahresbericht der Deutschen Mathematiker-Vereinigung* **107** (2005), 88–109.

[Dic85] Dickmann, Max: "The Infinitary Languages $\mathcal{L}_{\kappa,\lambda}$ and $\mathcal{L}_{\infty,\lambda}$," in [BarF85], 317–363.

[Din13] Dingler, Hugo: "Über die logischen Paradoxien der Mengenlehre und eine paradoxienfreie Mengendefinition," *Jahresbericht der Deutschen Mathematiker-Vereinigung* **22** (1913), 307–315.

[DreK97] Dreben, Burton and Kanamori, Akihiro: "Hilbert and Set Theory," *Synthese* **110** (1997), 77–125.

[Dru 86] Drüll, Dagmar: *Heidelberger Gelehrtenlexikon*, Vol. 2, Springer-Verlag: Berlin and Heidelberg 1986.

[DuB71] du Bois-Reymond, Emil: "Über die Grenzen des Naturerkennens," in Emil du Bois-Reymond, *Vorträge über Philosophie und Gesellschaft*, edited by Siegfried Wollgast, Meiner: Hamburg 1974, 54–77.

[DuB79a] du Bois-Reymond, Paul: "Erläuterungen zu den Anfangsgründen der Variationsrechnung," *Mathematische Annalen* **15** (1879), 283–314.

[DuB79b] du Bois-Reymond, Paul: "Fortsetzung der Erläuterungen zu den Anfangsgründen der Variationsrechnung," *Mathematische Annalen* **15** (1879), 564–576.

[Ebb03] Ebbinghaus, Heinz-Dieter: "Zermelo: Definiteness and the Universe of Definable Sets," *History and Philosophy of Science* **24** (2003), 197–219.

[Ebb04] Ebbinghaus, Heinz-Dieter: "Zermelo in the Mirror of the Baer Correspondence 1930–1931," *Historia Mathematica* **31** (2004), 76–86.

[Ebb05] Ebbinghaus, Heinz-Dieter: "Zermelo and the Heidelberg Congress 1904," manuscript, Freiburg 2005, to appear in *Historia Mathematica* **33**, 2006.

[Ebb06a] Ebbinghaus, Heinz-Dieter: "Zermelo: Boundary Numbers and Domains of Sets Continued," *History and Philosophy of Logic* **27** (2006), 285–306.

[Ebb06b] Ebbinghaus, Heinz-Dieter: "Löwenheim-Skolem Theorems," in [Jac06], 457–484.

[EE07] Ehrenfest, Paul and Ehrenfest, Tatyana: "Über zwei bekannte Einwände gegen das Boltzmannsche H-Theorem," *Physikalische Zeitschrift* **8** (1907), 311–314.

[EE11] Ehrenfest, Paul and Ehrenfest, Tatyana: "Begriffliche Grundlagen der statistischen Auffassung der Mechanik," in *Encyklopädie der mathematischen Wissenschaften mit Einschluss ihrer Anwendungen*, Vol. 4: *Mechanik*, Part 4. D. 32, Teubner: Leipzig, volume published 1904–1935, article published 1911.

[EE90] Ehrenfest, Paul and Ehrenfest, Tatyana: *The Conceptual Foundations of the Statistical Approach in Mechanics*, Dover: New York 1990.

[Elo78] Elo, Arpad E.: *The Rating of Chess Players Past and Present*, Arco Publishing: New York 1978.

[Enr22] Enriques, Federigo: *Per la storia della logica. I principii e l'ordine della scienza nel concetto dei pensatori matematici*, Zanichelli: Bologna 1922.

[Enr27] Enriques, Federigo: *Zur Geschichte der Logik. Grundlagen und Aufbau der Wissenschaft im Urteil der mathematischen Denker*, German translation by Ludwig Bieberbach, Teubner: Leipzig and Berlin 1927 (*Wissenschaft und Hypothese*; XXVI).

[Erd92] Erdmann, Benno: *Logik*, Vol. 1: *Logische Elementarlehre*, Niemeyer: Halle 1892.

[Euw29] Euwe, Max: "Mengentheoretische Betrachtungen über das Schachspiel," *Koninklijke Nederlandse Akademie van Wetenschappen, Proceedings of the Section of Sciences* **32** (1929), 633–642.

[Ewa96] Ewald, William (ed.): *From Kant to Hilbert: A Source Book in the Foundations of Mathematics*, Vol. II, Clarendon Press: Oxford 1996.

[FarZ27] Farber, Marvin and Zermelo, Ernst: *The Foundations of Logic: Studies concerning the Structure and Function of Logic*, Typescript, 1927, Marvin Farber Papers on Philosophy and Phenomenology 1920–1980; UBA, Box 28; pages 12, 13, 17, and 18 in Box 29.

[Fec75] Fechner, Gustav Theodor: *Kleine Schriften*, Breitkopf und Härtel: Leipzig 1875.

[Fef95] Feferman, Solomon: Introductory Note to [Goe33], in [Goe86], Vol. III, 36–44.

[Fef05] Feferman, Solomon: "Predicativity," in Shapiro, Stewart (ed.): *The Oxford Handbook of Philosophy of Mathematics and Logic*, Oxford University Press: Oxford 2005, 590-624.

[Fel79] Felgner, Ulrich (ed.): *Mengenlehre*, Wissenschaftliche Buchgesellschaft: Darmstadt 1979.

[Fel02a] Felgner, Ulrich: "Der Begriff der Funktion," in [Haus2002], 621–633.

[Fel02b] Felgner, Ulrich: "Ein Brief Gödels zum Fundierungsaxiom," in [Bul02], 205–212.

[Fel02c] Felgner, Ulrich: "Zur Geschichte des Mengenbegriffs," in [Bul02], 169–185.

[Fer56] Ferber, Christian v.: *Die Entwicklung des Lehrkörpers der deutschen Universitäten und Hochschulen 1864–1954*, Vandenhoeck & Ruprecht: Göttingen 1956 (*Untersuchungen zur Lage der deutschen Hochschullehrer*; III).

[Ferr99] Ferreirós, José: *Labyrinth of Thought. A History of Set Theory and its Role in Modern Mathematics*, Birkhäuser: Basel, Boston and Berlin 1999.

[Ferr01] Ferreirós, José: "The Road to Modern Logic – an Interpretation," *The Bulletin of Symbolic Logic* **7** (2001), 441–484.

[FloK06] Floyd, Juliet and Kanamori, Akihiro: "How Gödel Transformed Set Theory," *Notices of the American Mathematical Society* **53** (2006), 417–425.

[Fra19] Fraenkel, Abraham A.: *Einleitung in die Mengenlehre*, Springer-Verlag: Berlin 1919. 2nd enlarged edition 1923; 3rd enlarged edition 1928.

[Fra21] Fraenkel, Abraham A.: "Über die Zermelosche Begründung der Mengenlehre," *Jahresbericht der Deutschen Mathematiker-Vereinigung* **30** (1921), 2nd section, 97–98.

[Fra22a] Fraenkel, Abraham A.: "Der Begriff 'definit' und die Unabhängigkeit des Auswahlaxioms," *Die Preussische Akademie der Wissenschaften. Sitzungsberichte. Physikalisch-Mathematische Klasse* **1922**, 253–257. English translation in [vanH67], 284–289.

[Fra22b] Fraenkel, Abraham A.: "Zu den Grundlagen der Cantor-Zermeloschen Mengenlehre," *Mathematische Annalen* **86** (1922), 230–237.

[Fra23] Fraenkel, Abraham A.: "Die Axiome der Mengenlehre," *Scripta atque Bibliothecae Hierosolymitanarum* **1** (1923), 1–8.

[Fra24] Fraenkel, Abraham A.: "Ueber die gegenwärtige Grundlagenkrise der Mathematik," *Sitzungsberichte der Gesellschaft zur Beförderung der gesamten Naturwissenschaften zu Marburg* **1924**, 83–98.

[Fra25] Fraenkel, Abraham A.: "Untersuchungen über die Grundlagen der Mengenlehre," *Mathematische Zeitschrift* **22** (1925), 250–273.

[Fra27] Fraenkel, Abraham A.: *Zehn Vorlesungen über die Grundlegung der Mengenlehre*, Teubner: Leipzig and Berlin 1927.

[Fra30] Fraenkel, Abraham A.: "Georg Cantor," *Jahrebericht der Deutschen Mathematiker-Vereinigung* **39** (1930), 189–266.

[Fra32a] Fraenkel, Abraham A.: "Über die Axiome der Teilmengenbildung," in Saxer, Walter (ed.): *Verhandlungen des Internationalen Mathematiker-Kongresses Zürich 1932*, Vol. 2: *Sektionsvorträge*, Orell Füssli Verlag: Zurich and Berlin 1932, 341–342.

[Fra32b] Fraenkel, Abraham A.: "Das Leben Georg Cantors," in [Can32], 452–483.

[Fra66] Fraenkel, Abraham A.: "Logik und Mathematik," *Studium generale* **19** (1966), 127–135.

[Fra67] Fraenkel, Abraham A.: *Lebenskreise. Aus den Erinnerungen eines jüdischen Mathematikers*, Deutsche Verlags-Anstalt: Stuttgart 1967.

[FraB58] Fraenkel, Abraham A. and Bar-Hillel, Yehoshua: *Foundations of Set Theory*, North-Holland Publishing Company: Amsterdam 1958.

[Fran33] Frank, Philipp: "Die schnellste Flugverbindung zwischen zwei Punkten," *Zeitschrift für angewandte Mathematik und Mechanik* **13** (1933), 88–91.

[Fre1884] Frege, Gottlob: *Die Grundlagen der Arithmetik. Eine logisch mathematische Untersuchung über den Begriff der Zahl*, Koebner: Breslau 1884. English translation *The Foundations of Arithmetic*, Blackwell: Oxford 1974.

[Fre1893] Frege, Gottlob: *Grundgesetze der Arithmetik, begriffsschriftlich abgeleitet*, Vol. 1, Hermann Pohle: Jena 1893.

[Fre03] Frege, Gottlob: *Grundgesetze der Arithmetik, begriffsschriftlich abgeleitet*, Vol. 2, Hermann Pohle: Jena 1903.

[Fre76] Frege, Gottlob: *Wissenschaftlicher Briefwechsel*, edited by G. Gabriel et al., Felix Meiner: Hamburg 1976.

[Fre80] Frege, Gottlob: *Philosophical and Mathematical Correspondence*, edited by G. Gabriel et al., Basil Blackwell: Oxford 1980.

[Fre83] Frege, Gottlob: *Nachgelassene Schriften*, edited by Hans Hermes, Friedrich Kambartel, and Friedrich Kaulbach, 2nd revised edition, Meiner: Hamburg 1983. 1st edition 1969.

[Fre86] Frege, Gottlob: *Die Grundlagen der Arithmetik. Eine logisch mathematische Untersuchung über den Begriff der Zahl. Centenarausgabe*, edited by Christian Thiel, Meiner: Hamburg 1986.

[FreiS94] Frei, Günther and Stammbach, Urs: *Die Mathematiker an den Zürcher Hochschulen*, Birkhäuser Verlag: Basel, Boston and Berlin 1994.

[Frew81] Frewer, Magdalene: "Felix Bernstein," *Jahresbericht der Deutschen Mathematiker-Vereinigung* **83** (1981), 84–95.

[Fri30] Friedlaender, Salomon: *Der Philosoph Ernst Marcus als Nachfolger Kants. Leben und Lehre (3. IX. 1856–30. X. 1928). Ein Mahnruf*, Baedecker: Essen 1930.

[Gar92] Garciadiego Dantan, Alejandro R.: *Bertrand Russell and the Origins of the Set-theoretic "Paradoxes"*, Birkhäuser: Basel, Boston and Berlin 1992.

[Geo04] Georgiadou, Maria: *Constantin Carathéodory. Mathematics and Politics in Turbulent Times*, Springer-Verlag: Berlin et al. 2004.

[Ger55] Gericke, Helmuth: *Zur Geschichte der Mathematik an der Universität Freiburg i. Br.*, Eberhard Albert: Freiburg 1955 (*Beiträge zur Freiburger Wissenschafts- und Universitätsgeschichte*; 7).

[Gib02] Gibbs, Josiah Willard: *Elementary Principles in Statistical Mechanics, Developed with Especial Reference to the Rational Foundation of Thermodynamics*, Scribner: New York 1902.

[Gib05] Gibbs, Josiah Willard: *Elementare Grundlagen der statistischen Mechanik entwickelt besonders im Hinblick auf eine rationale Begründung der Thermodynamik*, German edition of [Gib02] by Ernst Zermelo, Johann Ambrosius Barth: Leipzig 1905.

[Gin57] Gini, Corrado: "Felix Bernstein 1878–1956," *Revue de l'Institut International de Statistique* **25** (1956), 185–186.

[Gla94] Glazebrook, Richard Tedley: *Light, an Elementary Textbook Theoretical and Practical*, Cambridge University Press: Cambridge 1894.

[Gla97] Glazebrook, Richard Tedley: *Das Licht. Grundriß der Optik für Studierende und Schüler*, German edition of [Gla94] by Ernst Zermelo, S. Calvary & Co.: Berlin 1897.

[Gli95] Glickman, Mark E.: "Chess Rating Systems," *American Chess Journal* **3** (1995), 59–102.

[GliJ99] Glickman, Mark E. and Jones, Albyn C.: "Rating the Chess Rating System," *Chance* **12** (1999), 21–28.

[Goe29] Gödel, Kurt: *Über die Vollständigkeit des Logikkalküls*, Ph. D. Thesis, University of Vienna. Reprinted and in English translation in [Goe86], 60–101.

[Goe30] Gödel, Kurt: "Die Vollständigkeit der Axiome des logischen Funktionenkalküls," *Monatshefte für Mathematik und Physik* **37** (1930), 349–360. Reprinted and in English translation in [Goe86], 102–123. English translation in *van Heijenoort 1967*, 582–591.

[Goe31a] Gödel, Kurt: "Über formal unentscheidbare Sätze der *Principia mathematica* und verwandter Systeme I," *Monatshefte für Mathematik und Physik* **38** (1931), 173–198. Reprinted and in English translation in [Goe86], 144–195; English translation in [vanH67], 596–616.

316   References

[Goe31b] Gödel, Kurt: Letter to E. Zermelo, 12 October 1931, UAF, C 129/36. Reprinted in [Gra79], 298–302, and with English translation in [Goe86], Vol. V, 422–429.

[Goe33] Gödel, Kurt: "The Present Situation in the Foundation of Mathematics," invited lecture delivered to the meeting of the American Mathematical Society, December 29–30, 1933 in Cambridge, Massachusetts. Printed in [Goe86], Vol. III, 45–53.

[Goe38] Gödel, Kurt: "The Consistency of the Axiom of Choice and of the Generalized Continuum-Hypothesis," *Proceedings of the National Academy of Sciences of the United States of America* **24** (1938), 556–557. Reprinted in [Goe86], Vol. II, 26–27.

[Goe39] Gödel, Kurt: *Vortrag Göttingen* (1939), lecture notes in shorthand, Gödel *Nachlass*. Transcription with English translation in [Goe86], Vol. III, 127–155.

[Goe40] Gödel, Kurt: *Lecture [on the] Consistency [of the] Continuum Hypothesis* (1940), lecture notes in shorthand, Gödel *Nachlass*. Transcription in [Goe86], Vol. III, 175–185.

[Goe47] Gödel, Kurt: "What is Cantor's Continuum Problem?," *American Mathematical Monthly* **54** (1947), 515–525. Reprinted in [Goe86], Vol. II, 176–187.

[Goe86] Gödel, Kurt: *Collected Works*, Vol. I, edited by Solomon Feferman et al., Oxford University Press: New York and Clarendon Press: Oxford; Vol. II: 1990, Vol. III: 1995, Vol. IV: 2003, Vol. V: 2003.

[Gol80] Goldstine, Herman H.: *A History of the Calculus of Variations From the 17th Through the 19th Century,* Springer-Verlag: New York, Heidelberg, and Berlin 1980.

[Gon41] Gonseth, Ferdinand (ed.): *Les entretiens de Zurich sur les fondements et la méthode des sciences mathématiques, 6–9 Décembre 1938. Exposés et discussions,* Leemann: Zürich 1941.

[Gra70] Grattan-Guinness, Ivor: "An Unpublished Paper by Georg Cantor: Principien einer Theorie der Ordnungstypen. Erste Mitteilung," *Acta Mathematica* **124** (1970), 65–107.

[Gra72] Grattan-Guinness, Ivor: "The Correspondence Between Georg Cantor and Philip Jourdain," *Jahresbericht der Deutschen Mathematiker-Vereinigung* **73** (1972–73), 111–130.

[Gra74] Grattan-Guinness, Ivor: "The Rediscovery of the Cantor-Dedekind-Correspondence," *Jahresbericht der Deutschen Mathematiker-Vereinigung* **76** (1974), 104–139.

[Gra77] Grattan-Guinness, Ivor: *Dear Russell – Dear Jourdain,* Duckworth: London 1977.

[Gra79] Grattan-Guinness, Ivor: "In Memoriam Kurt Gödel: His 1931 Correspondence with Zermelo on his Incompletability Theorem," *Historia Mathematica* **6** (1979), 294–304.

[Gra80] Grattan-Guinness, Ivor: *From the Calculus to Set Theory 1630–1910. An Introductory History,* Duckworth: London 1980.

[Gra00a] Grattan-Guinness, Ivor: *The Search for Mathematical Roots, 1870–1940. Logic, Set Theories and the Foundations of Mathematics from Cantor Through Russell to Gödel,* Princeton University Press: Princeton and Oxford.

[Gra00b] Grattan-Guinness, Ivor.: "A Sidewayslook at Hilbert's Twenty-three Problems of 1900," *Notices of the AMS* **47** (2000), 752–757.

[Gre04] Greene, Brian: *The Fabric of the Cosmos. Space, Time, and the Texture of Reality*, Alfred A. Knopf: New York 2004.

[Grel10] Grelling, Kurt: *Die Axiome der Arithmetik mit besonderer Berücksichtigung der Beziehungen zur Mengenlehre*, Ph. D. Thesis, Göttingen 1910.

[GrelN08] Grelling, Kurt and Nelson, Leonard: "Bemerkungen zu den Paradoxieen von Russell und Burali-Forti," *Abhandlungen der Fries'schen Schule* n.s. **2** (1908), issue 3, 301–334. Reprinted in [Nel59], 55–87, and in [Nel74], 95–127.

[Gro77a] Gröbli, Walter: *Specielle Probleme über die Bewegung geradliniger paralleler Wirbelfäden*, Zürich und Furrer: Zurich 1877

[Gro77b] Gröbli, Walter: "Specielle Probleme über die Bewegung geradliniger paralleler Wirbelfäden," *Vierteljahrsschrift der Nataurforschenden Gesellschaft in Zürich* **22** (1877), 37–81.

[Gro85] Gromeka, Ippolit Stepanovich: "On Vortex Motions of Liquid on a Sphere," in *Proceedings of the Physical-Mathematical Section of the Society on Natural Sciences, Kazan State University, Session 45, 13 April 1885*. Also in Gromeka, Ippolit Stepanovich: *Collected Papers*, Akademii Nauk: Moscow 1952, 185–205.

[Gru1981] Gruenberg, Karl W.: "Reinhold Baer," *Bulletin of the London Mathematical Society* **13** (1981), 339–361. Extended version *Illinois Journal of Mathematics* **47** (2003), 1–30.

[Gue02] Guerraggio, Angelo and Nastasi, Pietro (eds.): *Cantor e Richard Dedekind: lettere 1872–1899*, Springer-Verlag Italia: Milan 2002.

[Had95] Hadamard, Jacques: "La théorie des ensembles," *Revue Général des Sciences Pures et Appliquées* **16** (1895), 241–242.

[HahZ04] Hahn, Hans and Zermelo, Ernst: "Weiterentwicklung der Variationsrechnung in den letzten Jahren," *Encyklopädie der mathematischen Wissenschaften mit Einschluss ihrer Anwendungen*, Vol. 2: *Analysis*, Part 1. 1, 626–641, Teuber: Leipzig, volume published 1899–1916, article published 1904. French translation in [Lec13].

[Hal84] Hallett, Michael: *Cantorian Set Theory and Limitation of Size*, Clarendon Press: Oxford 1984 (*Oxford Logic Guides*; 10).

[Hal96] Hallett, Michael: "Ernst Friedrich Ferdinand Zermelo (1871–1953)," in [Ewa96], Vol. 2, 1208–1218.

[Ham05] Hamel, Georg: "Eine Basis aller Zahlen und die unstetigen Lösungen der Funktionalgleichung $f(x + y) = f(x) + f(y)$," *Mathematische Annalen* **60** (1905), 459–462.

[Ham23] Hamel, Georg: "Zum Gedächtnis an Hermann Amandus Schwarz," *Jahresbericht der Deutschen Mathematiker-Vereinigung* **32** (1923), 9–13.

[Han95] Hannequin, Arthur: *Essai critique sur l'hypothése des atomes dans la science contemporaine*, Masson: Paris 1895.

[Har15] Hartogs, Friedrich: "Über das Problem der Wohlordnung," *Mathematische Annalen* **76** (1915), 436–443

[Has27] Hasse, Helmut: "Über eindeutige Zerlegungen in Primelemente oder in Primhauptideale in Integritätsbereichen," *Journal für die reine und angewandte Mathematik* **159** (1927), 3–12.

[Hau33] Haupt, Otto: "Emil Hilb", *Jahresbericht der Deutschen Mathematiker-Vereinigung* **42** (1933), 183–198.

[Haus04] Hausdorff, Felix: "Der Potenzbegriff der Mengenlehre," *Jahresbericht der Deutschen Mathematiker-Vereinigung* **13** (1904), 569–571.

[Haus08]   Hausdorff, Felix: "Grundzüge einer Theorie der geordneten Mengen,"
           Mathematische Annalen **65** (1908), 435–505.

[Haus09]   Hausdorff, Felix: "Die Graduierung nach dem Endverlauf," *Abhandlungen
           der Königlich-Sächsischen Gesellschaft der Wissenschaften, mathematisch-
           physikalische Klasse* **31** (1909), 297–334.

[Haus14]   Hausdorff, Felix: *Grundzüge der Mengenlehre*, Verlag von Veit: Leipzig
           1914. Reprinted in [Haus2002], 91–576. 2nd and 3rd edition [Haus27].

[Haus27]   Hausdorff, Felix: *Mengenlehre*, Verlag de Gruyter: Berlin 1927, 2nd edition
           of [Haus14], ^{3}1935.

[Haus2002] Hausdorff, Felix: *Gesammelte Werke*, Vol. II: *Grundzüge der Mengen-
           lehre*," Springer-Verlag: Berlin et al. 2002.

[Hef52]   Heffter, Lothar: *Beglückte Rückschau auf neun Jahrzehnte. Ein Professoren-
           leben*, Schulz: Freiburg i. Br. 1952.

[Hei72]   Heine, Eduard: "Elemente der Functionenlehre," *Journal für die reine und
           angewandte Mathematik* **74** (1872), 172–188.

[Hein86]   Heinzmann, Gerhard: *Poincaré, Russell, Zermelo et Peano. Textes de la
           discussion (1906–1912) sur les fondements des mathématiques: des anti-
           nomies à la prédicativité*, Blanchard: Paris 1986 (*Bibliothèque scientifique
           Albert Blanchard*).

[Hel58]   Helmholtz, Hermann von: "Über Integrale der hydrodynamischen Gleichun-
           gen, welche den Wirbelbewegungen entsprechen," *Journal für die reine und
           angewandte Mathematik* **55** (1858), 25–55. Reprinted in *Wissenschaftliche
           Abhandlungen von Hermann Helmholtz*, Vol. I, Johann Ambrosius Barth:
           Leipzig 1882, 101–134.

[Hen50]   Henkin, Leon: "Completeness in the Theory of Types," *The Journal of
           Symbolic Logic* **15** (1950), 81–91.

[HerKK83] Hermes, Hans, Kambartel, Friedrich, and Kaulbach, Friedrich: "Ge-
           schichte des Frege-Nachlasses und Grundsätze seiner Edition," in [Fre83],
           XXXIV–XLI.

[Hert22]   Hertz, Paul: "Über Axiomensysteme für beliebige Satzsysteme, I. Teil:
           Sätze ersten Grades (Über die Axiomensysteme von der kleinsten Satz-
           zahl und den Begriff des idealen Elementes)," *Mathematische Annalen* **87**
           (1922), 246–269.

[Hert23]   Hertz, Paul: "Über Axiomensysteme für beliebige Satzsysteme, Teil II:
           Sätze höheren Grades," *Mathematische Annalen* **89** (1923), 76–102.

[Hert29]   Hertz, Paul: "Über Axiomensysteme für beliebige Satzsysteme," *Mathe-
           matische Annalen* **101** (1929), 457–514.

[Hes06]   Hessenberg, Gerhard: *Grundbegriffe der Mengenlehre*, Vandenhoeck &
           Ruprecht: Göttingen 1906.

[Hes08]   Hessenberg, Gerhard: "Willkürliche Schöpfungen des Verstandes?," *Jahres-
           bericht der Deutschen Mathematiker-Vereinigung* **17** (1908), 145–162.

[Hes10]   Hessenberg, Gerhard: "Kettentheorie und Wohlordnung," *Journal für die
           reine und angewandte Mathematik* **135** (1910), 81–133, 318.

[Hey56]   Heyting, Arend: *Intuitionism. An Introduction*, North-Holland Publishing
           Company; Amsterdam 1956.

[Hil1897] Hilbert, David: "Die Theorie der algebraischen Zahlkörper," *Jahresbericht
           der Deutschen Mathematiker-Vereinigung* **4** (1897), 175–546.

[Hil1899] Hilbert, David: *Grundlagen der Geometrie*, Teubner: Leipzig 1899, 14th
           edition [Hil1999], English translation [Hil02a].

[Hil00a] Hilbert, David: "Über den Zahlbegriff," *Jahresbericht der Deutschen Ma-thematiker-Vereinigung* **8** (1900), 180–184. English translation [Hil96a].

[Hil00b] Hilbert, David: "Mathematische Probleme. Vortrag, gehalten auf dem internationalen Mathematiker-Kongreß zu Paris 1900," *Nachrichten von der königlichen Gesellschaft der Wissenschaften zu Göttingen, Mathematisch-physikalische Klasse aus dem Jahre 1900* (1900), 253–297. Reprinted in [Hil35], 290–329. French translation [Hil00c]; English translation [Hil02b]; partial English translation [Hil96b].

[Hil00c] Hilbert, David: "Les principes fondamentaux de la géométrie. *Festschrift* publiée à l'occasion des fêtes pour l'inauguration du monument de Gauss-Weber à Göttingen. – Publiée par les soins du Comité des fêtes. Leipzig, Teubner, 1899" [translated by L. Laugel], *Annales scientifiques de l'École Normale Supérieure* (3) **17** (1900), 103–209.

[Hil02a] Hilbert, David: *The Foundations of Geometry*, Open Court: Chicago 1902. English translation of [Hil1899].

[Hil02b] Hilbert, David: "Mathematical Problems. Lecture Delivered Before the International Congress of Mathematicians at Paris 1900," *Bulletin of the American Mathematical Society* **8** (1902), 437–479.

[Hil02c] Hilbert, David: "Über den Satz von der Gleichheit der Basiswinkel im gleichschenkligen Dreieck," *Proceedings of the London Mathematical Society* **35** (1902/03), 50–68.

[Hil03] Hilbert, David: *Grundlagen der Geometrie*, 2nd edition, enlarged with five attachments, Teubner: Leipzig 1903.

[Hil05a] Hilbert, David: "Über die Grundlagen der Logik und der Arithmetik," in [Kra05], 174–185. English translation in *The Monist* **15** (1905), 338–352.

[Hil05b] Hilbert, David: *Logische Principien des mathematischen Denkens*, lecture course, sommer semester 1905, elaboration by Ernst Hellinger, Library of Mathematisches Seminar, University of Göttingen.

[Hil05c] Hilbert, David: *Logische Principien des mathematischen Denkens*, lecture course, summer semester 1905, elaboration by Max Born, Staats- und Universitätsbibliothek Göttingen, Cod. Ms. D. Hilbert 558a.

[Hil18] Hilbert, David: "Axiomatisches Denken," *Mathematische Annalen* **78** (1918), 405–415. Reprinted in [Hil35], 146–156; English translation [Hil96e].

[Hil20] Hilbert, David: *Probleme der Mathematischen Logik,* lecture notes by Moses Schönfinkel and Paul Bernays, summer semester 1920, Mathematical Institute Göttingen.

[Hil23] Hilbert, David: "Die logischen Grundlagen der Mathematik," *Mathematische Annalen* **88** (1923), 151–165. Reprinted in [Hil35], 178–191; English translation [Hil96c].

[Hil26] Hilbert, David: "Über das Unendliche," *Mathematische Annalen* **95** (1926), 161–190. English translation in [vanH67], 367–392.

[Hil28] Hilbert, David: "Die Grundlagen der Mathematik," *Abhandlungen aus dem mathematischen Seminar der Hamburgischen Universität* **6** (1928), 65–85. English translation in [vanH67], 464–479.

[Hil29] Hilbert, David: "Probleme der Grundlegung der Mathematik." *Atti del Congresso internazionale dei matematici; Bologna 3–10 settembre 1928*, Vol. 1, Zanichelli: Bologna 1929, 135–141. Reprinted in *Mathematische Annalen* **102** (1930), 1–9.

[Hil31] Hilbert, David: "Die Grundlegung der elementaren Zahlenlehre," *Mathematische Annalen* **104** (1931), 485–494. Reprinted in part in [Hil35], 192–195; English translation [Hil96d].

[Hil35] Hilbert, David: *Gesammelte Abhandlungen*, Vol. 3: *Analysis, Grundlagen der Mathematik, Physik, Verschiedenes, Lebensgeschichte*, Springer-Verlag: Berlin and Heidelberg 1935; 2nd edition Springer-Verlag: Berlin, Heidelberg, and New York 1970.

[Hil71] Hilbert, David: "Über meine Tätigkeit in Göttingen," in *Hilbert. Gedenkband*, edited by Kurt Reidemeister, Springer-Verlag: Berlin, Heidelberg, and New York 1971, 78–82.

[Hil96a] Hilbert, David: "On the Concept of Number," in [Ewa96], 1089–1095. English translation of [Hil00a].

[Hil96b] Hilbert, David: "*From* Mathematical Problems," in [Ewa96], 1096–1105. English translation of a part of [Hil00b].

[Hil96c] Hilbert, David: "The Logical Foundations of Mathematics," in [Ewa96], 1134–1148. English translation of [Hil23].

[Hil96d] Hilbert, David: "The Grounding of Elementary Number Theory," in [Ewa96], 1148–1157. English translation of [Hil31].

[Hil96e] Hilbert, David: "Axiomatic Thought," in [Ewa96], 1105–1115. English translation of [Hil18].

[Hil1999] Hilbert, David: *Grundlagen der Geometrie. Mit Supplementen von Paul Bernays*, edited by Michael Toepell, Teubner: Stuttgart and Leipzig 1999. 14th edition of [Hil1899].

[HilA28] Hilbert, David and Ackermann, Wilhelm: *Grundzüge der theoretischen Logik*, Springer-Verlag: Berlin 1928.

[HilB34] Hilbert, David and Bernays, Paul: *Grundlagen der Mathematik*, Vol. I (1934), Vol. II (1939), Springer-Verlag: Berlin and Heidelberg, ²1968/1970.

[HolZ85] Hollinger, Henry B. and Zenzen, Michael J.: *The Nature of Irreversibility. A Study of Its Dynamics and Physical Origins*, Reidel: Dordrecht et al. 1985 (*The University of Western Ontario Series in Philosophy of Science*; 28).

[Hus82] Husserl, Edmund: *Beiträge zur Variationsrechnung*, Ph. D. Thesis, University of Vienna 1882.

[Hus91a] Husserl, Edmund: *Philosophie der Arithmetik. Logische und psychologische Untersuchungen*, Vol. 1, C. E. M. Pfeffer (Robert Stricker): Halle a. S. 1891; critical edition *Husserliana*, Vol. 18, edited by E. Holenstein, Nijhoff: The Hague 1970.

[Hus91b] Husserl, Edmund: Review of [Schr90], *Göttingische gelehrte Anzeigen* (1891), 243–278; critical edition in [Hus1979], 3–43.

[Hus1956] Husserl, Edmund: *Husserliana*, Vol. 7 (1956) and Vol. 8 (1959), edited by R. Boehm, Nijhoff: The Hague.

[Hus1979] Husserl, Edmund: *Aufsätze und Rezensionen (1890–1910), mit ergänzenden Texten*, Husserliana, Vol. XXII, edited by B. Rang, Nijhoff: The Hague, Boston and London 1979.

[Jac06] Jacquette, Dale (ed.): *Handbook of the Philosophy of Science*, Vol. 5: *Philosophy of Logic*, Elsevier: Amsterdam et al. 2006

[Jec73] Jech, Thomas J.: *The Axiom of Choice*, North-Holland: Amsterdam and New York 1973.

[Jou04] Jourdain, Philip: "On the Transfinite Cardinal Numbers of Well-ordered Aggregates," *The London, Edinburgh, and Dublin Philosophical Magazine and Journal of Science*, ser. 6, **7** (1904), 61–75.

[Jou05] Jourdain, Philip: "On a Proof that Every Aggregate Can be Well-Ordered," *Mathematische Annalen* **60** (1905), 465–470.

[JunM86] Jungnickel, Christa and MacCormmach, Russell: *The Intellectual Mastery of Nature. Theoretical Physics from Ohm to Einstein*, Vol. 2: *The Now Mighty Theoretical Physics 1870–1925*, University of Chicago Press: Chicago and London 1986.

[Kac59] Kac, Mark: *Probability and Related Topics in Physical Sciences. Lectures in Applied Mathematics, Proceedings of the Summer Seminar, Boulder, Colorado, 1957*, American Mathematical Society: Providence, Rhode Island 1959.

[Kai88] Kaiser, Walter: "Die Struktur wissenschaftlicher Kontroversen," in [PosB88], 113–147.

[Kal28] Kalmár, László: "Zur Theorie der abstrakten Spiele," *Acta Literarum ac Scientiarum Regiae Universitatis Hungaricae Francisco-Josephinae, Sectio Scientiarum Mathematicarum* **4** (1928/29), 65–85.

[Kan94] Kanamori, Akihiro: *The Higher Infinite. Large Cardinals in Set Theory from Their Beginnings*, Springer-Verlag: Berlin et al. 1994.

[Kan96] Kanamori, Akihiro: "The Mathematical Development of Set Theory from Cantor to Cohen," *The Bulletin of Symbolic Logic* **2** (1996), 1–71.

[Kan97] Kanamori, Akihiro: "The Mathematical Import of Zermelo's Well-Ordering Theorem," *The Bulletin of Symbolic Logic* **3** (1997), 281–311.

[Kan03] Kanamori, Akihiro: "The Empty Set, the Singleton, and the Ordered Pair," *Bulletin of Symbolic Logic* **9** (2003), 273–298.

[Kan04] Kanamori, Akihiro: "Zermelo and Set Theory," *The Bulletin of Symbolic Logic* **10** (2004), 487–553.

[Kan06] Kanamori, Akihiro: "Gödel and Set Theory," manuscript, Boston 2006.

[Kar64] Karp, Carol R.: *Languages with Expressions of Infinite Length,* North-Holland Publishing Company: Amsterdam 1964.

[Kei77] Keisler, H. Jerome: "Fundamentals of Model Theory," in [Bar77], 47–103.

[Ker88] Kern, Bärbel: *Madame Doktorin Schlözer. Ein Frauenleben in den Widersprüchen der Aufklärung,* Beck: Munich 1988; 2nd edition 1990.

[Key04] Keyser, Cassius Jackson: "The Axiom of Infinity: A new Presupposition of Thought," *Hibbert Journal* **2** (1904), 532–552.

[Key05] Keyser, Cassius Jackson: "The Axiom of Infinity," *Hibbert Journal* **3** (1905), 380–383.

[Kli72] Kline, Morris: *Mathematical Thought from Ancient to Modern Times,* Oxford University Press: New York 1972.

[Kir76] Kirchhoff, Gustav Robert: *Vorlesungen über mathematische Physik: Mechanik,* Teubner: Leipzig 1876. 4th edition 1897 edited by Wilhelm Wien.

[KneK62] Kneale, William and Kneale, Martha: *The Development of Logic,* Clarendon Press: Oxford 1962.

[Knes1898] Kneser, Adolf: "Zur Variationsrechnung," *Mathematische Annalen* **50**, 1898, 27–50.

[Knes00] Kneser, Adolf: *Lehrbuch der Variationsrechnung,* Vieweg: Braunschweig 1900.

[Knes04] Kneser, Adolf: "Variationsrechnung," in *Encyklopädie der mathematischen Wissenschaften mit Einschluss ihrer Anwendungen*, Vol. 2: *Analysis*, Part 1. 1, 571–625, Teubner: Leipzig, volume published 1899–1916, article published 1904.

[Kob92a] Kobb, Gustaf: "Sur les maxima et les minima des intégrales doubles," *Acta Mathematica* **16** (1892), 65–140.

[Kob92b] Kobb, Gustaf: "Sur les maxima et les minima des intégrales doubles (second mémoire)," *Acta Mathematica* **17** (1892), 321–344.

[Koe02] Köhler, Eckehart et. al. (eds.): *Kurt Gödel. Wahrheit & Beweisbarkeit*, Volume 1: *Dokumente und historische Analysen*, öbv & hpt: Vienna 2002.

[Koed27] König, Dénes: "Über eine Schlussweise aus dem Endlichen ins Unendliche," *Acta Literarum ac Scientiarum Regiae Universitatis Hungaricae Francisco-Josephinae, Sectio Scientiarum Mathematicarum* **3** (1927), 121–130.

[Koej05a] König, Julius: "Zum Kontinuum-Problem," in [Kra05], 144–147.

[Koej05b] König, Julius: "Zum Kontinuum-Problem," *Mathematische Annalen* **60** (1905), 177–180.

[Koej05c] König, Julius: "Über die Grundlagen der Mengenlehre und das Kontinuumproblem," *Mathematische Annalen* **61** (1905), 156–160. English translation in [vanH67], 145–149.

[Koej07] König, Julius: "Über die Grundlagen der Mengenlehre und das Kontinuumproblem (Zweite Mitteilung)," *Mathematische Annalen* **63** (1907), 217–221.

[Kor11] Korselt, Alwin: "Über einen Beweis des Äquivalenzsatzes," *Mathematische Annalen* **70** (1911), 294–296.

[Kow50] Kowalewski, Gerhard: *Bestand und Wandel. Meine Lebenserinnerungen. Zugleich ein Beitrag zur neueren Geschichte der Mathematik*, Oldenbourg: München 1950.

[Kra05] Krazer, Adolf (ed.): *Verhandlungen des Dritten Internationalen Mathematiker-Kongresses in Heidelberg vom 8. bis 13. August 1904*, Teubner: Leipzig 1905.

[Kre80] Kreisel, Georg: "Kurt Gödel," *Biographical Memoirs of Fellows of the Royal Society* **26** (1980), 149–224; **27** (1981), 697; **28** (1982), 719.

[Kre87] Kreisel, Georg: "Gödel's Excursion into Intuitionistic Logic," in [WeinS87], 65–186.

[Krei83] Kreiser, Lothar: "Einleitung [zu: 'Nachschrift einer Vorlesung und Protokolle mathematischer Vorträge Freges']," in [Fre83], 327–346.

[Krei01] Kreiser, Lothar: *Gottlob Frege. Leben – Werk – Zeit*, Felix Meiner Verlag: Hamburg 2001.

[Kro87] Kronecker, Leopold: "Über den Zahlbegriff," *Journal für die reine und angewandte Mathematik* **101** (1887), 337–356. Reprinted in [Kro95], 249–274.

[Kro95] Kronecker, Leopold: *Leopold Kronecker's Werke*, Vol. 3.1, edited by Kurt Hensel, Teubner: Leipzig 1895.

[Kue76] Küssner, Martha: *Dorothea Schlözer. Ein Göttinger Leben*, Göttingen: Musterschmidt 1976.

[Kun98] Kunze, Konrad: *dtv-Atlas Namenkunde. Vor- und Familiennamen im deutschen Sprachgebiet*, Deutscher Taschenbuch-Verlag: Munich 1998.

[Lam79] Lamb, Horace: *A Treatise on the Mathematical Theory of the Motion of Fluids*, At the University Press: Cambridge 1879. German edition by J. C. B. Mohr under the title *Einleitung in die Hydrodynamic*, Akademische Verlagsbuchhandlung: Freiburg i. B. and Tübingen 1884.

[Lam95] Lamb, Horace: *Hydrodynamics,* enlarged and revised edition of [Lam79], At the University Press: Cambridge 1895.

[Lan14] Landau, Edmund: "Über Preisverteilung bei Spielturnieren," *Zeitschrift für Mathematik und Physik* **63** (1914), 192–202.

[Lan17] Landau, Edmund: "R. Dedekind. Gedächtnisrede, gehalten in der öffentlichen Sitzung der Königlichen Gesellschaft der Wissenschaften zu Göttingen am 12. Mai 1917," *Nachrichten der Königlichen Gesellschaft der Wissenschaften zu Göttingen. Geschäftliche Mitteilungen* **1917**, 51–70. Reprinted in [Lan86], 449–465.

[Lan86] Landau, Edmund: *Collected Works*, Vol. 6, edited by Paul T. Bateman et al., Thales Verlag: Essen 1986.

[Las05] Lasker, Emanuel: "Zur Theorie der Moduln und Ideale," *Mathematische Annalen* **60**, 20–116.

[Lec13] Lecat, Maurice: "Calcul des variations. Exposé, d'après les articles allemands de A. Kneser (Breslau), E. Zermelo (Zurich) et H. Hahn (Czernowitz)," in *Encyclopédie des Sciences Mathématiques pures et appliquées. Édition française*, Vol. 2, Part 6, Gauthier-Villar: Paris and Teubner: Leipzig, 1–288. German original [HahZ04].

[Levi31] Levi-Civita, Tullio: "Über Zermelos Luftfahrtproblem," *Zeitschrift für angewandte Mathematik und Mechanik* **11** (1931), 314–322.

[Levy60] Levy, Azriel: "Axiom Schemata of Strong Infinity in Axiomatic Set Theory," *Pacific Journal of Mathematics* **10** (1960), 223–238.

[Lia77a] Liard, Louis: "Un nouveau système de logique formelle M. Stanley Jevons," *Revue philosophique de la France et de l'Étranger* **3** (1877), 277–293.

[Lia77b] Liard, Louis: "La logique algébrique de Boole," *Revue philosophique de la France et de l'Étranger* **4** (1877), 285–317.

[Lia78] Liard, Louis: *Les logiciens anglais contemporains*, Germer Baillière: Paris 1878.

[Lia80] Liard, Louis: *Die neuere englische Logik*, Denicke: Leipzig 1880.

[Lin71] von Lindemann, Ferdinand: *Lebenserinnerungen*, manuscript without date, in typewritten form edited by I. Verholzer: Munich. Faculty of Mathematics and Physics, University of Freiburg 1971.

[Lind01] Lindley, David: *Boltzmann's Atoms*, The Free Press: New York *et al.* 2001.

[Link04] Link, Godehard (ed.): *One Hundred Years of Russell's Paradox. Mathematics, Logic, Philosophy*, Walter de Gruyter: Berlin and New York 2004.

[Loe15] Löwenheim, Leopold: "Über Möglichkeiten im Relativkalkül," *Mathematische Annalen* **76** (1915), 447–470. English translation in [vanH67], 232–251

[Lop66] Lopez-Escobar, Edgar G. K.: "On Defining Well-Orderings," *Fundamenta Mathematicae* **59** (1966), 13–21, 299–300.

[Los76] Loschmidt, Josef: "Über den Zustand des Wärmegleichgewichtes eines Systems von Körpern mit Rücksicht auf die Schwerkraft I," *Sitzungsberichte der Kaiserlichen Akademie der Wissenschaften in Wien, Mathematisch-Naturwissenschaftliche Classe*, 2. Abt. **73** (1876), 128–142.

[Mane70] Manegold, Karl-Heinz: *Universität, Technische Hochschule und Industrie. Ein Beitrag zur Emanzipation der Technik im 19. Jahrhundert unter besonderer Berücksichtigung der Bestrebungen Felix Kleins*, Duncker & Humblot: Berlin 1970 (*Schriften zur Wirtschafts- und Sozialgeschichte*; 16).

[Mani37] Manià, Basilio: "Sopra un problema di navigazione di Zermelo," *Mathematische Annalen* **113** (1937), 584–599.

[Mar93] Marchisotto, Elena Anne: "Mario Pieri and his Contributions to Geometry and the Foundations of Mathematics," *Historia Mathematica* **20** (1993), 285–303.

[Marc07] Marcus, Ernst: *Das Gesetz der Vernunft und die ethischen Strömungen der Gegenwart*, Menckhoff: Herford 1907. 2nd revised edition under the title *Der kategorische Imperativ. Eine gemeinverständliche Einführung in Kants Sittenlehre*, Reinhardt: München 1921.

[MariM84] Mariani, Mauro, Moriconi, Enrico: *Coerenza e completezza delle teorie elementari. La metateoria dei sistemi formali nella scuola hilbertiana*, ETS: Pisa 1984 (*Biblioteca di "Teoria"*; 3).

[Mars94] Marsch, Ulrich: *Notgemeinschaft der Deutschen Wissenschaft. Gründung und frühe Geschichte 1920–1925*, Lang: Frankfurt a. M. 1994.

[Martb89] Martin, Bernd: *Martin Heidegger und das 'Dritte Reich'. Ein Kompendium*, Wissenschaftliche Buchgesellschaft: Darmstadt 1989.

[Martg55] Martin, Gottfried: *Kant's Metaphysics and Theory of Science*, Manchester University Press: Manchester 1955. English translation of *Immanuel Kant. Ontologie und Wissenschaftstheorie*, Kölner Universitätsverlag: Cologne 1951.

[May99] Mayer, Adolph: "Über die Aufstellung der Differentialgleichungen für reibungslose Punktsysteme, die Bedingungsgleichungen unterworfen sind," *Berichte über die Verhandlungen der Königlich-Sächsischen Gesellschaft der Wissenschaften zu Leipzig* **51** (1899), 224–244.

[Mcc04] McCarty, David Charles: "David Hilbert und Paul du Bois-Rymond: Limits and Ideals," in [Link04], 517–532.

[Meh90] Mehrtens, Herbert: *Moderne – Sprache – Mathematik. Eine Geschichte des Streits um die Grundlagen der Disziplin und des Subjekts formaler Systeme*, Suhrkamp Verlag: Frankfurt a. M. 1990.

[Mes83] Meschkowski, Herbert: *Georg Cantor. Leben, Werk und Wirkung*, Bibliographisches Institut: Mannheim, Vienna and Zurich.

[Mir17] Mirimanoff, Dimitry: "Les antinomies de Russell et de Burali-Forti et le problème fondamental de la théorie des ensembles," *L'Enseignement Mathématique* **19** (1917), 37–52.

[Mis31] von Mises, Richard: "Zum Navigationsproblem der Luftfahrt," *Zeitschrift für angewandte Mathememathik und Mechanik* **11** (1931), 373–381.

[Moo75] Moore, Gregory H.: "A Prospective Biography of Ernst Zermelo (1871–1953)," *Historia Mathematica* **2** (1975), 62–63.

[Moo76] Moore, Gregory H.: "E. Zermelo, A. E. Harward, and the Axiomatization of Set Theory," *Historia Mathematica* **3** (1976), 206–209.

[Moo78] Moore, Gregory H.: "The Origins of Zermelo's Axiomatization of Set Theory," *Journal of Philosophical Logic* **7** (1978), 307–329.

[Moo80] Moore, Gregory H.: "Beyond First-order Logic: The Historical Interplay between Mathematical Logic and Axiomatic Set Theory," *History and Philosophy of Logic* **1** (1980), 95–137.

[Moo82] Moore, Gregory H.: *Zermelo's Axiom of Choice. Its Origins, Development and Influence*, Springer-Verlag: New York, Heidelberg, and Berlin 1982 (*Studies in the History of Mathematics and Physical Sciences*; 8).

[Moo87] Moore, Gregory H.: "A House Divided Against Itself: The Emergence of First-Order Logic as the Base for Mathematics," in Esther R. Phillips (ed.): *Studies in the History of Mathematics*, The Mathematical Association of America: Washington, D. C. 1987, 98–136.

[Moo89] Moore, Gregory H.: "Towards a History of Cantor's Continuum Problem," in [RowM89], 79–121.

[Moo90]   Moore, Gregory H.: "Bernays, Paul Isaac," *Dictionary of Scientific Biography*, edited by Frederic L. Holmes, Vol. 17, Supplement II, Charles Scribner's Sons: New York 1990, 75–78.

[Moo97]   Moore, Gregory H.: "The Prehistory of Infinitary Logic: 1885–1955," in Maria Luisa Dalla Chiara et al. (eds.): *Structures and Norms in Science.* Kluwer Academic Publishers: Dordrecht et al. 1997, 105–123.

[Moo02a] Moore, Gregory H.: "Hilbert on the Infinite: The Role of Set Theory in the Evolution of Hilbert's Thought," *Historia Mathematica* **29** (2002), 40–64.

[Moo02b] Moore, Gregory H.: "Die Kontroverse zwischen Gödel und Zermelo," in [Bul02], 55–64.

[MooG81] Moore, Gregory H., Garciadiego, Alejandro R.: "Burali-Forti's Paradox: A Reappraisal of Its Origins," *Historia Mathematica* **8** (1981), 319–350.

[Mos72]   Moss, J. Michael B.: "Some B. Russell's Sprouts (1903–1908)," in Hodges, Wilfrid (ed.): *Conference in Mathematical Logic London '70*, Springer-Verlag: Berlin, Heidelberg, and New York 1972 (*Lecture Notes in Mathematics*; 255), 211–250.

[Nad85]   Nadel, Mark: "$\mathcal{L}_{\omega_\infty,\omega}$ and Admissible Fragments," in [BarF85], 271–316.

[Nel07]   Nelson, Leonard: Review of Ernst Mach, *Erkenntnis und Irrtum*, Leipzig [2]1905, *Göttingische gelehrte Anzeigen* **169** (1907), 636–657.

[Nel08]   Nelson, Leonard: "Ist metaphysikfreie Naturwissenschaft möglich?," *Abhandlungen der Fries'schen Schule* n.s. **2**, No. 3 (1908), 241–299.

[Nel28]   Nelson, Leonard: "Kritische Philosophie und mathematische Axiomatik," *Unterrichtsblätter für Mathemaik und Naturwissenschaften* **34** (1928), 108–142. Reprinted in [Nel59], 89–124 and in [Nel74], 187–220.

[Nel59]   Nelson, Leonard: *Beiträge zur Philosophie der Logik und Mathematik, mit einführenden und ergänzenden Bemerkungen von Wilhelm Ackermann, Paul Bernays, David Hilbert* †, Verlag öffentliches Leben: Frankfurt a.M. 1959.

[Nel73]   Nelson, Leonard: *Geschichte und Kritik der Erkenntnistheorie*, Meiner: Hamburg 1973 (*Gesammelte Schriften in neun Bänden*, edited by Paul Bernays et al., Vol. 2).

[Nel74]   Nelson, Leonard: *Die kritische Methode in ihrer Bedeutung für die Wissenschaft*, Meiner: Hamburg 1974 (*Gesammelte Schriften in neun Bänden*, edited by Paul Bernays et al., Vol. 3).

[Neu23]   von Neumann, John: "Zur Einführung der transfiniten Zahlen," *Acta litterarum ac scientiarum Regiae Universitatis Hungaricae Francisco-Josephinae, Sectio scientiarum mathematicarum* **1** (1923), 199–208. Reprinted in [Neu61], 24–33; English translation in [vanH67], 346–354.

[Neu25]   von Neumann, John: "Eine Axiomatisierung der Mengenlehre," *Journal für die reine und angewandte Mathematik* **154** (1925), 219–240. Reprinted in [Neu61], 35–56. English translation in [vanH67], 393–413.

[Neu28a] von Neumann, John: "Über die Definition durch transfinite Induktion und verwandte Fragen der allgemeinen Mengenlehre," *Mathematische Annalen* **99** (1928), 373–391. Reprinted in *von Neumann 1961*, 339–422

[Neu28b] von Neumann, John: "Zur Theorie der Gesellschaftsspiele," *Mathematische Annalen* **100** (1928), 295–320.

[Neu29]   von Neumann, John: "Über eine Widerspruchsfreiheitsfrage der axiomatischen Mengenlehre," *Journal für die reine und angewandte Mathematik* **160** (1929), 227–241. Reprinted in [Neu61], 494–508.

[Neu61]   von Neumann, John: *Collected Works*, Vol. 1, edited by Abraham Haskel Taub, Pergamon Press: Oxford et al. 1961.

[NeuM53] von Neumann, John and Morgenstern, Oskar: *Theory of Games and Economic Behavior*, Princeton University Press: Princeton 1953.

[New01] Newton, Paul K.: *The N-Vortex Problem. Analytical Techniques*, Springer-Verlag: New York, Heidelberg, and Berlin 2001.

[Noe16] Noether, Emmy: "Die allgemeinsten Bereiche aus ganzen transzendenten Zahlen," *Mathematische Annalen* **77** (1916), 103–128.

[NoeC37] Noether, Emmy and Cavaillès, Jean (eds.): *Briefwechsel Cantor–Dedekind*, Hermann: Paris 1937.

[Ott88] Ott, Hugo: *Martin Heidegger. Unterwegs zu seiner Biographie*, Campus Verlag: Frankfurt and New York 1988.

[Pan32] Pannwitz, Erika: Review of [Can32], *Jahrbuch über die Fortschritte der Mathematik* **58** (1932), 43–44 [JFM 58.0043.01].

[Par87] Parsons, Charles Dacre: "Developing Arithmetic in Set Theory Without Infinity: Some Historical Remarks," *History and Philosophy of Logic* **8** (1987), 201–213.

[Pea90] Peano, Guiseppe: "Démonstration de l'intégrabilité des équations différentielles ordinaires," *Mathematische Annalen* **37** (1890), 182–228.

[Pea91] Peano, Giuseppe: "Formole di logica matematica," *Rivista di matematica* **1** (1891), 24–31, 182–184.

[Pea95] Peano, Giuseppe: *Formulaire de mathématiques*, Vol. 1–5, Bocca: Torino 1895–1908. Reprint [Pea60].

[Pea97] Peano, Giuseppe: *Formulaire de mathématiques*, Vol. 2.1: *Logique mathématique*, Bocca: Turin 1897.

[Pea06a] Peano, Guiseppe: "Super theorema de Cantor-Bernstein," *Rendiconti del Circulo Matematico di Palermo* **21** (1906), 360–366, and *Rivista de matematica* **8** (1908), 136-143.

[Pea06b] Peano, Guiseppe: "Additione," *Rivista de matematica* **8** (1906), 143–157. English translation in [Pea73], 206–218.

[Pea60] Peano, Guiseppe: *Formulario Mathematico*, Cremonese: Rom 1960. Reprint of [Pea95].

[Pea73] Peano, Guiseppe: *Selected Works of Guiseppe Peano*, translated and edited by Hubert C. Kennedy, University of Toronto Press: Toronto and Buffalo 1973.

[Pec88] Peckhaus, Volker: "Historiographie wissenschaftlicher Disziplinen als Kombination von Problem- und Sozialgeschichtsschreibung: Formale Logik im Deutschland des ausgehenden 19. Jahrhunderts," in [PosB88], 177–215.

[Pec89] Peckhaus, Volker: "Die Institutionalisierung der Mathematischen Logik in Deutschland," *XVIIIth International Congress of History of Science. General Theme: Science and Political Order. Wissenschaft und Staat. 1st–9th August 1989 Hamburg-München. Abstracts*, edited by Fritz Krafft and Christoph J. Scriba, ICHS: Hamburg and München 1989, E1.4.

[Pec90a] Peckhaus, Volker: " 'Ich habe mich wohl gehütet, alle Patronen auf einmal zu verschießen'. Ernst Zermelo in Göttingen," *History and Philosophy of Logic* **11** (1990), 19–58.

[Pec90b] Peckhaus, Volker: *Hilbertprogramm und Kritische Philosophie. Das Göttinger Modell interdisziplinärer Zusammenarbeit zwischen Mathematik und Philosophie*, Vandenhoeck & Ruprecht: Göttingen 1990 (*Studien zur Wissenschafts-, Sozial- und Bildungsgeschichte der Mathematik*; 7).

[Pec92] Peckhaus, Volker: "Hilbert, Zermelo und die Institutionalisierung der mathematischen Logik in Deutschland," *Berichte zur Wissenschaftsgeschichte* **15** (1992), 27–38

[Pec95] Peckhaus, Volker: "The Genesis of Grelling's Paradox," in *Logik und Mathematik. Frege-Kolloquium Jena 1993*, edited by Ingolf Max/Werner Stelzner, Walter de Gruyter: Berlin and New York 1995 (*Perspectives in Analytical Philosophy*, 5), 269–280.

[Pec96] Peckhaus, Volker: "The Axiomatic Method and Ernst Schröder's Algebraic Approach to Logic," *Philosophia Scientiae* (Nancy) **1**/3 (1996), 1–15.

[Pec97] Peckhaus, Volker: *Logik, Mathesis universalis und allgemeine Wissenschaft. Leibniz und die Wiederentdeckung der formalen Logik im 19. Jahrhundert*, Akademie Verlag: Berlin 1997 (*Logica nova*).

[Pec99] Peckhaus, Volker: "19th Century Logic Between Philosophy and Mathematics," *Bulletin of Symbolic Logic* **5** (1999), 433–450.

[Pec02] Peckhaus, Volker: "Regressive Analysis," in Uwe Meixner and Albert Neven (eds.): *Philosophiegeschichte und logische Analyse. Logical Analysis and History of Philosophy*, Vol. 5, Mentis: Paderborn 2002.

[Pec04a] Peckhaus, Volker: "Schröder's Logic," in Dov M. Gabbay and John Woods (eds.), *Handbook of the History of Logic, Vol. 3: The Rise of Modern Logic: From Leibniz to Frege*, Elsevier North Holland: Amsterdam et al. 2004, 557–609.

[Pec04b] Peckhaus, Volker: "Paradoxes in Göttingen," in [Link04], 501–515.

[Pec04c] Peckhaus, Volker: "Garantiert Widerspruchsfreiheit Existenz?", in Bente Christiansen and Uwe Scheffler (eds.), *Was folgt? Themen zu Wessel*, Logos Verlag: Berlin 2004 (*Logische Philosophie*), 111–127.

[Pec04d] Peckhaus, Volker: "Die Zeitschrift für die Grundlagen der gesamten Mathematik (1908)," unpublished manuscript, Paderborn 2004.

[Pec05a] Peckhaus, Volker (ed.): *Oskar Becker und die Philosophie der Mathematik*, Wilhelm Fink Verlag: Munich 2005.

[Pec05b] Peckhaus, Volker: "Impliziert Widerspruchsfreiheit Existenz? Oskar Beckers Kritik am formalistischen Existenzbegriff," in [Pec05a], 77–99.

[Pec05c] Peckhaus, Volker: "Becker und Zermelo," in [Pec05a], 279–297.

[PecK02] Peckhaus, Volker and Kahle, Reinhard: "Hilbert's Paradox," *Historia Mathematica* **29** (2002), 157–175.

[Ped79] Pedlosky, Joseph: *Geophysical Fluid Dynamics*, Springer-Verlag: New York, Heidelberg, and Berlin 1979.

[Pei85] Peirce, Charles Sanders: "On the Algebra of Logic: A Contribution to the Philosophy of Notation," *American Journal of Mathematics* **7** (1885), 180–202.

[Pet03] Peters, Martin: *Altes Reich und Europa. Der Historiker, Statistiker und Publizist August Ludwig von Schlözer (1735–1839)*, Münster et al.: LIT 2003 (also Ph.D. thesis Marburg 2000).

[Pin69] Pinl, Maximilian: "Kollegen in einer dunklen Zeit," first part, *Jahresbericht der Deutschen Mathematiker-Vereinigung* **71** (1969), 167–228.

[Pla00] Planck, Max: "Zur Theorie des Gesetzes der Energieverteilung im Normalspektrum," *Verhandlungen der Deutschen Physikalischen Gesellschaft* **2** (1900), 237–245.

[Pla08] Planck, Max: "Die Einheit des physikalischen Weltbildes," lecture delivered to the Faculty of Science of the Student's Corporation of the University of Leiden on 9 December 1908, in [Pla49a], 28–51.

[Pla14] Planck, Max: "Dynamische und statistische Gesetzmäßigkeit," lecture delivered at the ceremony to the memory of the founder of the Friedrich-Wilhelms-Universität Berlin on 3 August 1914, in [Pla49a], 81–94.

[Pla33] Planck, Max: *Wege zur physikalischen Erkenntnis*, S. Hirzel Verlag: Stuttgart 1933.

[Pla49a] Planck, Max: *Vorträge und Erinnerungen*, S. Hirzel Verlag: Stuttgart. 5th extended edition of [Pla33].

[Pla49b] Planck, Max: "Persönliche Erinnerungen aus alten Zeiten," in [Pla49a], 1–14.

[Poi90] Poincaré, Henri: "Sur le problème de trois corps et le équations de dynamique", *Acta Mathematica* **13** (1890), 1–270. Reprinted in *Oeuvres de Henri Poincaré*, Vol. 7, Paris: Gauthier-Villars 1952, 265–479.

[Poi93a] Poincaré, Henri: "Le mécanisme et l'expérience," *Revue de métamathematique et de morale* **1** (1893), 534–537. English translation in [Bru66], 203–207.

[Poi93b] Poincaré, Henri: *Théorie des Tourbillons*, George Carré: Paris 1893.

[Poi06a] Poincaré, Henri: "Les mathématiques et la logique," *Revue de métamathematique et de morale* **14** (1906), 294–317. English translation in [Ewa96], 1052–1071.

[Poi06b] Poincaré, Henri: Letter to E. Zermelo of 16 June 1906, UAF, C 129/90. Reprinted in [Hein86], 105.

[Poi10a] Poincaré, Henri: *Sechs Vorträge über ausgewählte Gegenstände aus der reinen Mathematik und mathematischen Physik*, Teuber: Leipzig and Berlin 1910 (*Mathematische Vorlesungen an der Universität Göttingen*; 4).

[Poi10b] Poincaré, Henri: "Über transfinite Zahlen," in [Poi10a], 43–48. Reprinted in *Oeuvres de Henri Poincaré*, Vol. 11, Paris: Gauthier-Villars 1956, 120–124. English translation in [Ewa96], 1071–1074.

[PosB88] Poser, Hans and Burrichter, Clemens (eds.): *Die geschichtliche Perspektive in den Disziplinen der Wissenschaftsforschung. Kolloquium an der TU Berlin, Oktober 1988*, Universitätsbibliothek der TU Berlin: Berlin 1988 (*TUB-Dokumentation Kongresse und Tagungen*; 39).

[Pur02] Purkert, Walter: "Grundzüge der Mengenlehre – Historische Einführung," in [Haus2002], 1–89.

[PurI87] Purkert, Walter and Ilgauds, Hans Joachim: *Georg Cantor 1845–1918*, Birkhäuser: Basel, Boston, Stuttgart 1987 (*Vita Mathematica*; 1).

[Qui66] Quine, William Van Orman: *The Ways of Paradox*, Random House: New York 1966.

[RanT81] Rang, Bernhard and Thomas, Wolfgang: "Zermelo's Discovery of the 'Russell Paradox'," *Historia Mathematica* **8** (1981), 15–22.

[Rei70] Reid, Constance: *Hilbert. With an Appreciation of Hilbert's Mathematical Work by Hermann Weyl*, Springer-Verlag: New York, Heidelberg, and Berlin 1970.

[Rem95] Remmert, Volker: "Zur Mathematikgeschichte in Freiburg. Alfred Loewy (1873–1935): Jähes Ende späten Glanzes," *Freiburger Universitätsblätter* **129** (1995), 81–102.

[Rem97] Remmert, Volker: "Alfred Loewy (1873–1935) in Freiburg," *Mitteilungen der Deutschen Mathematiker-Vereinigung* (1997), No. 3, 17–22.

[Rem99a] Remmert, Volker: "Vom Umgang mit der Macht: Das Freiburger Mathematische Institut im 'Dritten Reich'," *Zeitschrift für Sozialgeschichte des 20. und 21. Jahrhunderts* **14**, No 2 (1999), 56–85.

[Rem99b] Remmert, Volker: "Mathematicians' Community: Gustav Doetsch and Wilhelm Süss," *Revue d'Histoire des Mathématiques* **5** (1999), 7–59.

[Rem99c] Remmert, Volker: "Griff aus dem Elfenbeinturm. Mathematik, Macht und Nationalsozialismus: das Beispiel Freiburg," *DMV-Mitteilungen* (1999), No. 3, 13–24.

[Rem00] Remmert, Volker: "Mathematical Publishing in the Third Reich: Springer-Verlag and the Deutsche Mathematiker-Vereinigung," *Mathematical Intelligencer* **22**, No. 3 (2000), 22–30.

[Rem01] Remmert, Volker: "Das Mathematische Institut der Universität Freiburg (1900–1950)," in Michael Toepell (ed.): *Mathematik im Wandel. Anregungen zum fachübergreifenden Mathematikunterricht*, Vol. 2, Verlag Franzbecker: Hildesheim and Berlin 2001, 374–392.

[Rem02] Remmert, Volker: "Ungleiche Partner in der Mathematik im 'Dritten Reich': Heinrich Behnke und Wilhelm Süss," *Mathematische Semesterberichte* **49** (2002), 11–27.

[Ric05] Richard, Jules Antoine: "Les principes des mathématiques et le problème des ensembles," *Revue Général des Sciences Pures et Appliquées* **16** (1905), 541–543. English translation in [vanH67], 142–144.

[Ric07] Richard, Jules Antoine: "Sur une paradoxe de la theorie des ensembles et sur l'axiom de Zermelo," *L'Enseignement Mathématiques* **9** (1907), 94–98.

[RieZ09] Riesenfeld, Ernst and Zermelo, Ernst: "Die Einstellung der Grenzkonzentration an der Trennungsfläche zweier Lösungsmittel," *Physikalische Zeitschrift* **10** (1909), 958–961.

[Row89] Rowe, David E.: "Felix Klein, David Hilbert, and the Göttingen Mathematical Tradition," in Kathryn Olesko (ed.), *Science in Germany. The Intersection of Institutional and Intellectual Issues*, The University of Chicago Press: Chicago, IL, 1989 (*Osiris*, Vol. 5).

[RowM89] Rowe, David E. and McCleary, John (eds.): *The History of Modern Mathematics, Vol. I: Ideas and Their Reception, Proceedings of the Symposium on the History of Modern Mathematics, Vassar College, Poughkeepsie, New York, June 20–24, 1989*, Academic Press: Boston, San Diego, New York et al. 1989.

[Rup35] Ruprecht, Wilhelm: *Väter und Söhne. Zwei Jahrhunderte Buchhändler in einer deutschen Universitätsstadt*, Vandenhoeck & Ruprecht: Göttingen 1935.

[Rus02] Russell, Bertrand: Letter to G. Frege of 16 June 1902, Collection Darmstädter, Berlin, Signatur H 1897. English translation in [vanH67], 124–125.

[Rus03] Russell, Bertrand: *The Principles of Mathematics*, Vol. 1, The University Press: Cambridge 1903.

[Rus06] Russell, Bertrand: "On Some Difficulties in the Theory of Transfinite Numbers and Order Types," *Proceedings of the London Mathematical Society*, 2nd ser., **4** (1906), 29–53. Reprinted in [Hein86], 54–78.

[Rus08] Russell, Bertrand: "Mathematical Logic as Based on the Theory of Types," *American Journal of Mathematics* **30** (1908), 222–262. Reprinted in [vanH67], 150–182, and in [Hein86], 200–223.

[Scha87] Schappacher, Norbert: "Das Mathematische Institut der Universität Göttingen 1929–1950," in [BecDW87], 345–373.

[Schn62] Schneeburger, Guido: *Nachlese zu Heidegger – Dokumente zu seinem Leben*, Guido Schneeburger: Bern 1962.

[Scho00] Schoenflies, Arthur: "Die Entwickelung der Lehre von den Punktmannig-faltigkeiten. Bericht, erstattet der Deutschen Mathematiker-Vereinigung," *Jahresbericht der Deutschen Mathematiker-Vereinigung* **8**, no. 2, 1–250.

[Scho05] Schoenflies, Arthur: "Über wohlgeordnete Mengen," *Mathematische An-nalen* **60** (1905), 181–186.

[Scho06] Schoenflies, Arthur: "Über die logischen Paradoxien der Mengenlehre," *Jahresbericht der Deutschen Mathematiker-Vereinigung* **15**, 177–184.

[Scho11] Schoenflies, Arthur: "Über die Stellung der Definition in der Axiomatik," *Jahresbericht der Deutschen Mathematiker-Vereinigung* **20** (1911), 222–255.

[Scho22] Schoenflies, Arthur: "Zur Erinnerung an Georg Cantor," *Jahresbericht der Deutschen Mathematiker-Vereinigung* **31** (1922), 97–106.

[Schol31] Scholz, Arnold: "Zermelos neue Theorie der Mengenbereiche," *Jahres-bericht der Deutschen Mathematiker-Vereinigung* **40** (1931), 2nd section, 42–43.

[Schol39] Scholz, Arnold: *Einführung in die Zahlentheorie*, Walter de Gruyter: Berlin 1939 (*Sammlung Göschen*; 1131).

[Schol55] Scholz, Arnold: *Einführung in die Zahlentheorie*, revised and edited by Bruno Schoeneberg, Walter de Gruyter: Berlin 1955 (*Sammlung Göschen*; 1131).

[Scholh31] Scholz, Heinrich: *Geschichte der Logik*, Junker und Dünnhaupt: Berlin 1931.

[Scholh61] Scholz, Heinrich: *Mathesis Universalis. Abhandlungen zur Philosophie als strenger Wissenschaft*, edited by Hans Hermes, Friedrich Kambartel, and Joachim Ritter, Benno Schwabe: Basel and Stuttgart 1961.

[Schr77] Schröder, Ernst: *Der Operationskreis des Logikkalkuls*, Teubner: Leipzig 1977.

[Schr90] Schröder, Ernst: *Vorlesungen über die Algebra der Logik (Exakte Logik)*, Vol. 1, Teubner: Leipzig 1890.

[Schr91] Schröder, Ernst: *Vorlesungen über die Algebra der Logik (Exakte Logik)*, Vol. 2.1, Teubner: Leipzig 1891.

[Schr95a] Schröder, Ernst: *Vorlesungen über die Algebra der Logik (Exakte Logik)*, Vol. 3.1: *Algebra und Logik der Relative*, Teubner: Leipzig 1895.

[Schr95b] Schröder, Ernst: "Note über die Algebra der binären Relative," *Mathe-matische Annalen* **46** (1895), 144–158.

[Schr98a] Schröder, Ernst: "Über Pasigraphie, ihren gegenwärtigen Stand und die pasigraphische Bewegung in Italien," in *Verhandlungen des Ersten Interna-tionalen Mathematiker-Kongresses in Zürich vom 9. bis 11. August 1897*, edited by Ferdinand Rudio, Teubner: Leipzig 1898, 147–162.

[Schr98b] Schröder, Ernst: "On Pasigraphy. Its Present State and the Pasigraphic Movement in Italy," *The Monist* **9** (1898–99), 44–62, Corrigenda 320.

[Schr98c] Schröder, Ernst: "Ueber zwei Definitionen der Endlichkeit und G. Can-tor'sche Sätze," *Nova Acta Academiae Caesareae Leopoldino-Caroliae Ger-manicae Naturae Curiosum* **71** (1898), 303–362.

[Schr98d] Schröder, Ernst: "Die selbständige Definition der Mächtigkeiten 0, 1, 2, 3 und die explicite Gleichzahligkeitsbedingung," *Nova acta Academiae Cae-sareae Leopoldino-Caroliae Germanicae Naturae Curiosum* **71** (1898), 365–376.

[Schr05] Schröder, Ernst: *Vorlesungen über die Algebra der Logik (Exakte Logik)*, Vol. 2.2, edited by Karl Eugen Müller, Leipzig 1905.

[SchrH02] Schröder-Heister, Peter: "Resolution and the Origin of Structural Reasoning. Early Proof-Theoretic Ideas of Hertz and Gentzen," *The Bulletin of Symbolic Logic* **8** (2002), 246–265.

[Schu89] Schubring, Gert: "Pure and Applied Mathematics in Divergent Institutional Settings in Germany: The Role and Impact of Felix Klein," in David E. Rowe and John McCleary (eds.), *The History of Modern Mathematics*, Vol. II: *Institutions and Applications*, Academic Press: Boston et al. 1989, 171–207.

[SchwW01] Schwalbe, Ulrich and Walker, Paul: "Zermelo and the Early History of Game Theory," *Games and Economic Behavior* **34** (2001), 123–137.

[Sco74] Scott, Dana: "Axiomatizing Set Theory," in Scott, Dana and Jech, Thomas J. (eds.): *Axiomatic Set Theory*, American Mathematical Society: Providence, R. I. 1974 (*Proceedings of Symposia in Pure Mathematics*, Vol. 2), 207–214.

[Seg03] Segal, Sanford L.: *Mathematicians under the Nazis*, Princeton University Press: Princeton and Oxford 2003.

[Ser99] Serret, Joseph Alfred: *Lehrbuch der Differential- und Integralrechnung*, German edited by Axel Harnack, 2nd rev. edition, Vol. 2: *Integralrechnung*, edited by Georg Bohlmann with assistance of Heinrich Liebmann and Ernst Zermelo, Teubner: Leipzig 1899.

[Ser04] Serret, Joseph Alfred: *Lehrbuch der Differential- und Integralrechnung*, German edition by Axel Harnack, 2nd revised edition, Vol. 3: *Differentialgleichungen und Variationsrechnung*, edited by Georg Bohlmann and Ernst Zermelo, Teubner: Leipzig 1904.

[Sha06] Shapiro, Stewart: "We Hold These Truths to Be Self-Evident: But What Do We Mean by That?," manuscript, The Ohio State University 2006.

[She52] Shepherdson, John C.: "Inner Models for Set Theory – Part II," *The Journal of Symbolic Logic* **17** (1952), 225–237.

[Sho77] Shoenfield, Joseph R.: "Axioms of Set Theory," in [Bar77], 321–344.

[Sie18] Sierpiński, Wacław Franciszek: "L'axiome de M. Zermelo et son rôle dans la théorie des ensembles et l'analyse," *Bulletin de l'Académie des Sciences de Cracovie, Classe des Sciences Mathematiques-Sciences Naturelles*, Serie A, **1918** (1918), 97–152.

[Skl93] Sklar, Lawrence: *Physics and Chance: Philosophical Issues in the Foundations of Statistical Mechanics*, Cambridge University Press: Cambridge 1993.

[Sko20] Skolem, Thoralf: "Logisch-kombinatorische Untersuchungen über die Erfüllbarkeit und Beweisbarkeit mathematischer Sätze nebst einem Theoreme über dichte Mengen," in *Skrifter, Vitenskapsakademiet i Kristiania* **4** (1920), 1–36. Reprinted in [Sko70], 103–136; English translation of the first section in [vanH67], 252-263.

[Sko23] Skolem, Thoralf: "Einige Bemerkungen zur axiomatischen Begründung der Mengenlehre," *Wissenschaftliche Vorträge gehalten auf dem Fünften Kongress der Skandinavischen Mathematiker in Helsingfors vom 4. bis 7. Juli 1922*, Akad. Buchhandlung: Helsingfors 1923, 217–232. Reprinted in [Sko70], 137–152; English translation in [vanH67], 290–301.

[Sko29] Skolem, Thoralf: "Über einige Grundlagenfragen der Mathematik," *Norske Videnskaps-Akademi i Oslo. Matematik-naturvidenskabelig Klasse* **4** (1929), 1–49. Reprinted in [Sko70], 227–273.

[Sko30a] Skolem, Thoralf: "Einige Bemerkungen zu der Abhandlung von E. Zermelo: 'Über die Definitheit in der Axiomatik'," *Fundamenta Mathematicae* **15** (1930), 337–341. Reprinted in [Sko70], 275–279.

[Sko30b] Skolem, Thoralf: "Über die Grundlagendiskussionen in der Mathematik," in *Den syvende Skandinaviske Matematikerkongress i Oslo 19–22 August 1929*, Broggers: Oslo 1930, 3–21. Reprinted in [Sko70], 207–225.

[Sko58] Skolem, Thoralf: "Une relativisation des notions mathématiques fondamentales," in *Le raisonnement en mathématiques fondamentales. Colloques Internationaux des Centre National de la Recherche Scientifique*, Editions du Centre National de la Recherche Scientifique: Paris 1958, 13–18. Reprinted in [Sko70], 633–638.

[Sko70] Skolem, Thoralf: *Selected Works in Logic*, edited by Jens Erik Fenstad, Universitetsforlaget: Oslo, Bergen, Tromsö 1970.

[Sta04] Stammbach, Urs: *Die Zürcher Vorlesung von Issai Schur über Darstellungstheorie*, ETH-Bibliothek: Zurich 2004.

[Ste10] Steinitz, Ernst: "Algebraische Theorie der Körper," *Journal für die reine und angewandte Mathematik* **137** (1910), 167–309.

[StrR83] Strauss, Herbert A. and Röder, Werner (eds.): *International Biographical Dictionary of Central European Emigrés 1933–1945*, Vol. 2: *The Arts, Sciences, and Literature*, Part 1, Saur: München et al. 1983.

[Stu65] Stüper, Josef: *Automatische Automobilgetriebe*, Springer-Verlag: Vienna and New York 1965.

[Tai98] Tait, William W.: "Zermelo's Conception of Set Theory and Reflection Principles," in Matthias Schirn (ed.): *The Philosophy of Mathematics Today*, Clarendon Press: Oxford 1998, 469–483.

[Tak78] Takeuti, Gaisi: *Two Applications of Logic to Mathematics*, Princeton University Press: Princeton 1978.

[Tar24] Tarski, Alfred: "Sur les ensembles finis," *Fundamenta Mathematicae* **6** (1924), 45–95.

[Tau52] Taussky-Todd, Olga: "Arnold Scholz zum Gedächtnis," *Mathematische Nachrichten* **7** (1952), 379–386.

[Tau87] Taussky-Todd, Olga: "Remembrances of Kurt Gödel," in Rudolf Gödel et al., *Gödel remembered. Salzburg 10–12 July 1983*, Bibliopolis: Naples 1987 *History of Logic*; IV), 29–41.

[Tay93] Taylor, R. Gregory: "Zermelo, Reductionism, and the Philosophy of Mathematics," *Notre Dame Journal Formal Logic* **34** (1993), 539–563.

[Tay02] Taylor, R. Gregory: "Zermelo's Cantorian Theory of Systems of Infinitely Long Propositions," *The Bulletin of Symbolic Logic* **8** (2002), 478–515.

[Tay04] Taylor, R. Gregory: "Zermelo's Analysis of Generality," manuscript, Manhattan College 2004.

[Thi87] Thiel, Christian: "Scrutinizing an Alleged Dichotomy in the History of Mathematical Logic," in *8th International Congress of Logic, Methodology, and Philosophy of Science. Abstracts*, Vol. 3, Moskau 1987, 254–255.

[Thie06] Thiele, Rüdiger: *Von der Bernoullischen Brachistochrone zum Kalibratorkonzept. Ein historischer Abriss zur Entstehung der Feldtheorie in der Variationsrechnung*, Brepols: Turnhout 2006.

[Tie29] Tietjens, Oskar G.: *Hydro- und Aeromechanik nach Vorlesungen von L. Prandtl*, Vol. 1: *Gleichgewicht und reibungslose Bewegung*, Springer-Verlag: Berlin 1929.

[Tie31] Tietjens, Oskar G.: *Hydro- und Aeromechanik nach Vorlesungen von L. Prandtl*, Vol. 2: *Bewegung reibender Flüssigkeiten und technische Anwendungen*, Springer-Verlag: Berlin 1931.

[Tie57] Tietjens, Oskar G.: *Fundamentals of Hydro and Aeromechanics: Based on Lectures of L. Prandtl*, Dover: New York 1957.

[Tob81] Tobies, Renate: *Felix Klein*, Teubner: Leipzig 1981 (*Biographien hervorragender Naturwissenschaftler, Techniker und Mediziner*; 50).

[Tob87] Tobies, Renate: "Zur Berufungspolitik Felix Kleins – Grundsätzliche Ansichten," *NTM. Schriftenreihe für Geschichte der Naturwissenschaften, Technik und Medizin* **24**, No. 2 (1987), 43–52.

[Toe86] Toepell, Michael: *Über die Entstehung von Hilberts "Grundlagen der Geometrie"*, Vandenhoeck & Ruprecht: Göttingen 1986 (*Studien zur Wissenschafts-, Sozial- und Bildungsgeschichte der Mathematik*; 2).

[Toe99] Toepell, Michael: "Zur Entstehung und Weiterentwicklung von David Hilberts 'Grundlagen der Geometrie'," in [Hil1999], 283-324.

[TroD88] Troelstra, Anne S. and van Dalen, Dirk: *Constructivism in Mathematics*, Vol. I, North-Holland: Amsterdam et al. 1988.

[Uff04] Uffink, Jos: "Boltzmann's Work in Statistical Physics," in Edward N. Zalta (ed.): *The Stanford Encyclopedia of Philosophy*, http: //plato.stanford.edu/ entries/statphys-Boltzmann/ (2004).

[Uff07] Uffink, Jos: "Compendium of the Foundations of Classical Statistical Physics," in Butterfield, Jeremy and Earman, John (eds.), *Handbook of the Philosophy of Science, Philosophy of Physics, Part B*, Elsevier: Amsterdam et al. 2007, 923–1074.

[vanD05] van Dalen, Dirk: *Mystic, Geometer, and Intuitionist. The Life of L. E. J. Brouwer 1881-1966, Vol. 2: Hope and Disillusion*, Clarendon Press: Oxford 2005.

[vanDE00] van Dalen, Dirk and Ebbinghaus, Heinz-Dieter: "Zermelo and the Skolem Paradox," *The Bulletin of Symbolic Logic* **6** (2000), 145–161.

[vanH67] van Heijenoort, Jean: *From Frege to Gödel. A Source Book in Mathematical Logic, 1879-1931*, Harvard University Press: Cambridge, Mass. 1967.

[vanR76] van Rootselaar, Bob: "Zermelo, Ernst Friedrich Ferdinand," *Dictionary of Scientific Biography*, edited by Charles Coulston Gillispie, Vol. 14, Scribner: New York 1976, 613–616.

[Ver91] Veronese, Giuseppe: *Fondamenti di geometria*, Tip. del Seminario: Padua 1891.

[Ver94] Veronese, Giuseppe: *Grundzüge der Geometrie von mehreren Dimensionen und mehreren Arten gradliniger Einheiten in elementarer Form entwickelt. Mit Genehmigung des Verfassers nach einer neuen Bearbeitung des Originals*, translated by Adolf Schepp, Teubner: Leipzig 1894.

[Vil02] Vilkko, Risto: *A Hundred Years of Logical Investigations. Reform Efforts of Logic in Germany 1781-1879*, Mentis: Paderborn 2002.

[Vol95] Vollrath, Hans-Joachim: "Emil Hilb (1882–1929)", in Peter Baumgart (ed.), *Lebensbilder bedeutender Würzburger Professoren*, Degener: Neustadt/Aisch 1995, 320–338.

[Wan49] Wang, Hao: "On Zermelo's and von Neumann's axioms for set theory," *Proceedings of the National Academy of Sciences (U. S. A.)* **35** (1949), 150–155.

[Wan57] Wang, Hao: "The Axiomatization of Arithmetic," *The Journal of Symbolic Logic* **22** (1957), 145–158.

[Wan70] Wang, Hao: "A Survey of Skolem's Work in Logic," in [Sko70], 17–52.

[Wan81a] Wang, Hao: *Popular Lectures on Mathematical Logic*, Science Press: Beijing 1981.

[Wan81b] Wang, Hao: "Some Facts about Kurt Gödel," *The Journal of Symbolic Logic* **46** (1981), 653–659.

[Wan87] Wang, Hao: *Reflections on Kurt Gödel*, The MIT Press: Cambridge, Mass. 1987.

[WangT97] Wangerin, Albert and Taschenberg, Otto (eds.): *Verhandlungen der Gesellschaft Deutscher Naturforscher und Ärzte. 68. Versammlung zu Frankfurt a. M., 12.–26. September 1896*, Vol. 2.1, Teubner: Leipzig 1897.

[Wap05] Wapnewski, Peter: *Mit dem anderen Auge. Erinnerungen 1922–1959*, Berlin Verlag: Berlin 2005.

[Wei27] Weierstraß, Karl: *Mathematische Werke*, Vol. 7: *Vorlesungen über Variationsrechnung*, edited by Rudolf Rothe, Akademische Verlagsgesellschaft: Leipzig 1927.

[WeinS87] Weingartner, Paul and Schmetterer, Leopold (eds.): *Gödel Remembered. Papers from the 1983 Gödel Symposium Held in Salzburg, July 10–12, 1983*, Bibliopolis: Naples 1987.

[Weis34] Weiss, Paul: Review of [Can32], *The Philosophical Review* **43** (1934), 214–215.

[Wey10] Weyl, Hermann: "Über die Definition der mathematischen Grundbegriffe," *Mathematisch-naturwissenschaftliche Blätter* **7** (1910), 93–95, 109–113.

[Wey18] Weyl, Hermann: *Das Kontinuum*, Verlag von Veit: Leipzig 1918. Reprinted in [Wey66].

[Wey21] Weyl, Hermann: "Über die neue Grundlagenkrise der Mathematik," *Mathematische Zeitschrift* **10** (1921), 39–70.

[Wey44] Weyl, Hermann: "David Hilbert and His Mathematical Work," *Bulletin of the American Mathematical Society* **50** (1944), 612–654. Reprinted in [Rei70], 245–283.

[Wey66] Weyl, Hermann: *Das Kontinuum und andere Monographien*, Chelsea: New York 1966

[Whi02] Whitehead, Alfred North: "On Cardinal Numbers," *American Journal of Mathematics* **24** (1902), 367–394.

[WhiR10] Whitehead, Alfred North and Russell, Bertrand: *Principia Mathematica*, Cambridge University Press: Cambridge, Vol. I (1910), Vol. II (1912), Vol. III (1913), 2nd edition 1925–1927.

[Wit51] Witt, Ernst: "Beweisstudien zum Satz von E. Zorn," *Mathematische Nachrichten* **4** (1951), 434–438.

[Witt21] Wittgenstein, Ludwig: "Logisch-philosophische Abhandlung," *Annalen der Naturphilosophie* **14** (1921), 185–262.

[Witt22] Wittgenstein, Ludwig: *Tractatus logico-philosophicus*, Routledge & Kegan Paul: New York and London 1922.

[Wri32] Wrinch, Dorothy Maud: Review of [Can32], *The Mathematical Gazette* **17** (1932), 332.

[YouY06] Young, Grace Chisholm and Young, William Henry: *The Theory of Sets of Points*, At the University Press: Cambridge 1906.

[Zac06] Zach, Richard: "Hilbert's Program Then and Now," in [Jac06], 393–429.

[Zeh89] Zeh, Heinz-Dieter: *The Physical Basis of the Direction of Time*, Springer-Verlag: Berlin et al. 1989

[Zer94] Zermelo, Ernst: *Untersuchungen zur Variations-Rechnung*, Ph. D. Thesis, Berlin 1894.

[Zer96a] Zermelo, Ernst: "Ueber einen Satz der Dynamik und die mechanische Wärmetheorie," *Annalen der Physik und Chemie* n. s. **57** (1896), 485–494. Reprinted in [Bru70], 264–275; English translation in [Bru66], 208–217.

[Zer96b] Zermelo, Ernst: "Ueber mechanische Erklärungen irreversibler Vorgänge. Eine Antwort auf Hrn. Boltzmann's 'Entgegnung'," *Annalen der Physik und Chemie* n. s. **59** (1896), 793–801. Reprinted in [Bru70], 290–300; English translation in [Bru66], 229–237.

[Zer99a] Zermelo, Ernst: "Über die Bewegung eines Punktsystems bei Bedingungsungleichungen," *Nachrichten der Königlichen Gesellschaft der Wissenschaften zu Göttingen, Mathematisch-physikalische Klasse* **1899**, 306–310.

[Zer99b] Zermelo, Ernst: "Wie bewegt sich ein unausdehnbarer materieller Faden unter dem Einfluß von Kräften mit dem Potentiale $W(x, y, z)$?," manuscript, 3 pp., SUB Göttingen, Cod. Ms. D. Hilbert 722.

[Zer00a] Zermelo, Ernst: "Über die Anwendung der Wahrscheinlichkeitsrechnung auf dynamische Systeme," *Physikalische Zeitschrift* **1** (1899–1900), 317–320.

[Zer00b] Zermelo, Ernst: *Mengenlehre*, Lecture notes, University of Göttingen, winter semester 1900/01; UAF, C 129/150.

[Zer02a] Zermelo, Ernst: "Ueber die Addition transfiniter Cardinalzahlen," *Nachrichten von der Königl. Gesellschaft der Wissenschaften zu Göttingen. Mathematisch-physikalische Klasse aus dem Jahre 1901*, Göttingen 1902, 34–38.

[Zer02b] Zermelo, Ernst: "Hydrodynamische Untersuchungen über die Wirbelbewegungen in einer Kugelfläche," *Zeitschrift für Mathematik und Physik* **47** (1902), 201–237.

[Zer02c] Zermelo, Ernst: "Zur Theorie der kürzesten Linien," *Jahresbericht der Deutschen Mathematiker-Vereinigung* **11** (1902), 184–187.

[Zer03] Zermelo, Ernst: "Über die Herleitung der Differentialgleichung bei Variationsproblemen," *Mathematische Annalen* **58** (1903), 558–564.

[Zer04] Zermelo, Ernst: "Beweis, daß jede Menge wohlgeordnet werden kann," *Mathematische Annalen* **59** (1904), 514–516. English translation [Zer67a].

[Zer06a] Zermelo, Ernst: Review of [Gib02] and [Gib05], *Jahresbericht der Deutschen Mathematiker-Vereinigung* **15** (1906), 232–242.

[Zer06b] Zermelo, Ernst: *Mengenlehre und Zahlbegriff*, lecture notes, University of Göttingen, summer semester 1906; UAF, C 129/151.

[Zer08a] Zermelo, Ernst: "Neuer Beweis für die Möglichkeit einer Wohlordnung," *Mathematische Annalen* **65** (1908), 107–128. Reprinted in [Hein86], 107–128, and in [Fel79], 105–126; English translation [Zer67b].

[Zer08b] Zermelo, Ernst: "Untersuchungen über die Grundlagen der Mengenlehre. I," *Mathematische Annalen* **65** (1908), 261–281. Reprinted in [Hein86], 261–281, and in [Fel79], 28–48; English translation [Zer67c].

[Zer08c] Zermelo, Ernst: *Mathematische Logik. Sommer-Semester 1908*, lecture notes; UAF, C 129/152.

[Zer08d] Zermelo, Ernst: *Mathematische Logik. Vorlesungen gehalten von Prof. Dr. E. Zermelo zu Göttingen im S. S. 1908*, lecture notes by Kurt Grelling; UAF, C 129/224 (Part I) and C 129/215 (Part II).

[Zer08e] Zermelo, Ernst: Letter to Georg Cantor of 24 July 1908, reprinted in Herbert Meschkowski (ed.), *Probleme des Unendlichen. Werk und Leben Georg Cantors*, Friedr. Vieweg & Sohn: Braunschweig 1967, 267–268.

[Zer09a] Zermelo, Ernst: "Sur les ensembles finis et le principe de l'induction complète," *Acta Mathematica* **32** (1909), 185–193. Reprinted in [Hein86], 148–156.

[Zer09b] Zermelo, Ernst: "Ueber die Grundlagen der Arithmetik," in *Atti del IV Congresso Internazionale dei Matematici (Roma, 6–11 Aprile 1908)*, edited by Guido Castelnuovo, Vol. 2: *Comunicazioni delle sezioni I e II*, Tipogr. della R. Accad. dei Lincei: Roma 1909, 8–11.

[Zer13] Zermelo, Ernst: "Über eine Anwendung der Mengenlehre auf die Theorie des Schachspiels," in Ernest William Hobson and Augustus Edward Hough Love (eds.): *Proceedings of the 5th International Congress of Mathematicians, Cambridge 1912*, Vol. 2, Cambridge University Press: Cambridge 1913, 501–504. English translation in [SchwW01], 133–137.

[Zer14] Zermelo, Ernst: "Über ganze transzendente Zahlen," *Mathematische Annalen* **75** (1914), 434–442.

[Zer15] Zermelo, Ernst: Notes on ordinals and on a 'Neue Theorie der Wohlordnung' in a writing pad, ca. 1915; UAF, C 129/261.

[Zer21] Zermelo, Ernst: "Thesen über das Unendliche in der Mathematik," manuscript, 1 p., ca. 1921; UAF, C 129/278. Published in [vanDE00], 158–159; English translation ibid., 148–149, and [Tay02], 479–482.

[Zer27a] Zermelo, Ernst: "Über das Maß und die Diskrepanz von Punktmengen," *Journal für die reine und angewandte Mathematik* **158** (1927), 154–167.

[Zer27b] Zermelo, Ernst: Addition to [Koed27] in [Koed27], 129–130.

[Zer27c] Zermelo, Ernst: *Allgemeine Mengenlehre und Punktmengen*, lecture notes, winter semester 1927/28; UAF, C 129/157.

[Zer28] Zermelo, Ernst: "Die Berechnung der Turnier-Ergebnisse als ein Maximumproblem der Wahrscheinlichkeitsrechnung," *Mathematische Zeitschrift* **29** (1928), 436–460.

[Zer29a] Zermelo, Ernst: "Über den Begriff der Definitheit in der Axiomatik," *Fundamenta Mathematicae* **14** (1929), 339–344.

[Zer29b] Zermelo, Ernst: "Vortrags-Themata für Warschau 1929," notes of six lectures, manuscript, 9 pp.; UAF, C 129/288. Notes of 1st and 3rd lecture published in [Moo80], 134–136, notes of 6th lecture in [Ebb03], 213–214.

[Zer30a] Zermelo, Ernst: "Über Grenzzahlen und Mengenbereiche. Neue Untersuchungen über die Grundlagen der Mengenlehre," *Fundamenta mathematicae* **16** (1930), 29–47. English translation in [Ewa96], 1219–1233.

[Zer30b] Zermelo, Ernst: "Über die logische Form der mathematischen Theorien," *Annales de la Société Polonaise de Mathématique* **9** (1930), 187.

[Zer30c] Zermelo, Ernst: "Über die Navigation in der Luft als Problem der Variationsrechnung," *Jahresbericht der Deutschen Mathematiker-Vereinigung* **39** (1930), 2nd sect., 44–48.

[Zer30d] Zermelo, Ernst: "Bericht an die Notgemeinschaft der Deutschen Wissenschaft über meine Forschungen betreffend die *Grundlagen der Mathematik*," manuscript, 5 pp. with 2 pp. appendices, dated 3 December 1930; UAF, C 129/140. Published in [Moo80], 130–134.

[Zer30e] Zermelo, Ernst: "Über Grenzzahlen und Mengenbereiche. Neue Untersuchungen über die Grundlagen der Mengenlehre," manuscript, 22 pp.; UAF, C 129/216.

[Zer30f] Zermelo, Ernst: "Über das mengentheoretische Modell," manuscript, 5 pp.; UAF, C 129/292. Copy of first 4 pp., dated September 1930 in UAF, C 129/212. Published in [Ebb06a], 298–302.

[Zer30g] Zermelo, Ernst: "Aus Homers Odyssee. V 149–225," *Wiener Blätter für die Freunde der Antike* **6** (1930), 93–94.

[Zer31a] Zermelo, Ernst: "Über das Navigationsproblem bei ruhender oder veränderlicher Windverteilung," *Zeitschrift für angewandte Mathematik und Mechanik* **11** (1931), 114–124.

[Zer31b] Zermelo, Ernst: Letter to Reinhold Baer of 7 October 1931. Reprinted with English translation in [WeinS87], 43–48.

[Zer31c] Zermelo, Ernst: Letter to K. Gödel of 21 September 1931; UAF, C 129/36. Facsimile and English translation in [Daw85]. Published with English translation in [Goe86], Vol. V, 420–423.

[Zer31d] Zermelo, Ernst: Letter to K. Gödel of 29 October 1931; UAF, C 129/36. Reprinted in [Gra79], 302–303, and with English translation in [Goe86], Vol. V, 430–431.

[Zer31e] Zermelo, Ernst: Eight notes without general title, 10 pp.; undated, ca. 1931; UAF, C 129/236. Seven notes published in [Ebb06a], 302–306, remaining note cf. [Zer31f]

[Zer31f] Zermelo, Ernst: "Sätze über *geschlossene Bereiche*," manuscript, 2 pp; undated, ca. 1931; UAF, Zermelo *Nachlass*, part of [Zer31e]. Published in [Ebb03], 214–215.

[Zer32a] Zermelo, Ernst: "Über Stufen der Quantifikation und die Logik des Unendlichen," *Jahresbericht der Deutschen Mathematiker-Vereinigung* **41** (1932), 2nd sect., 85–88.

[Zer32b] Zermelo, Ernst: "Über mathematische Systeme und die Logik des Unendlichen," *Forschungen und Fortschritte* **8** (1932), 6–7.

[Zer32c] Zermelo, Ernst: "Mengenlehre 1932," manuscript, 5 pp.; UAF, C 129/290. Partially published in [Ebb03], 215–217.

[Zer32d] Zermelo, Ernst: "Vorwort," in [Can32], iii–v.

[Zer33a] Zermelo, Ernst: "Über die Bruchlinien zentrierter Ovale. Wie zerbricht ein Stück Zucker?," *Zeitschrift für angewandte Mathematik und Mechanik* **13** (1933), 168–170.

[Zer33b] Zermelo, Ernst: "Die unbegrenzte Zahlenreihe und die exorbitanten Zahlen," 1 p., November 1933; UAF, C 129/239. Published in [Ebb06a], 306.

[Zer34] Zermelo, Ernst: "Elementare Betrachtungen zur Theorie der Primzahlen," *Nachrichten von der Gesellschaft der Wissenschaften zu Göttingen, Fachgruppe I; Mathematik*, n,. s. **1** (1934), 43–46.

[Zer35] Zermelo, Ernst: "Grundlagen einer allgemeinen Theorie der mathematischen Satzsysteme. Erste Mitteilung)," *Fundamenta mathematicae* **25** (1935), 136–146.

[Zer37] Zermelo, Ernst: "Der Relativismus in der Mengenlehre und der sogenannte Skolemsche Satz," manuscript, dated 4 October 1937, 3 pp.; UAF, C 129/217. Published in [vanDE00], 159–160.

[Zer41] Zermelo, Ernst: Letter to Paul Bernays of 1 October 1941, ETH-Bibliothek Zürich, Nachlaß Bernays, Hs. 975.5259. Published in [Pec90b], 118–119.

[Zer67a] Zermelo, Ernst: "Proof that Every Set Can Be Well-Ordered," in [vanH67], 139–141.

[Zer67b] Zermelo, Ernst: "A New Proof of the Possibility of a Well-Ordering," in [vanH67], 183–198.

[Zer67c] Zermelo, Ernst: "Investigations in the Foundations of Set Theory I," in [vanH67], 199–215.

[Zert75] Zermelo, Theodor: *August Ludwig Schlözer, ein Publicist im alten Reich*, W. Weber: Berlin 1875.

[Zie68] Zierold, Kurt: *Forschungsförderung in drei Epochen. Deutsche Forschungsgemeinschaft. Geschichte, Arbeitsweise, Kommentar*, Steiner: Wiesbaden 1968.

[Zit31] Zita, Kurt: *Beiträge zu einem Variationsproblem von Zermelo*, Ph. D. Thesis, University of Breslau 1931.

[Zor35] Zorn, Max: "A Remark on Method in Transfinite Algebra," *Bulletin of the American Mathematical Society* **41** (1935), 667–670.

[ZoreR24] Zoretti, Ludovic and Rosenthal, Artur: "Neuere Untersuchungen über Funktionen reeller Veränderlichen. Die Punktmengen," in *Enzyklopädie der mathematischen Wissenschaften mit Einschluss ihrer Anwendungen*, Vol. 2: *Analysis*, Part 3. 2, 855–1030, Teubner: Leipzig, volume published 1899–1924, article published 1924.

Index

CPSIA information can be obtained at www.ICGtesting.com
Printed in the USA
LVOW10*1828040314

376006LV00010B/332/P